MATH USER'S HANDBOOK

hot words *hot* topics

Glencoe
McGraw-Hill

New York, New York
Columbus, Ohio
Chicago, Illinois
Peoria, Illinois
Woodland Hills, California

The McGraw·Hill Companies

Send all inquiries to:
Glencoe/McGraw-Hill
8787 Orion Place
Columbus, OH 43240-4027

ISBN 0-07-860160-6 *Quick Review Math Handbook, Course 3*

Printed in the United States of America.
1 2 3 4 5 6 7 8 9 10 027 10 09 08 07 06 05 04 03

HANDBOOK AT A GLANCE

Introduction xiv

PART ONE

Hot Words 2

Glossary 4
Formulas 62
Symbols 64
Patterns 65

PART TWO

Hot Topics 68

1 Numbers and Computation 70
2 Fractions, Decimals, and Percents 98
3 Powers and Roots 164
4 Data, Statistics, and Probability 192
5 Logic 254
6 Algebra 272
7 Geometry 338
8 Measurement 404
9 Tools 430

PART THREE

Hot Solutions and Index 460

HANDBOOK

CONTENTS

Introduction xiv

Descriptions of features show you how to use this handbook.

PART ONE

Hot Words 2

Glossary **4**

Definitions for boldfaced words and other key mathematical terms in the Hot Topics section

Formulas **62**

Explanations of commonly used formulas

Symbols **64**

Mathematical symbols with their meanings

Patterns **65**

A collection of common and significant patterns that are woven through mathematics

PART TWO

Hot Topics 68

A reference to key topics spread over nine areas of mathematics

1 NUMBERS AND COMPUTATION 70

What Do You Already Know? 72

1•1 Place Value of Whole Numbers
Understanding Our Number System 74
Using Expanded Notation 75
Comparing and Ordering Numbers 75
Using Approximations 76
Exercises 77

1•2 Properties
Commutative and Associative Properties 78
Properties of One and Zero 79
Distributive Property 79
Exercises 81

1•3 Order of Operations
Understanding the Order of Operations 82
Exercises 83

1•4 Factors and Multiples
Factors 84
Divisibility Rules 86
Prime and Composite Numbers 87
Multiples and Least Common Multiples 90
Exercises 91

1•5 Integer Operations
Positive and Negative Integers 92
Opposites of Integers and Absolute Value 92
Adding and Subtracting Integers 93
Multiplying and Dividing Integers 94
Exercises 95

What Have You Learned? 96

2 FRACTIONS, DECIMALS, AND PERCENTS 98

What Do You Already Know? 100

2•1 Fractions and Equivalent Fractions
Naming Fractions 102
Methods for Finding Equivalent Fractions 104
Writing Fractions in Lowest Terms 106
Writing Improper Fractions and Mixed Numbers 109
Exercises 111

2•2 Comparing and Ordering Fractions
Comparing Fractions 112
Ordering Fractions 114
Exercises 115

2•3 Addition and Subtraction of Fractions
Adding and Subtracting Fractions with Like Denominators 116
Adding and Subtracting Fractions with Unlike Denominators 117
Adding and Subtracting Mixed Numbers 118
Exercises 121

2•4 Multiplication and Division of Fractions
Multiplying Fractions 122
Dividing Fractions 124
Exercises 125

2•5 Naming and Ordering Decimals
Decimal Place Value: Tenths and Hundredths 126
Decimal Place Value: Thousandths 127
Naming Decimals Greater Than and Less Than One 128
Comparing Decimals 129
Exercises 131

2•6 Decimal Operations
Adding and Subtracting Decimals 132
Multiplying Decimals 133
Dividing Decimals 136
Exercises 139

2•7 Meaning of Percent
Naming Percents 140
Understanding the Meaning of Percent 140
Exercises 143

2•8 Using and Finding Percents
Finding a Percent of a Number 144
Finding Percent and Base 146
Percent of Increase or Decrease 147
Discounts and Sale Prices 149
Finding Simple Interest 151
Exercises 153

2•9 Fraction, Decimal, and Percent Relationships
Percents and Fractions 154
Percents and Decimals 156
Fractions and Decimals 159
Exercises 161

What Have You Learned? 162

3 POWERS AND ROOTS 164

What Do You Already Know? 166

3•1 Powers and Exponents
Exponents 168
Evaluating the Square of a Number 169
Evaluating the Cube of a Number 170
Evaluating Higher Powers 172
Powers of Ten 173
Using a Calculator to Evaluate Powers 174
Exercises 175

3•2 Square and Cube Roots
Square Roots 176
Cube Roots 179
Exercises 181

3•3 Scientific Notation
Using Scientific Notation 182
Converting from Scientific Notation to Standard Form 185
Exercises 187

3•4 Laws of Exponents
Exponents Within the Order of Operations 188
Exercises 189

What Have You Learned? 190

4 DATA, STATISTICS, AND PROBABILITY 192

What Do You Already Know? 194

4•1 Collecting Data
Surveys 196
Random Samples 197
Questionnaires 198
Compiling Data 199
Exercises 201

4•2 Displaying Data

Interpret and Create a Table 202
Interpret a Box Plot 203
Interpret and Create a Circle Graph 204
Interpret and Create a Frequency Graph 206
Interpret a Line Graph 207
Interpret a Stem-and-Leaf Plot 208
Interpret and Create a Bar Graph 209
Interpret a Double-Bar Graph 210
Interpret and Create a Histogram 211
Exercises 213

4•3 Analyzing Data

Scatter Plots 214
Correlation 216
Line of Best Fit 218
Distribution of Data 219
Exercises 221

4•4 Statistics

Mean 222
Median 224
Mode 226
Range 228
Weighted Averages 229
Exercises 231

4•5 Combinations and Permutations

Tree Diagrams 232
Permutations 235
Combinations 236
Exercises 239

4•6 Probability

Experimental Probability 240
Theoretical Probability 241
Strip Graphs 244
Outcome Grids 246
Probability Line 247
Dependent and Independent Events 249
Sampling 250
Exercises 251

What Have You Learned? 252

5 LOGIC 254

What Do You Already Know? 256

5•1 If/Then Statements
Conditional Statements? 258
Converse of a Conditional 259
Negations and the Inverse of a Conditional 260
Contrapositive of a Conditional 261
Exercises 263

5•2 Counterexamples
Counterexamples 264
Exercises 265

5•3 Sets
Sets and Subsets 266
Union of Sets 266
Intersection of Sets 267
Venn Diagrams 267
Exercises 269

What Have You Learned? 270

6 ALGEBRA 272

What Do You Already Know? 274

6•1 Writing Expressions and Equations
Expressions 276
Writing Expressions Involving Addition 276
Writing Expressions Involving Subtraction 277
Writing Expressions Involving Multiplication 278
Writing Expressions Involving Division 279
Writing Expressions Involving Two Operations 280
Equations 282
Exercises 283

6•2 Simplifying Expressions
Terms 284
The Commutative Property of Addition and Multiplication 284
The Associative Property of Addition and Multiplication 285
The Distributive Property 286
Equivalent Expressions 287
Like Terms 289
Simplifying Expressions 290
Exercises 291

6•3 Evaluating Expressions and Formulas
Evaluating Expressions 292
Evaluating Formulas 292
Exercises 295

6•4 Solving Linear Equations
Additive Inverses 296
True or False Equations 296
The Solution of an Equation 297
Equivalent Equations 298
Solving Equations 298
Solving Equations Requiring Two Operations 301
Solving Equations with the Variable on Both Sides 302
Equations Involving the Distributive Property 304
Solving for a Variable in a Formula 305
Exercises 307

6•5 Ratio and Proportion
Ratio 308
Proportions 308
Using Proportions to Solve Problems 310
Exercises 311

6•6 Inequalities
Showing Inequalities 312
Solving Inequalities 313
Exercises 315

6•7 Graphing on the Coordinate Plane
Axes and Quadrants 316
Writing an Ordered Pair 317
Locating Points on the Coordinate Plane 318
The Graph of an Equation with Two Variables 319
Exercises 323

6•8 Slope and Intercept
Slope 324
Calculating the Slope of a Line 325
The y-Intercept 328
Using the Slope and y-Intercept to Graph a Line 329
Slope-Intercept Form 330
Writing Equations in Slope-Intercept Form 330
Writing the Equation of a Line 332
Exercises 335

What Have You Learned? 336

7 GEOMETRY 338

What Do You Already Know? 340

7•1 Naming & Classifying Angles & Triangles
Points, Lines, and Rays 342
Naming Angles 343
Measuring Angles 344
Classifying Angles 345
Classifying Triangles 346
The Triangle Inequality 348
Exercises 349

7•2 Naming & Classifying Polygons & Polyhedrons
Quadrilaterals 350
Angles of a Quadrilateral 350
Types of Quadrilaterals 351
Polygons 353
Angles of a Polygon 355
Polyhedrons 356
Exercises 358

7•3 Symmetry & Transformations
Reflections 360
Reflection Symmetry 362
Rotations 363
Translations 364
Exercises 365

7•4 Perimeter
Perimeter of a Polygon 366
Perimeter of a Rectangle 367
Perimeter of a Right Triangle 369
Exercises 370

7•5 Area
What Is Area? 372
Estimating Area 372
Area of a Rectangle 373
Area of a Parallelogram 374
Area of a Triangle 375
Area of a Trapezoid 376
Exercises 377

7•6 Surface Area
Surface Area of a Rectangular Prism 378
Surface Area of Other Solids 379
Exercises 381

7•7 Volume
What Is Volume? 382
Volume of a Prism 383
Volume of a Cylinder 384
Volume of a Pyramid and a Cone 384
Exercises 387

7•8 Circles
Parts of a Circle 388
Circumference 389
Central Angles 390
Area of a Circle 392
Exercises 393

7•9 Pythagorean Theorem
Right Triangles 394
The Pythagorean Theorem 395
Pythagorean Triples 396
Exercises 397

7•10 Tangent Ratio
Sides and Angles in a Right Triangle 398
Tangent of an Angle 398
Tangent Table 400
Exercises 401

What Have You Learned? 402

8 MEASUREMENT 404

What Do You Already Know? 406

8•1 Systems of Measurement
The Metric and Customary Systems 408
Accuracy 410
Exercises 411

8•2 Length and Distance
About What Length? 412
Metric and Customary Units 413
Conversions Between Systems 414
Exercises 415

8•3 Area, Volume, and Capacity
Area 416
Volume 417
Capacity 418
Exercises 419

8•4 Mass and Weight
Mass and Weight 420
Exercises 421

8•5 Time
Working with Time Units 422
Exercises 423

8•6 Size and Scale
Similar Figures 424
Scale Factors 425
Scale Factors and Area 426
Exercises 427

What Have You Learned? 428

9 TOOLS 430

What Do You Already Know? 432

9•1 Four-Function Calculator
Basic Operations 435
Memory 436
Special Keys 437
Exercises 439

9•2 Scientific Calculator
Frequently Used Functions 441
Tangent 442
Exercises 443

9•3 Geometry Tools
Ruler 444
Protractor 445
Compass 446
Construction Problem 448
Exercises 450

9•4 Spreadsheets
What Is a Spreadsheet? 452
Spreadsheet Formulas 453
Fill Down and Fill Right 454
Spreadsheet Graphs 456
Exercises 457

What Have You Learned? 458

PART THREE
Hot Solutions and Index 460

HANDBOOK

INTRODUCTION

Why use this handbook?

You will use this mathematics handbook to refresh your memory of concepts and skills.

What are Hot Words and how do you find them?

The Hot Words section includes a glossary of terms, a collection of common or significant mathematical patterns, and lists of symbols and formulas in alphabetical order. Many entries in the glossary will refer you to chapters and topics in the Hot Topics section for more detailed information.

4 absolute value

hot words A

absolute value a number's distance from zero on the number line *see 1•5 Integer Operations*

Example:

−2 is 2 units from 0

−5 −4 −3 −2 −1 0 1 2 3 4 5

the absolute value of −2 is 2 or $|-2| = 2$

accuracy the exactness of a number. For example, a number such as 62.42812 might be rounded off to three decimal places (62.428), to two decimal places (62.43), to one decimal place (62.4), or to the nearest whole number (62). The first answer is more accurate than the second, the second more accurate than the third, and so on. *see 2•5 Naming and Ordering Decimals, 8•1 Systems of Measurement*

actual size the true size of an object represented by a scale model or drawing *see 8•6 Size and Scale*

acute angle any angle that measures less than 90° *see 7•1 Naming and Classifying Angles and Triangles*

Example:

A B C

∠ABC is an acute angle

$0° < m\angle ABC < 90°$

...triangle a triangle in which all angles measure less than ... *...d Classifying Angles and Triangles*

HOT WORDS

What are Hot Topics and how do you use them?

The Hot Topics section consists of nine chapters. Each chapter has several topics that give you to-the-point explanations of key mathematical concepts. Each topic includes one or more concepts. After each concept is a Check It Out section, which gives you a few problems to check your understanding of the concept. At the end of each topic, there is an exercise set.

There are problems and a vocabulary list at the beginning and end of each chapter to help you preview and review what you know.

What are Hot Solutions?

The Hot Solutions section gives you easy-to-locate answers to Check It Out and What Do You Already Know? problems.

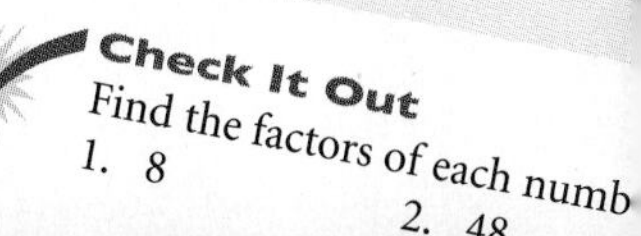

PART **ONE**

hot words

The Hot Words section includes a glossary of terms, a collection of common or significant mathematical patterns, and lists of symbols and formulas. Many entries in the glossary will refer to chapters and topics in the Hot Topics section.

Glossary	4	Symbols	64
Formulas	62	Patterns	65

absolute value a number's distance from zero on the number line *see 1•5 Integer Operations*

Example:

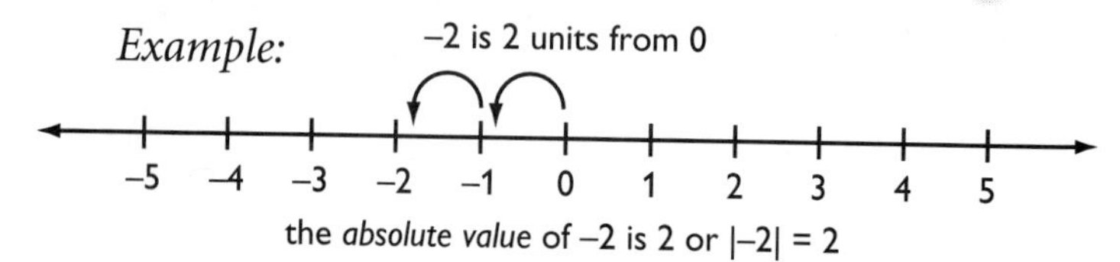

the *absolute value* of −2 is 2 or $|-2| = 2$

accuracy the exactness of a number. For example, a number such as 62.42812 might be rounded off to three decimal places (62.428), to two decimal places (62.43), to one decimal place (62.4), or to the nearest whole number (62). The first answer is more accurate than the second, the second more accurate than the third, and so on. *see 2•5 Naming and Ordering Decimals, 8•1 Systems of Measurement*

actual size the true size of an object represented by a scale model or drawing *see 8•6 Size and Scale*

acute angle any angle that measures less than 90° *see 7•1 Naming and Classifying Angles and Triangles*

Example:

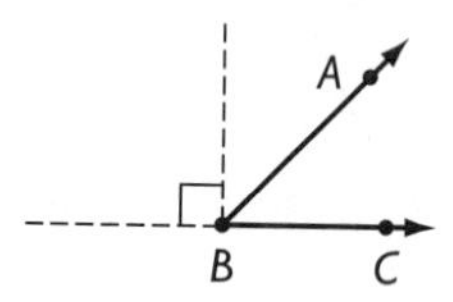

$\angle ABC$ is an *acute angle*

$0^\circ < m\angle ABC < 90^\circ$

acute triangle a triangle in which all angles measure less than 90° *see 7•1 Naming and Classifying Angles and Triangles*

Example:

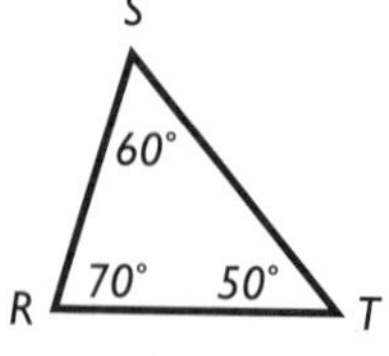

$\triangle RST$ is an *acute triangle*

addition the operation used to combine numbers into a sum *see 1•3 Order of Operations, 1•5 Integer Operations, 2•3 Addition and Subtraction of Fractions, 2•6 Decimal Operations, 3•4 Laws of Exponents, 9•1 Four Function Calculator, 9•2 Scientific Calculator, 9•4 Spreadsheets*

additive inverse a number that when added to a given number results in a sum of zero

Example: $(+3) + (-3) = 0$
(-3) is the *additive inverse* of 3

additive property the mathematical rule that states that if the same number is added to each side of an equation, the expressions remain equal *see 1•2 Properties*

additive system a mathematical system in which the values of individual symbols are added together to determine the value of a sequence of symbols

Examples: The Roman numeral system, which uses symbols such as I, V, D, and M, is a well-known additive system.

This is another example of an additive system:

▽▽□

If □ equals 1 and ▽ equals 7,
then ▽▽□ equals 7 + 7 + 1 = 15

algebra a branch of mathematics in which symbols are used to represent numbers and express mathematical relationships *see Chapter 6 Algebra*

algorithm a specific step-by-step procedure for any mathematical operation *see 2•3 Addition and Subtraction of Fractions, 2•4 Multiplication and Division of Fractions, 2•6 Decimal Operations*

altitude the perpendicular distance from the base of a shape to the vertex. *Altitude* indicates the height of a shape.

Example:

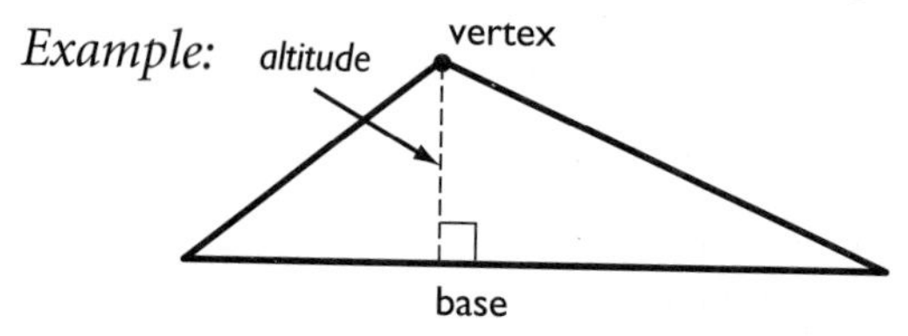

angle two rays that meet at a common endpoint
see 7•1 Naming and Classifying Angles and Triangles

Example:

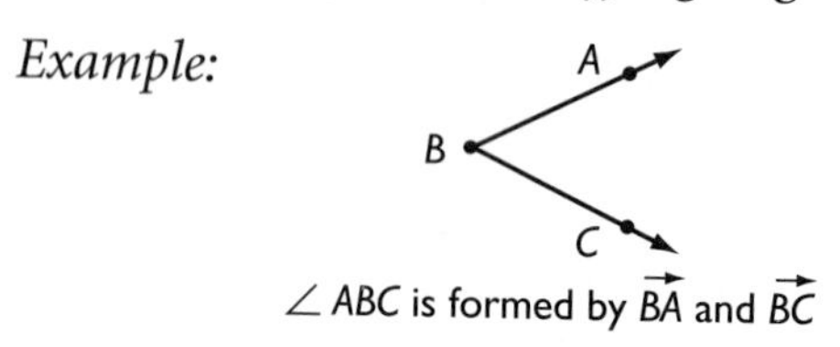

angle of elevation an angle formed by an upward line of sight and the horizontal

Example:

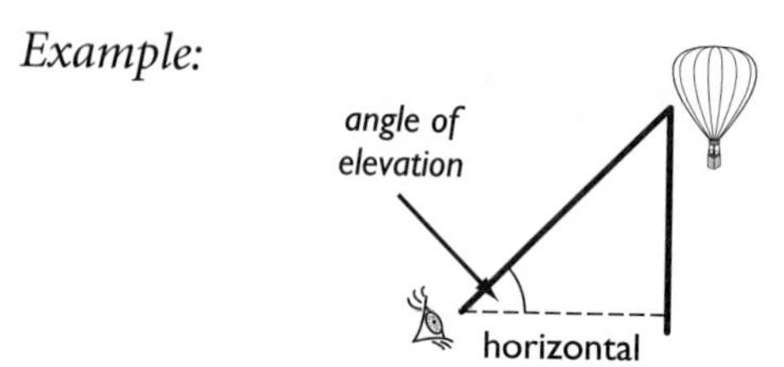

apothem a perpendicular line from the center of a regular polygon to any one of its sides

Example:

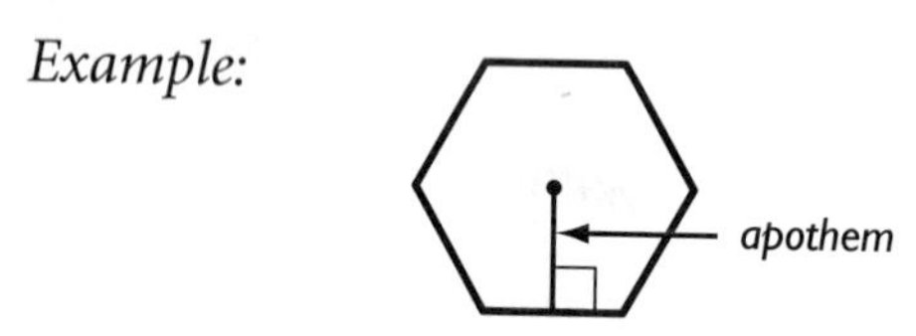

approximation an estimate of a mathematical value that is not exact but close enough to be of use

Arabic numerals (or Hindu-Arabic numerals) the number symbols we presently use {0, 1, 2, 3, 4, 5, 6, 7, 8, 9}

arc a section of a circle *see 7•8 Circles*

Example:

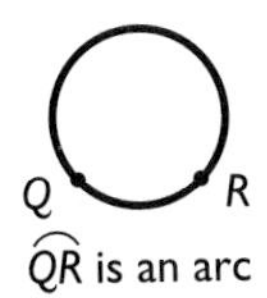

$\overset{\frown}{QR}$ is an arc

area the size of a surface, usually expressed in square units *see 7•5 Area, 7•6 Surface Area, 7•8 Circles, 8•3 Area, Volume, and Capacity*

Example:

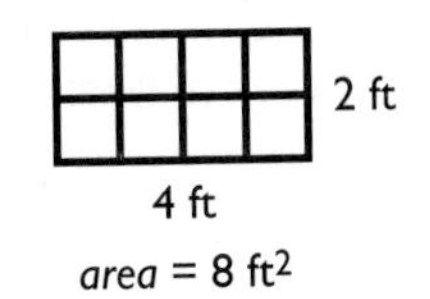

area = 8 ft^2

arithmetic expression a mathematical relationship expressed as a number, or two or more numbers with operation symbols
see 6•1 Writing Expressions and Equations

arithmetic sequence a mathematical progression in which the difference between any two consecutive numbers in the sequence is the same *see page 65*

Example: 2, 6, 10, 14, 18, 22, 26
the common difference of this *arithmetic sequence* is 4

associative property a rule that states that the sum or product of a set of numbers is the same, no matter how the numbers are grouped

Examples: $(x + y) + z = x + (y + z)$
$x \times (y \times z) = (x \times y) \times z$

average the sum of a set of values divided by the number of values *see 4•4 Statistics*

Example: the *average* of 3, 4, 7, and 10 is
$(3 + 4 + 7 + 10) \div 4 = 6$

average speed the average rate at which an object moves

HOT WORDS

axis (pl. *axes*) [1] one of the reference lines by which a point on a coordinate graph may be located; [2] the imaginary line about which an object may be said to be symmetrical (*axis* of symmetry); [3] the line about which an object may revolve (*axis* of rotation) *see 6•7 Graphing on the Coordinate Plane, 7•3 Symmetry and Transformations*

bar graph a way of displaying data using horizontal or vertical bars *see 4•2 Displaying Data*

base [1] the side or face on which a three-dimensional shape stands; [2] the number of characters a number system contains *see 1•1 Place Value of Whole Numbers, 7•6 Surface Area, 7•7 Volume*

base-ten system the number system containing ten single-digit symbols {0, 1, 2, 3, 4, 5, 6, 7, 8, and 9} in which the numeral 10 represents the quantity ten *see 1•1 Place Value of Whole Numbers, 2•5 Naming and Ordering Decimals*

base-two system the number system containing two single-digit symbols {0 and 1} in which 10 represents the quantity two *see binary system*

benchmark a point of reference from which measurements can be made *see 2•7 Naming Percents*

best chance in a set of values, the event most likely to occur *see 4•6 Probability*

bimodal distribution a statistical model that has two different peaks of frequency distribution *see 4•3 Analyzing Data*

binary system the base two number system, in which combinations of the digits 1 and 0 represent different numbers, or values

binomial an algebraic expression that has two terms

Examples: $x^2 + y$; $x + 1$; $a - 2b$

box plot a diagram, constructed from a set of numerical data, that shows a box indicating the middle 50% of the ranked statistics, as well as the maximum, minimum, and medium statistics *see 4•2 Displaying Data*

broken-line graph a type of line graph used to show change over a period of time *see 4•2 Displaying Data*

Example:

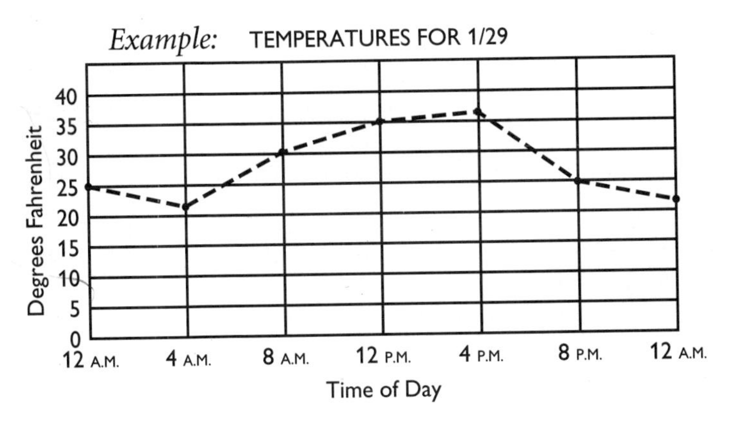

budget a spending plan based on an estimate of income and expenses *see 9•4 Spreadsheets*

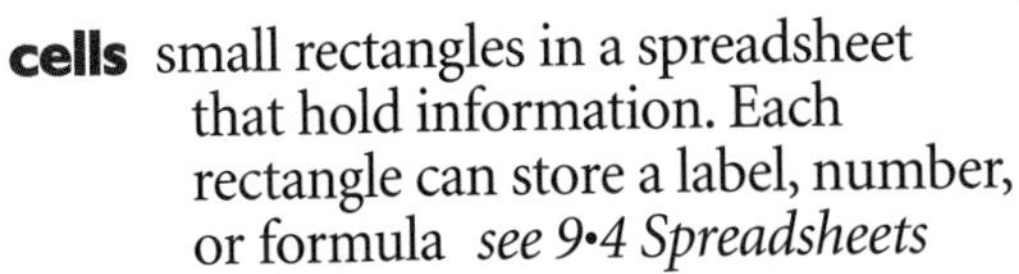

cells small rectangles in a spreadsheet that hold information. Each rectangle can store a label, number, or formula *see 9•4 Spreadsheets*

center of the circle the point from which all points on a circle are equidistant *see 7•8 Circles*

chance the probability or likelihood of an occurrence, often expressed as a fraction, decimal, percentage, or ratio *see 4•6 Probability*

circle a perfectly round shape with all points equidistant from a fixed point, or center *see 7•8 Circles*

Example:

center

a circle

circle graph (pie chart) a way of displaying statistical data by dividing a circle into proportionally-sized "slices" *see 4•2 Displaying Data*

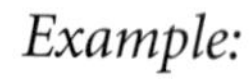

Example:

FAVORITE PRIMARY COLOR

circumference the distance around a circle, calculated by multiplying the diameter by the value pi *see 7•8 Circles*

classification the grouping of elements into separate classes or sets *see 5•3 Sets*

collinear a set of points that lie on the same line

Example:

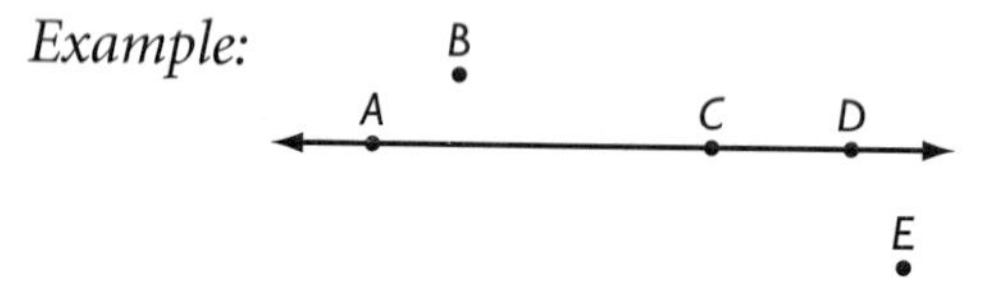

points *A*, *C*, and *D* are *collinear*

columns vertical lists of numbers or terms; in spreadsheets, the names of cells in a column all beginning with the same letter {A1, A2, A3, A4, . . .} *see 9•4 Spreadsheets*

combination a selection of elements from a larger set in which the order does not matter *see 4•5 Combinations and Permutations*

Example: 456, 564, and 654 are one *combination* of three digits from 4567

common denominator a whole number that is the denominator for all members of a group of fractions *see 2•3 Addition and Subtraction of Fractions*

Example: the fractions $\frac{5}{8}$ and $\frac{7}{8}$ have a *common denominator* of 8

common difference the difference between any two consecutive terms in an arithmetic sequence *see arithmetic sequence*

common factor a whole number that is a factor of each number in a set of numbers
see 1•4 Factors and Multiples

Example: 5 is a *common factor* of 10, 15, 25, and 100

common ratio the ratio of any term in a geometric sequence to the term that precedes it *see geometric sequence*

commutative property the mathematical rule that states that for any numbers x and y
$x + y = y + x$ and
$xy = yx$
see 1•2 Properties

composite number a number exactly divisible by at least one whole number other than itself and 1
see 1•4 Factors and Multiples

concave polygon a polygon that has an interior angle greater than 180°

Example:

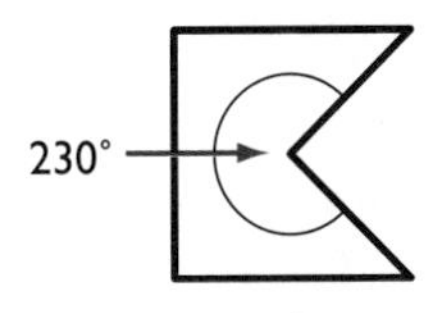

a *concave polygon*

conditional a statement that something is true or will be true provided that something else is also true
see contrapositive, converse, 5•1 If/Then Statements

Example: if a polygon has three sides, then it is a triangle

cone a solid consisting of a circular base and one vertex

Example:

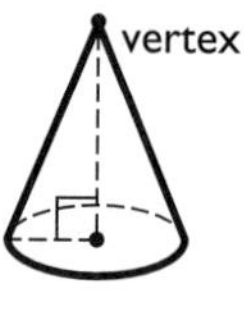

a cone

congruent figures figures that have the same size and shape. The symbol ≅ is used to indicate congruence.

Example:

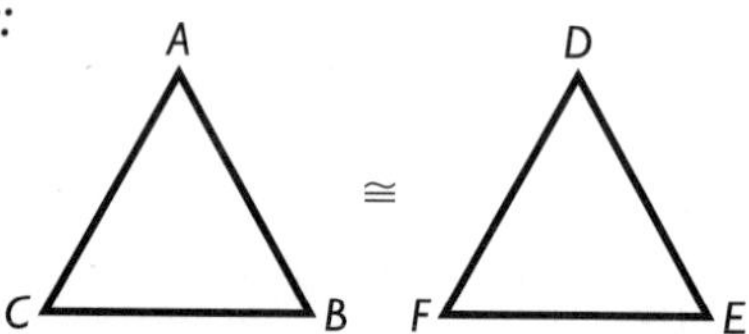

triangles *ABC* and *DEF* are congruent

conic section the curved shape that results when a conical surface is intersected by a plane

Example:

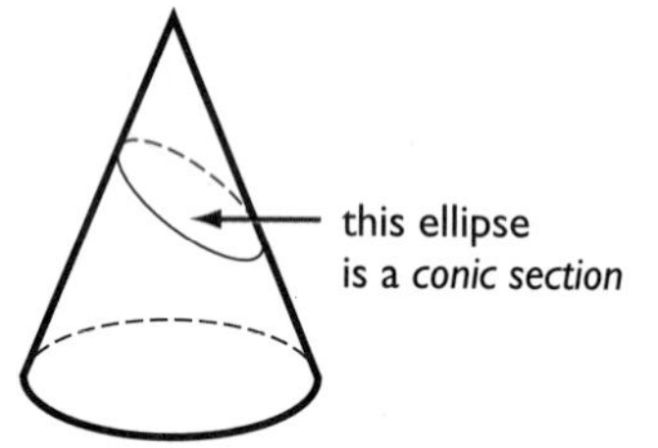

continuous data that relate to a complete range of values on the number line

Example: the possible sizes of apples are *continuous* data

contrapositive a logical equivalent of a given conditional statement, often expressed in negative terms
see 5•1 If/Then Statements

Example: "if *x*, then *y*" is a conditional statement; "if not *y*, then not *x*" is the *contrapositive* statement

HOT WORDS

convenience sampling a sample obtained by surveying people on the street, at a mall, or in another convenient way as opposed to a random sample *see 4•1 Collecting Data*

converse a conditional statement in which terms are expressed in reverse order *see 5•1 If/Then Statements*

Example: "if x, then y" is a conditional statement; "if y, then x" is the *converse* statement

convex polygon a polygon that has no interior angle greater than 180°
see 7•2 Naming and Classifying Polygons and Polyhedrons

Example:

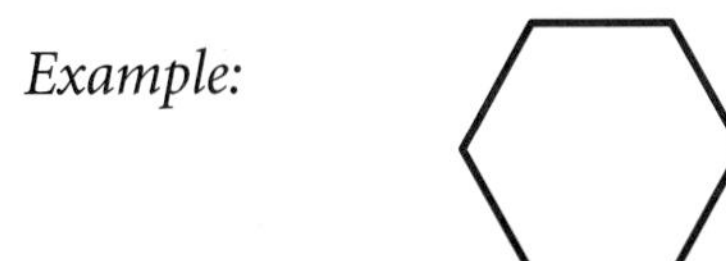

a regular hexagon is a *convex polygon*

coordinate graph the representation of points in space in relation to reference lines—usually, a horizontal x-axis and a vertical y-axis
see coordinates, 6•7 Graphing on the Coordinate Plane

coordinates an ordered pair of numbers that describes a point on a coordinate graph. The first number in the pair represents the point's distance from the origin (0, 0) along the x-axis, and the second represents its distance from the origin along the y-axis. *see ordered pairs, 6•7 Graphing on the Coordinate Plane, 6•8 Slope and Intercept*

Example:

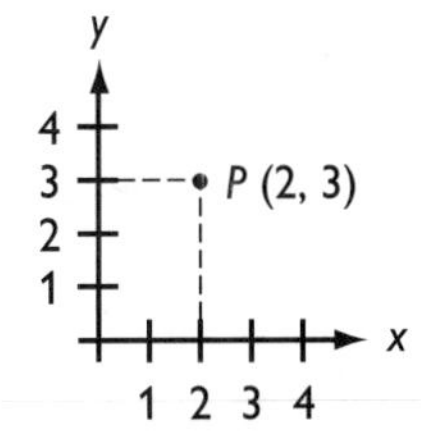

point P has *coordinates* (2, 3)

coplanar points or lines lying in the same plane

correlation the way in which a change in one variable corresponds to a change in another

cost an amount paid or required in payment

cost estimate an approximate amount to be paid or to be required in payment

counterexample a specific example that proves a general mathematical statement to be false
see 5•2 Counterexamples

counting numbers the set of numbers used to count objects; therefore, only those numbers that are whole and positive {1, 2, 3, 4. . .}
see positive integers

cross product a method used to solve proportions and test whether ratios are equal: $\frac{a}{b} = \frac{c}{d}$ if $ad = bc$
see 6•5 Ratio and Proportion

cross section the figure formed by the intersection of a solid and a plane

Example:

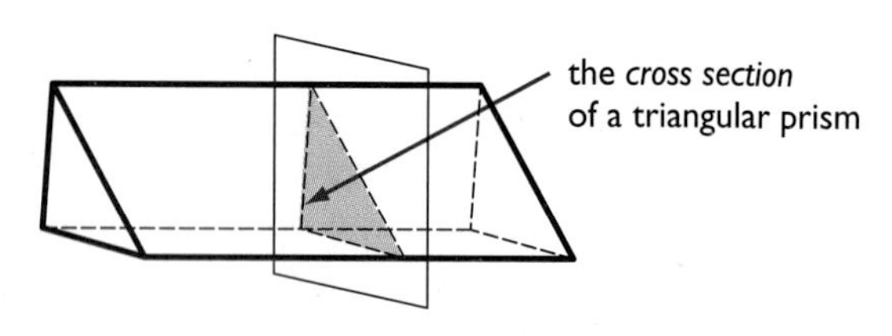

cube (n.) a solid figure with six square faces
see 7•2 Naming and Classifying Polygons and Polyhedrons

Example:

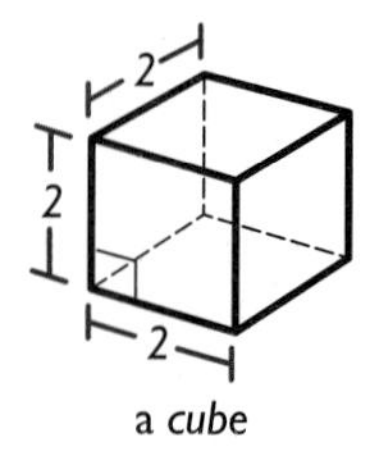

a cube

cube (v.) to multiply a number by itself and then by itself again *see 3•1 Powers and Exponents*

Example: $2^3 = 2 \times 2 \times 2 = 8$

cube root the number that must be multiplied by itself and then by itself again to produce a given number *see 3•2 Square and Cube Roots*

Example: $\sqrt[3]{8} = 2$

cubic centimeter the amount contained in a cube with edges that are 1 cm in length *see 7•7 Volume*

cubic foot the amount contained in a cube with edges that are 1 foot in length *see 7•7 Volume*

cubic inch the amount contained in a cube with edges that are 1 inch in length *see 7•7 Volume*

cubic meter the amount contained in a cube with edges that are 1 meter in length *see 7•7 Volume*

customary system units of measurement used in the United States to measure length in inches, feet, yards, and miles; capacity in cups, pints, quarts, and gallons; weight in ounces, pounds, and tons; and temperature in degrees Fahrenheit *see English system, 8•1 Systems of Measurement*

cylinder a solid shape with parallel circular bases *see 7•6 Surface Area*

Example:

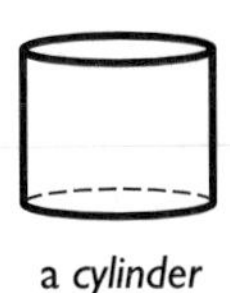

a *cylinder*

decagon a plane polygon with ten angles and ten sides

decimal system the most commonly used number system, in which whole numbers and fractions are represented using base ten
see 2•5 Naming and Ordering Decimals

Example: decimal numbers include 1230, 1.23, 0.23, and -123

degree [1] (algebraic) the exponent of a single variable in a simple algebraic term; [2] (algebraic) the sum of the exponents of all the variables in a more complex algebraic term; [3] (algebraic) the highest degree of any term in an equation; [4] (geometric) a unit of measurement of an angle or arc, represented by the symbol °
see [1] 3•1 Powers and Exponents, 3•4 Laws of Exponents, [4] 7•1 Naming and Classifying Angles and Triangles, 7•8 Circles, 9•2 Scientific Calculator

Examples: [1] In the term $2x^4y^3z^2$, x has a *degree* of 4, y has a *degree* of 3, and z has a *degree* of 2.
[2] The term $2x^4y^3z^2$ as a whole has a *degree* of $4 + 3 + 2 = 9$.
[3] The equation $x^3 = 3x^2 + x$ is an equation of the third *degree.*
[4] An acute angle is an angle that measures less than 90°.

denominator the bottom number in a fraction
see 2•1 Fractions and Equivalent Fractions

Example: for $\frac{a}{b}$, b is the *denominator*

dependent events a group of happenings, each of which affects the probability of the occurrence of the others
see 4•6 Probability

diagonal a line segment that connects one vertex to another (but not one next to it) on a polygon
see 7•2 Naming and Classifying Polygons and Polyhedrons

Example:

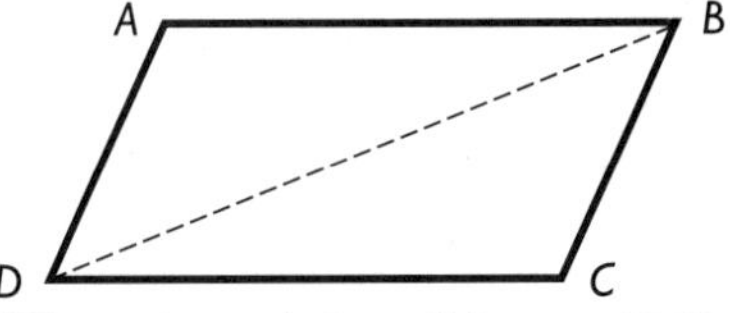

$\overline{BD}$ is a *diagonal* of parallelogram *ABCD*

diameter a line segment that passes through the center of a circle and divides it in half *see 7•8 Circles*

Example:

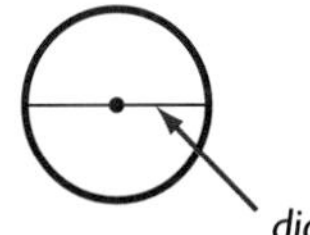

difference the result obtained when one number is subtracted from another

dimension the number of measures needed to describe a figure geometrically

Examples: A point has 0 *dimensions.*
A line or curve has 1 *dimension.*
A plane figure has 2 *dimensions.*
A solid figure has 3 *dimensions.*

direct correlation the relationship between two or more elements that increase and decrease together *see 4•3 Analyzing Data*

Example: At an hourly pay rate, an increase in the number of hours you work means an increase in the amount you get paid, while a decrease in the number of hours you work means a decrease in the amount you get paid.

discount a deduction made from the regular price of a product or service *see 2•8 Using and Finding Percents*

discrete data that can be described by whole numbers or fractional values. The opposite of *discrete* data is continuous data.

Example: the number of oranges on a tree is *discrete* data

distance the length of the shortest line segment between two points, lines, planes, and so forth
see 8•2 Length and Distance

distance-from graph a coordinate graph that shows distance from a specified point as a function of time

distribution the frequency pattern for a set of data
see 4•3 Analyzing Data

division the operation in which a dividend is divided by a divisor to obtain a quotient *see 1•5 Integer Operations, 2•4 Multiplication and Division of Fractions, 2•6 Decimal Operations, 3•1 Powers and Exponents*

Example:

$12 \div 3 = 4$

12 ← dividend
3 ← divisor
4 ← quotient

double-bar graph a graphical display that uses paired horizontal or vertical bars to show a relationship between data *see 4•2 Displaying Data*

Example:

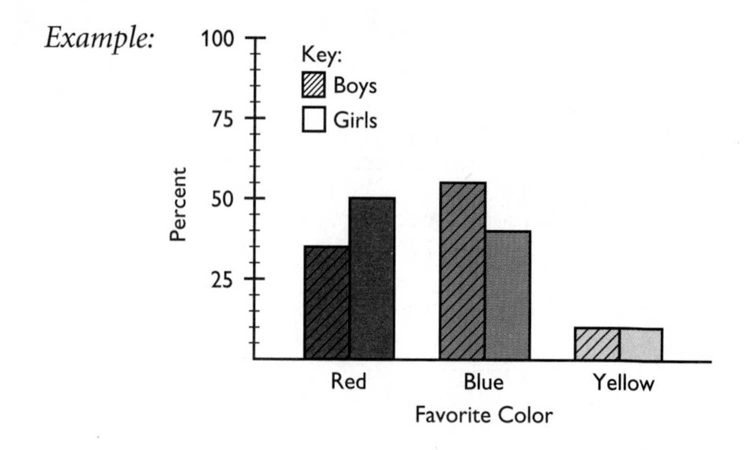

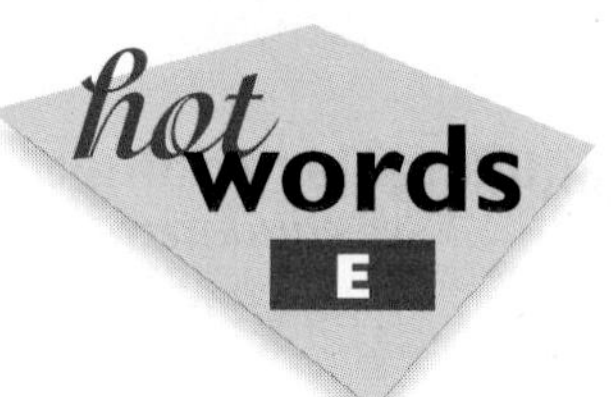

edge a line along which two planes of a solid figure meet *see 7•2 Naming and Classifying Polygons and Polyhedrons*

ellipse a figure shaped like an oval

Example:

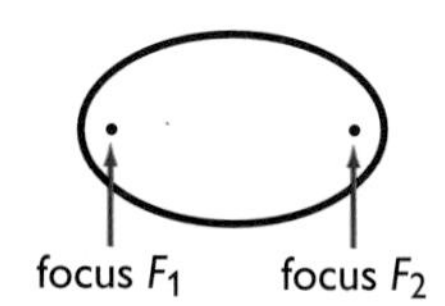

English system units of measurement used in the United States that measure length in inches, feet, yards, and miles; capacity in cups, pints, quarts, and gallons; weight in ounces, pounds, and tons; and temperature in degrees Fahrenheit *see customary system*

equal angles angles that measure the same number of degrees *see 7•1 Naming and Classifying Angles and Triangles*

equally likely describes outcomes or events that have the same chance of occurring *see 4•6 Probability*

equally unlikely describes outcomes or events that have the same chance of not occurring *see 4•6 Probability*

equation a mathematical sentence stating that two expressions are equal *see 6•1 Writing Expressions and Equations, 6•8 Slope and Intercept*

Example: $3 \times (7 + 8) = 9 \times 5$

equiangular having more than one angle, each of which is the same size *see 7•1 Naming and Classifying Angles and Triangles, 7•2 Naming and Classifying Polygons and Polyhedrons*

equiangular triangle a triangle in which each angle is 60°
see equilateral triangle, 7•1 Naming and Classifying Angles and Triangles

equilateral a shape having more than two sides, each of which is the same length

equilateral triangle a triangle in which each side is of equal length *see equiangular triangle, 7•1 Naming and Classifying Angles and Triangles*

Example:

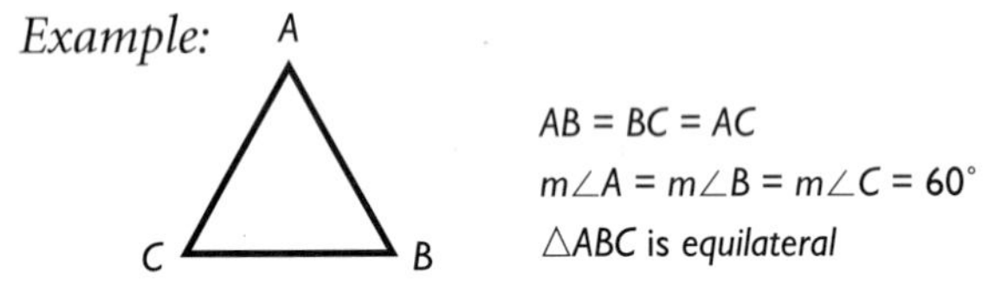

equivalent equal in value

equivalent expressions expressions that always result in the same number, or have the same mathematical meaning for all replacement values of their variables
see 6•2 Simplifying Expressions

Examples: $\frac{9}{3} + 2 = 10 - 5$

$2x + 3x = 5x$

equivalent fractions fractions that represent the same quotient but have different numerators and denominators
see 2•1 Fractions and Equivalent Fractions

Example: $\frac{5}{6} = \frac{15}{18}$

equivalent ratios ratios that are equal
see 6•5 Ratio and Proportion

Example: $\frac{5}{4} = \frac{10}{8}$; 5:4 = 10:8

estimate an approximation or rough calculation

even number any whole number that is a multiple of 2
{2, 4, 6, 8, 10, 12, . . .}

HOT WORDS

event any happening to which probabilities can be assigned *see 4•6 Probability*

expanded notation a method of writing a number that highlights the value of each digit *see 1•1 Place Value of Whole Numbers*

Example: $867 = 800 + 60 + 7$

expense an amount of money paid; cost

experimental probability a ratio that shows the total number of times the favorable outcome happened to the total number of times the experiment was done *see 4•6 Probability*

exponent a numeral that indicates how many times a number or expression is to be multiplied by itself *see 1•3 Order of Operations, 3•1 Powers and Exponents, 3•3 Scientific Notation, 3•4 Laws of Exponents*

Example: in the equation $2^3 = 8$, the *exponent* is 3

expression a mathematical combination of numbers, variables, and operations; e.g., $6x + y^2$ *see 6•1 Writing Expressions and Equations, 6•2 Simplifying Expressions, 6•3 Evaluating Expressions and Formulas*

face a two-dimensional side of a three-dimensional figure *see 7•2 Naming and Classifying Polygons and Polyhedrons, 7•6 Surface Area*

factor a number or expression that is multiplied by another to yield a product *see 1•4 Factors and Multiples*

Example: 3 and 11 are *factors* of 33

factor pair two unique numbers multiplied together to yield a product, such as $2 \times 3 = 6$ *see 1•4 Factors and Multiples*

factorial represented by the symbol !, the product of all the whole numbers between 1 and a given positive whole number *see 4•5 Combinations and Permutations*

Example: $5! = 1 \times 2 \times 3 \times 4 \times 5 = 120$

fair describes a situation in which the theoretical probability of each outcome is equal
see 4•6 Probability

Fibonacci numbers *see page 65*

flat distribution a frequency graph that shows little difference between responses *see 4•3 Analyzing Data*

Example:

flat distribution

Number of Occurrences

	X				X
X	X	X	X	X	X
X	X	X	X	X	X
X	X	X	X	X	X
1	2	3	4	5	6

Die roll

flip to "turn over" a shape
see reflection, 7•3 Symmetry and Transformations

Example:

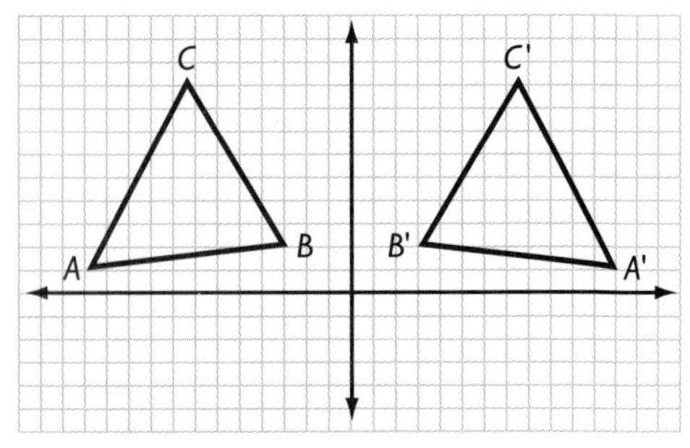

$\triangle A'B'C'$ is a flip of $\triangle ABC$

forecast to predict a trend, based on statistical data
see 4•3 Analyzing Data

formula an equation that shows the relationship between two or more quantities; a calculation performed by spreadsheet *see pages 62–63, 6•3 Evaluating Expressions and Formulas, 9•4 Spreadsheets*

Example: $A = \pi r^2$ is the *formula* for calculating the area of a circle; A2 × B2 is a spreadsheet *formula*

fraction a number representing some part of a whole; a quotient in the form $\frac{a}{b}$
see 2•1 Fractions and Equivalent Fractions

frequency graph a graph that shows similarities among the results so one can quickly tell what is typical and what is unusual *see 4•2 Displaying Data*

function assigns exactly one output value to each input value

Example: You are driving at 50 mi/hr. There is a relationship between the amount of time you drive and the distance you will travel. You say that the distance is a *function* of the time.

geometric sequence a sequence in which the ratio between any two consecutive terms is the same
see common ratio and page 65

Example: 1, 4, 16, 64, 256, . . .
the common ratio of this *geometric sequence* is 4

geometry the branch of mathematics concerned with the properties of figures
see Chapter 7 Geometry, 9•3 Geometry Tools

gram a metric unit used to measure mass
see 8•3 Area, Volume, and Capacity

greatest common factor (GCF) the greatest number that is a factor of two or more numbers
see 1•4 Factors and Multiples

Example: 30, 60, 75
the *greatest common factor* is 15

growth model a description of the way data change over time

hot words
H

harmonic sequence *see page 65*

height the distance from the base to the top of a figure

heptagon a polygon that has seven sides

Example:

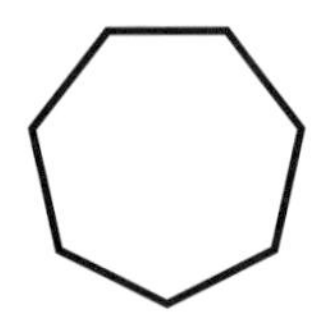

a *heptagon*

hexagon a polygon that has six sides

Example:

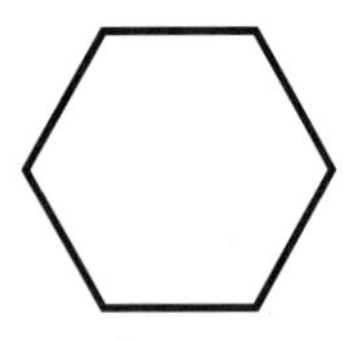

a *hexagon*

hexagonal prism a prism that has two hexagonal bases and six rectangular sides

Example:

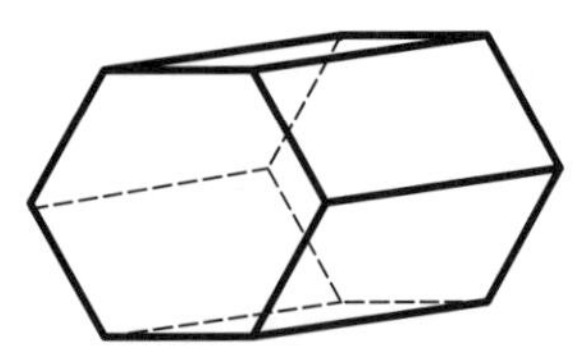

a *hexagonal prism*

hexahedron a polyhedron that has six faces

Example:

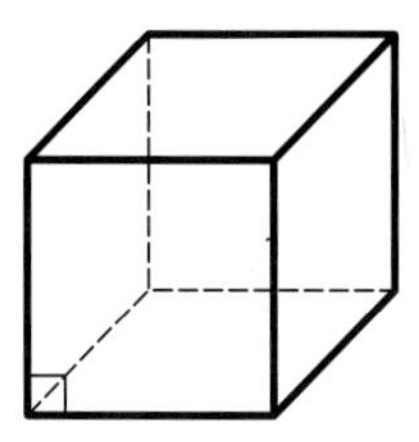

a cube is a *hexahedron*

histogram a graph in which statistical data is represented by blocks of proportionately-sized areas
see 4•2 Displaying Data

horizontal a flat, level line or plane

hyperbola the curve of an inverse variation function, such as $y = \frac{1}{x}$, is a hyperbola

Example:

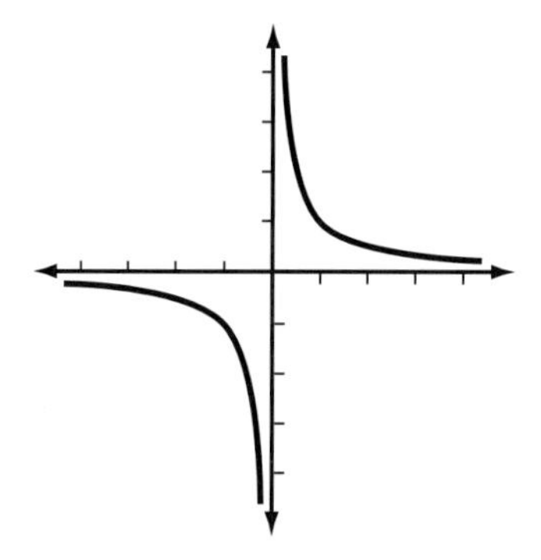

hypotenuse the side of a right triangle, opposite the right angle
see 7•1 Naming and Classifying Angles and Triangles

Example:

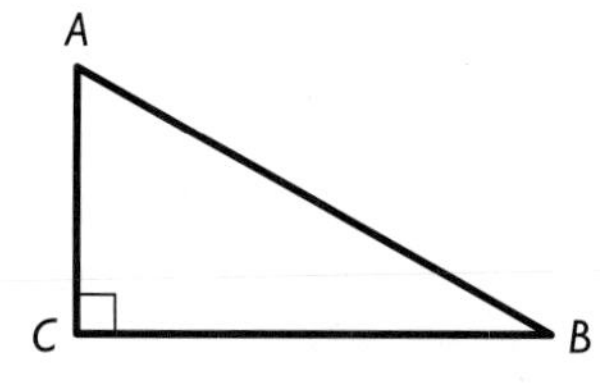

side $\overline{AB}$ is the *hypotenuse* of this right triangle

improper fraction a fraction in which the numerator is greater than the denominator *see 2•1 Fractions and Equivalent Fractions*

Examples: $\frac{21}{4}, \frac{4}{3}, \frac{2}{1}$

income the amount of money received for labor, services, or the sale of goods or property

independent event an event in which the outcome does not influence the outcome of other events *see 4•6 Probability*

inequality a statement that uses the symbols > (greater than), < (less than), ≥ (greater than or equal to), and ≤ (less than or equal to) to indicate that one quantity is greater than or less than another *see 6•6 Inequalities*

Examples: $5 > 3$; $\frac{4}{5} < \frac{5}{4}$; $2(5 - x) > 3 + 1$

infinite, nonrepeating decimal irrational numbers, such as π and $\sqrt{2}$, that are decimals with digits that continue indefinitely but do not repeat

inscribed figure a figure that is enclosed by another figure as shown below

Examples:

a triangle *inscribed* in a circle

a circle *inscribed* in a triangle

integers the set of all whole numbers and their additive inverses $\{\ldots -5, -4, -3, -2, -1, 0, 1, 2, 3, 4, 5 \ldots\}$

intercept [1] the cutting of a line, curve, or surface by another line, curve, or surface; [2] the point at which a line or curve cuts across a given axis *see 6•8 Slope and Intercept*

intersection the set of elements that belong to each of two overlapping sets *see 5•3 Sets*

Example:

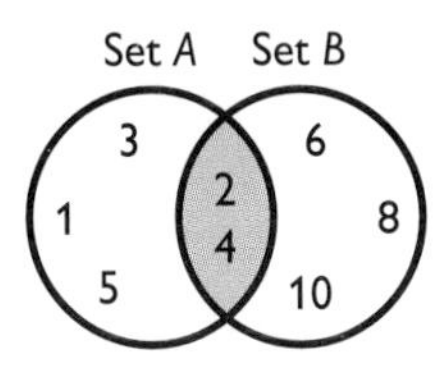

the shaded area is the *intersection* of set A (numbers 1 through 5) and set B (even numbers to 10)

inverse negation of the *if* idea and the *then* idea of a conditional statement *see 5•1 If/Then Statements*

inverse operations operations that undo each other

Examples: Addition and subtraction are inverse operations: $5 + 4 = 9$ and $9 - 4 = 5$. Multiplication and division are inverse operations: $5 \times 4 = 20$ and $20 \div 4 = 5$.

irrational numbers the set of all numbers that cannot be expressed as finite or repeating decimals

Examples: $\sqrt{2}$ (1.414214 . . .) and π (3.141592 . . .) are *irrational numbers*

isometric drawing a two-dimensional representation of a three-dimensional object in which parallel edges are drawn as parallel lines

Example:

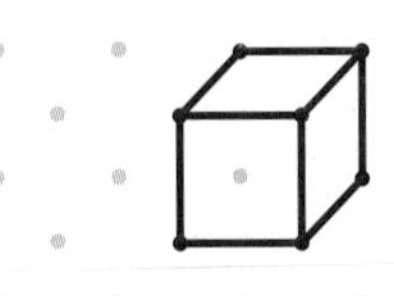

isosceles trapezoid a trapezoid in which the two nonparallel sides are of equal length

Example:

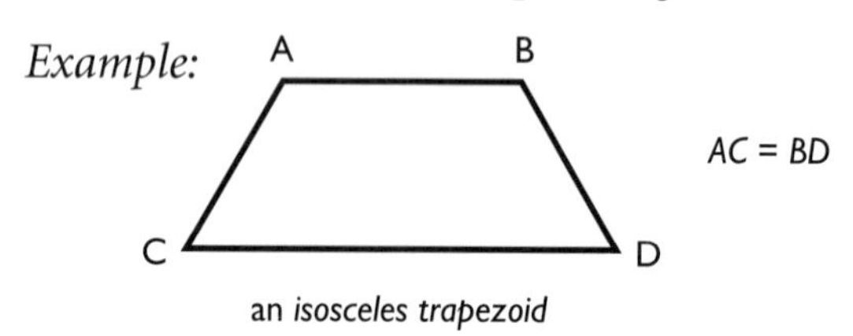

an *isosceles trapezoid*

isosceles triangle a triangle with at least two sides of equal length

Example:

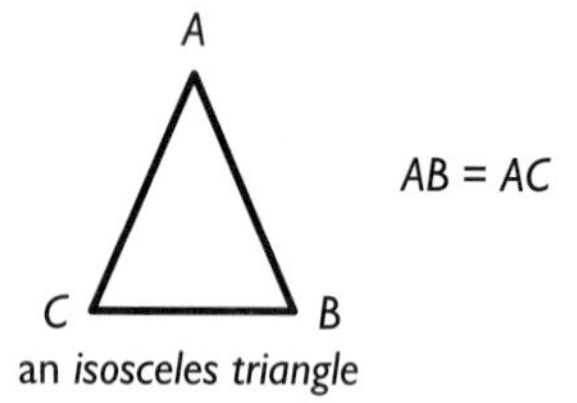

an *isosceles triangle*

hot words L

law of large numbers when you experiment by doing something over and over, you get closer and closer to what things "should" be theoretically. For example, when you repeatedly throw a die, the proportion of 1's that you throw will get closer to $\frac{1}{6}$ (which is the theoretical proportion of 1's in a batch of throws).

leaf the unit-digit of an item of numerical data between 1 and 99

least common denominator (LCD) the least common multiple of the denominators of two or more fractions

Example: 12 is the *least common denominator* of $\frac{1}{3}$, $\frac{2}{4}$, and $\frac{3}{6}$

least common multiple (LCM) the smallest nonzero whole number that is a multiple of two or more whole numbers

Example: the *least common multiple* of 3, 9, and 12 is 36

legs of a triangle the sides adjacent to the right angle of a right triangle

Example:

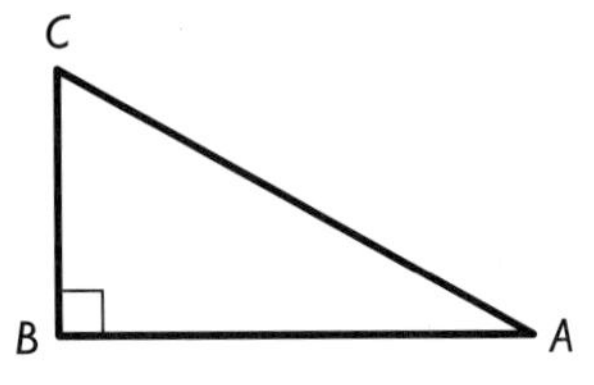

$\overline{AB}$ and $\overline{BC}$ are the *legs of triangle ABC*

length a measure of the distance of an object from end to end
see 8•2 Length and Distance

like terms terms that include the same variables raised to the same powers. *Like terms* can be combined.
see 6•2 Simplifying Expressions

Example: $5x^2$ and $6x^2$ are like terms; $3xy$ and $3zy$ are not like terms

likelihood the chance of a particular outcome occurring
see 4•6 Probability

line a connected set of points extending forever in both directions
see 7•1 Naming and Classifying Angles and Triangles

line graph data displayed visually to show change over time *see 4•2 Displaying Data*

Example:

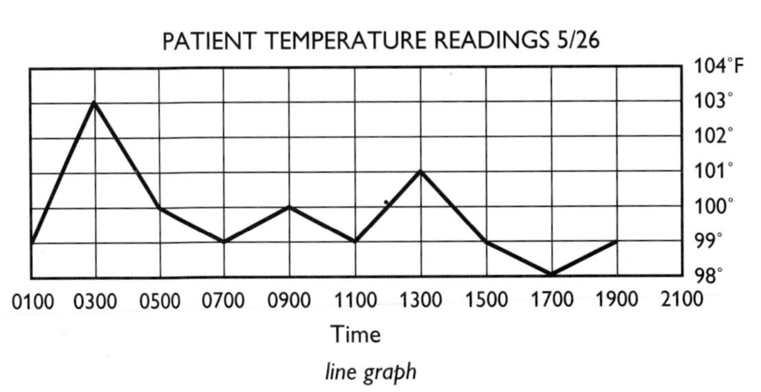

line graph

line of best fit on a scatter plot, a line drawn as near as possible to the various points so as to best represent the trend being graphed

Example:

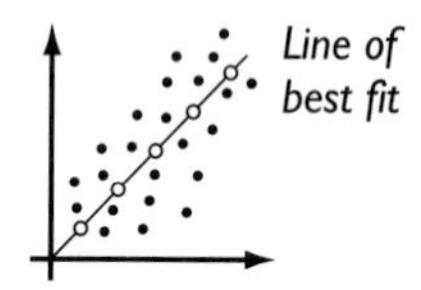

line of symmetry a line along which a figure can be folded so that the two resulting halves match

Example:

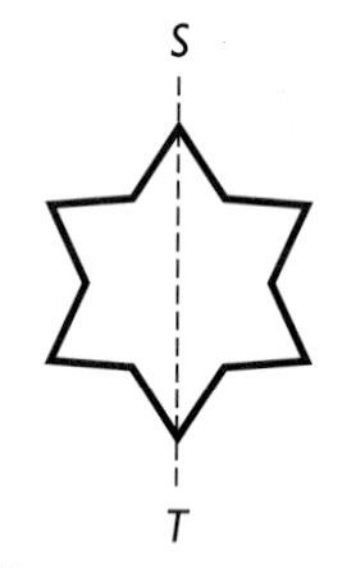

$\overleftrightarrow{ST}$ is a *line of symmetry*

line segment a section of a line running between two points

Example: A •———• B

$\overline{AB}$ is a *line segment*

linear equation an equation with two variables (x and y) that takes the general form $y = mx + b$, where m is the slope of the line and b is the y-intercept
see 6•4 Solving Linear Equations

linear measure the measure of the distance between two points on a line

liter a basic metric unit of capacity
see 8•3 Area, Volume, and Capacity

logic the mathematical principles that use existing theorems to prove new ones *see Chapter 5 Logic*

loss an amount of money that is lost

lowest common multiple the smallest number that is a multiple of all the numbers in a given set; same as least common multiple

Example: for 6, 9, and 18, 18 is the *lowest common multiple*

Lucas numbers *see page 66*

magic square *see page 66*

mathematical argument a series of logical steps a person might follow to determine whether a statement is correct

maximum value the greatest value of a function or a set of numbers

mean the quotient obtained when the sum of the numbers in a set is divided by the number of addends *see average, 4•4 Statistics*

Example: the *mean* of 3, 4, 7, and 10 is $(3 + 4 + 7 + 10) \div 4 = 6$

measurement units standard measures, such as the meter, the liter, and the gram, or the foot, the quart, and the pound *see 8•1 Systems of Measurement*

median the middle number in an ordered set of numbers *see 4•4 Statistics*

Example: 1, 3, 9, 16, 22, 25, 27
16 is the *median*

meter the basic metric unit of length

metric system a decimal system of weights and measurements based on the meter as its unit of length, the kilogram as its unit of mass, and the liter as its unit of capacity *see 8•1 Systems of Measurement*

midpoint the point on a line segment that divides it into two equal segments

Example:

A —— M —— B

$AM = MB$

M is the *midpoint* of $\overline{AB}$

minimum value the least value of a function or a set of numbers

mixed number a number composed of a whole number and a fraction *see 2•3 Addition and Subtraction of Fractions*

Example: $5\frac{1}{4}$

mode the number or element that occurs most frequently in a set of data *see 4•4 Statistics*

Example: 1, 1, 1, 2, 2, 3, 5, 5, 6, 6, 6, 6, 8
6 is the *mode*

monomial an algebraic expression consisting of a single term. $5x^3y$, xy, and $2y$ are three *monomials.*

multiple the product of a given number and an integer *see 1•4 Factors and Multiples*

Examples: 8 is a *multiple* of 4
3.6 is a *multiple* of 1.2

multiplication one of the four basic arithmetical operations, involving the repeated addition of numbers

multiplication growth number a number that when used to multiply a given number a given number of times results in a given goal number

Example: grow 10 into 40 in two steps by multiplying $(10 \times 2 \times 2 = 40)$
2 is the *multiplication growth number*

multiplicative inverse the number for any given number that will yield 1 when the two are multiplied, same as reciprocal

Example: $10 \times \frac{1}{10} = 1$
$\frac{1}{10}$ is the *multiplicative inverse* of 10

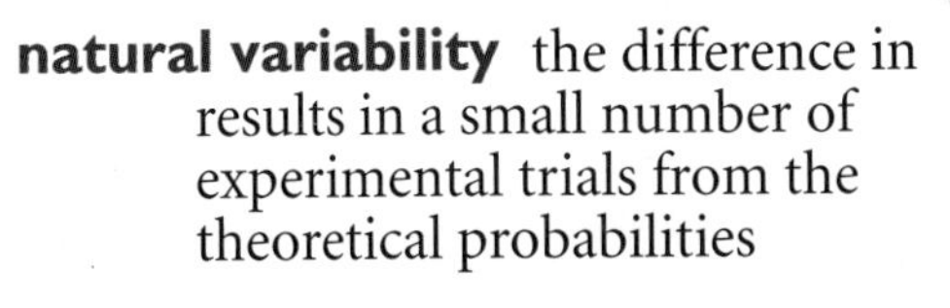

natural variability the difference in results in a small number of experimental trials from the theoretical probabilities

negative integers the set of all integers that are less than zero

Examples: $-1, -2, -3, -4, -5, \ldots$

negative numbers the set of all real numbers that are less than zero

Examples: $-1, -1.36, -\sqrt{2}, -\pi$

net a two-dimensional plan that can be folded to make a three-dimensional model of a solid

Example:

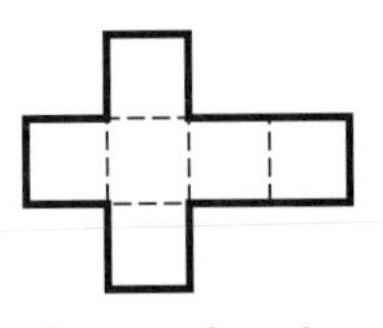

the *net* of a cube

nonagon a polygon that has nine sides

Example:

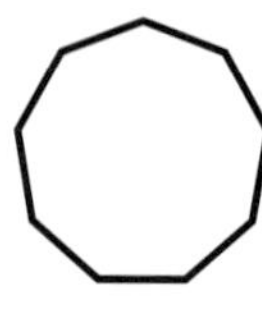

a *nonagon*

noncollinear not lying on the same straight line

noncoplanar not lying on the same plane

normal distribution represented by a bell curve, the most common distribution of most qualities across a given population
see 4•3 Analyzing Data

Example:

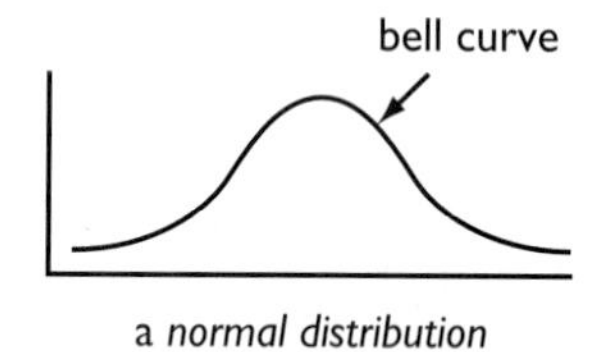

a *normal distribution*

number line a line showing numbers at regular intervals on which any real number can be indicated

Example:

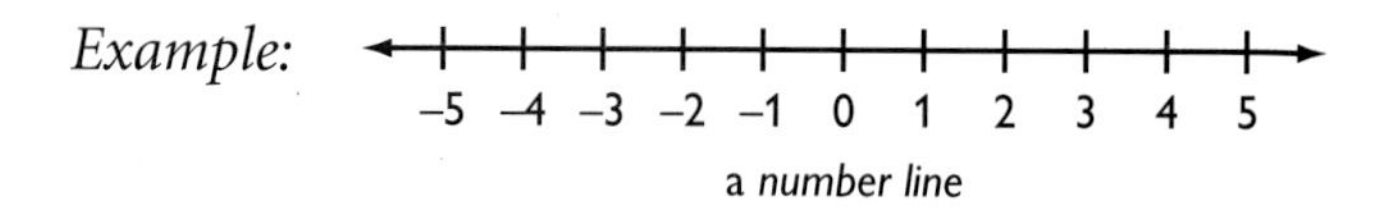

a *number line*

number symbols the symbols used in counting and measuring

Examples: $1, -\frac{1}{4}, 5, \sqrt{2}, -\pi$

number system a method of writing numbers. The Arabic *number system* is most commonly used today.

numerator the top number in a fraction. In the fraction $\frac{a}{b}$, a is the *numerator*. *see 2•1 Fractions and Equivalent Fractions*

obtuse angle any angle that measures more than 90° but less than 180°

Example:

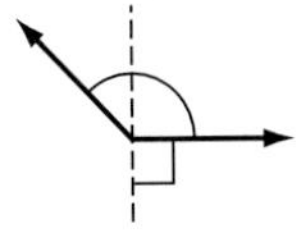

an *obtuse angle*

obtuse triangle a triangle that has one obtuse angle

Example:

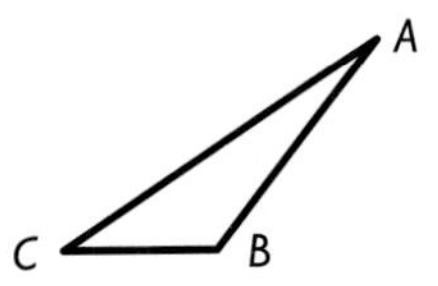

$\triangle ABC$ is an *obtuse triangle*

octagon a polygon that has eight sides

Example:

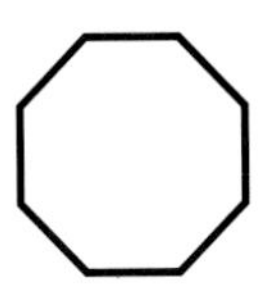

an *octagon*

octagonal prism a prism that has two octagonal bases and eight rectangular faces

Example:

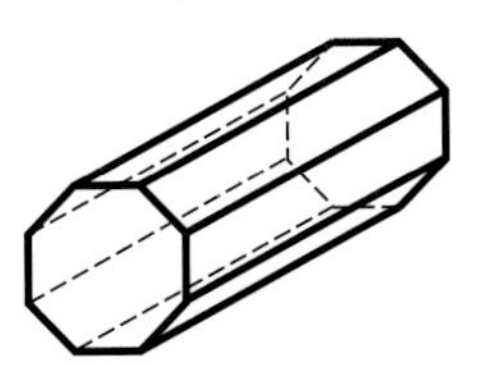

an *octagonal prism*

odd numbers the set of all integers that are not multiples of 2

odds against the ratio of the number of unfavorable outcomes to the number of favorable outcomes
see 4•6 Probability

odds for the ratio of the number of favorable outcomes to the number of unfavorable outcomes
see 4•6 Probability

one-dimensional having only one measurable quality

Example: a line and a curve are *one-dimensional*

operations arithmetical actions performed on numbers, matrices, or vectors

opposite angle in a triangle, a side and an angle are said to be opposite if the side is not used to form the angle

Example:

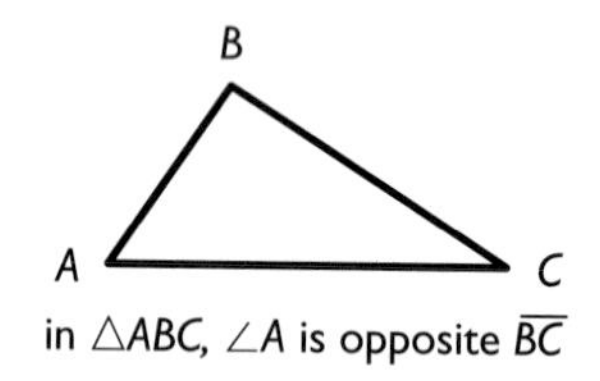

in $\triangle ABC$, $\angle A$ is opposite $\overline{BC}$

order of operations to find the answer to an equation, follow this four step process: 1) do all operations with parentheses first; 2) simplify all numbers with exponents; 3) multiply and divide in order from left to right; 4) add and subtract in order from left to right *see 1•3 Order of Operations*

ordered pair two numbers that tell the *x*-coordinate and *y*-coordinate of a point
see 6•7 Graphing on the Coordinate Plane

Example: The coordinates (3, 4) are an *ordered pair*. The *x*-coordinate is 3, and the *y*-coordinate is 4.

origin the point (0, 0) on a coordinate graph where the *x*-axis and the *y*-axis intersect

orthogonal drawing always shows three views of an object—top, side, and front. The views are drawn straight-on.

Example:

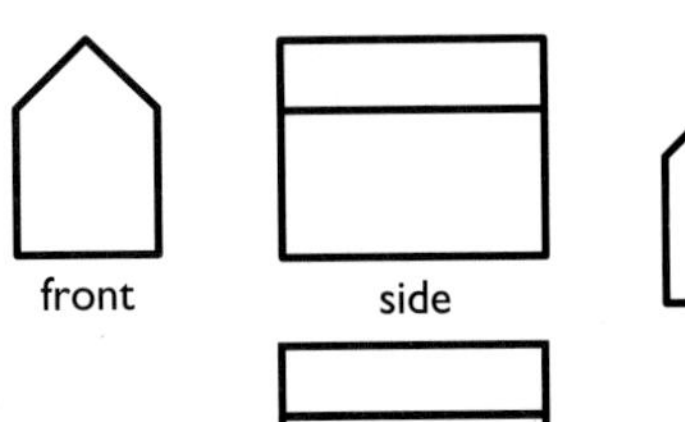

outcome a possible result in a probability experiment

outcome grid a visual model for analyzing and representing theoretical probabilities that shows all the possible outcomes of two independent events *see 4•6 Probability*

Example:

A grid used to find the sample space for rolling a pair of dice. The outcomes are written as ordered pairs.

	1	2	3	4	5	6
1	(1, 1)	(2, 1)	(3, 1)	(4, 1)	(5, 1)	(6, 1)
2	(1, 2)	(2, 2)	(3, 2)	(4, 2)	(5, 2)	(6, 2)
3	(1, 3)	(2, 3)	(3, 3)	(4, 3)	(5, 3)	(6, 3)
4	(1, 4)	(2, 4)	(3, 4)	(4, 4)	(5, 4)	(6, 4)
5	(1, 5)	(2, 5)	(3, 5)	(4, 5)	(5, 5)	(6, 5)
6	(1, 6)	(2, 6)	(3, 6)	(4, 6)	(5, 6)	(6, 6)

There are 36 possible outcomes.

hot words P

parabola the curve formed by a quadratic equation such as $y = x^2$

Example:

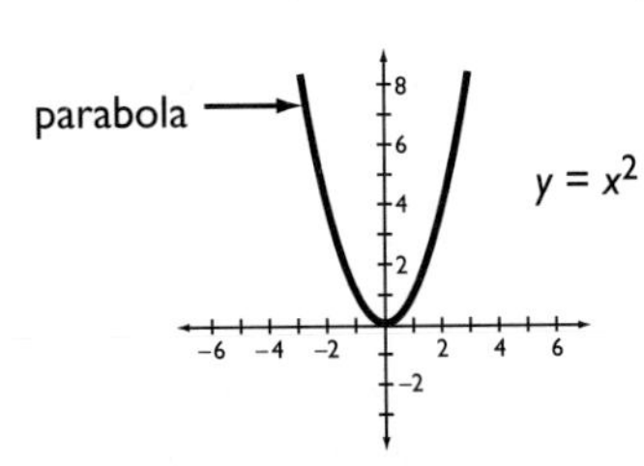

HOT WORDS

parallel straight lines or planes that remain a constant distance from each other and never intersect, represented by the symbol ||

Example:

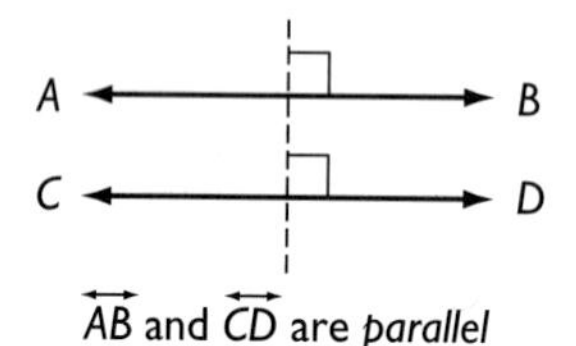

$\overleftrightarrow{AB}$ and $\overleftrightarrow{CD}$ are *parallel*

parallelogram a quadrilateral with two pairs of parallel sides

Example:

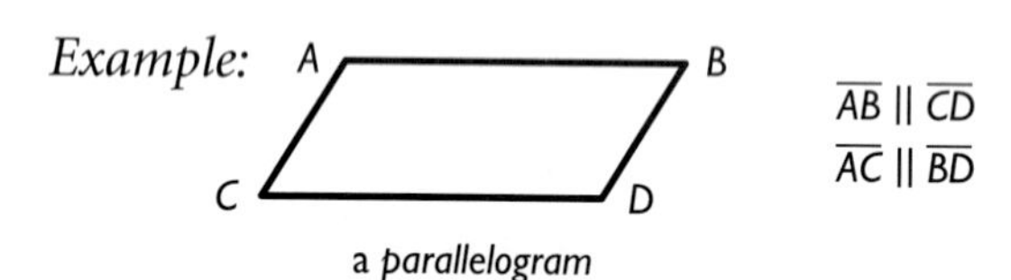

a *parallelogram*

parentheses the enclosing symbols (), which indicate that the terms within are a unit; for example, $(2 + 4) \div 2 = 3$

pattern a regular, repeating design or sequence of shapes or numbers *see Patterns, pages 65–67*

PEMDAS a reminder for the order of operations: 1) do all operations within **p**arentheses first; 2) simplify all numbers with **e**xponents; 3) **m**ultiply and **d**ivide in order from left to right; 4) **a**dd and **s**ubtract in order from left to right

pentagon a polygon that has five sides

Example:

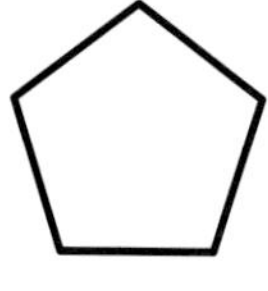

a *pentagon*

percent a number expressed in relation to 100, represented by the symbol % *see 2•7 Meaning of Percent*

Example: 76 out of 100 students use computers
76 *percent* of students use computers

percent grade the ratio of the rise to the run of a hill, ramp, or incline expressed as a percent

Example:

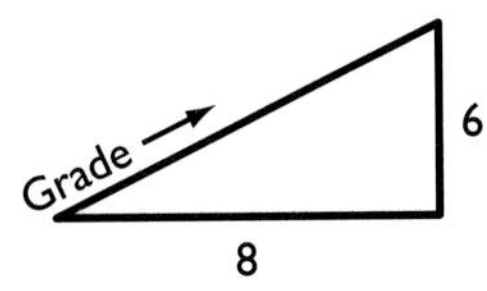

percent grade = 75% ($\frac{6}{8}$)

perfect cube a number that is the cube of an integer. For example, 27 is a *perfect cube* since $27 = 3^3$.

perfect number an integer that is equal to the sum of all its positive whole number divisors, excluding the number itself

Example: $1 \times 2 \times 3 = 6$ and $1 + 2 + 3 = 6$
6 is a *perfect number*

perfect square a number that is the square of an integer. For example, 25 is a *perfect square* since $25 = 5^2$.

perimeter the distance around the outside of a closed figure
see 7•4 Perimeter

Example:

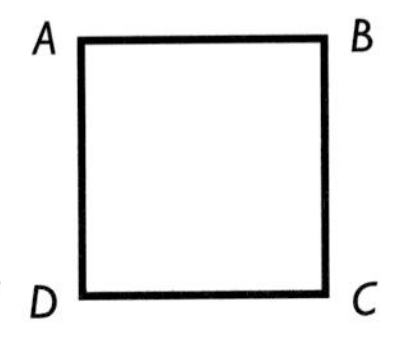

AB + BC + CD + DA = *perimeter*

permutation a possible arrangement of a group of objects. The number of possible arrangements of *n* objects is expressed by the term n!
see factorial, 4•5 Combinations and Permutations

perpendicular two lines or planes that intersect to form a right angle

Example:

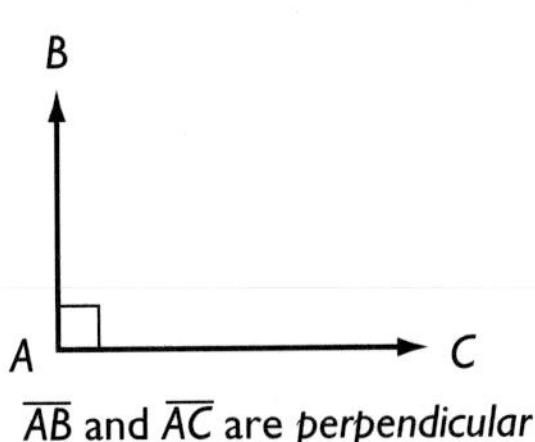

$\overline{AB}$ and $\overline{AC}$ are *perpendicular*

HOT WORDS

pi the ratio of a circle's circumference to its diameter. *Pi* is shown by the symbol π, and is approximately equal to 3.14. *see 7•8 Circles*

picture graph a graph that uses pictures or symbols to represent numbers

place value the value given to a place a digit may occupy in a numeral *see 1•1 Place Value of Whole Numbers*

place-value system a number system in which values are given to the places digits occupy in the numeral. In the decimal system, the value of each place is 10 times the value of the place to its right.
see 1•1 Place Value of Whole Numbers

point one of four undefined terms in geometry used to define all other terms. A *point* has no size.
see 6•7 Graphing on the Coordinate Plane

polygon a simple, closed plane figure, having three or more line segments as sides
see 7•2 Naming and Classifying Polygons and Polyhedrons

Examples:

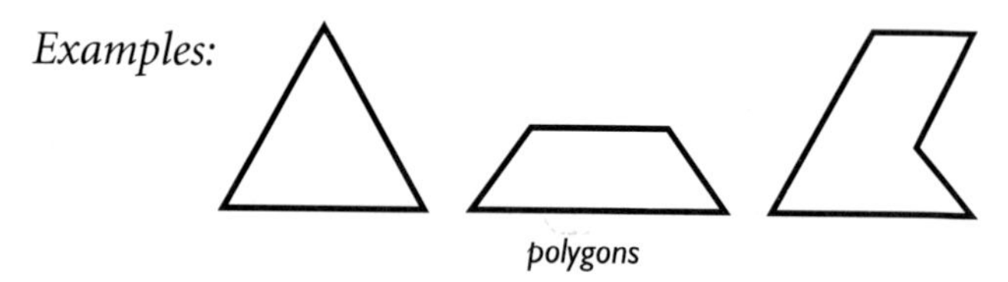

polygons

polyhedron a solid geometrical figure that has four or more plane faces
see 7•2 Naming and Classifying Polygons and Polyhedrons

Examples:

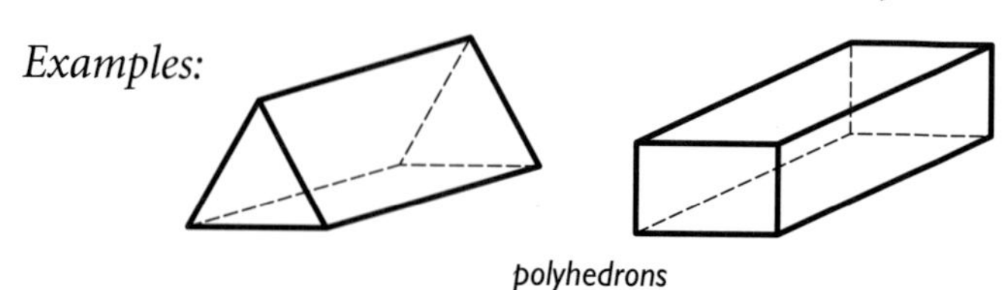

polyhedrons

population the universal set from which a sample of statistical data is selected

positive integers the set of all positive whole numbers $\{1, 2, 3, 4, 5, \ldots\}$ *see counting numbers*

positive numbers the set of all numbers that are greater than zero

Examples: $1, 1.36, \sqrt{2}, \pi$

power represented by the exponent *n*, to which a number is raised by multiplying itself *n* times *see 3•1 Powers and Exponents*

Example: 7 raised to the fourth *power*
$7^4 = 7 \times 7 \times 7 \times 7 = 2{,}401$

predict to anticipate a trend by studying statistical data *see trend, 4•3 Analyzing Data*

price the amount of money or goods asked for or given in exchange for something else

prime factorization the expression of a composite number as a product of its prime factors *see 1•4 Factors and Multiples*

Examples: $504 = 2^3 \times 3^2 \times 7$
$30 = 2 \times 3 \times 5$

prime number a whole number greater than 1 whose only factors are 1 and itself *see 1•4 Factors and Multiples*

Examples: 2, 3, 5, 7, 11

prism a solid figure that has two parallel, congruent polygonal faces (called *bases*) *see 7•2 Naming and Classifying Polygons and Polyhedrons*

Examples:

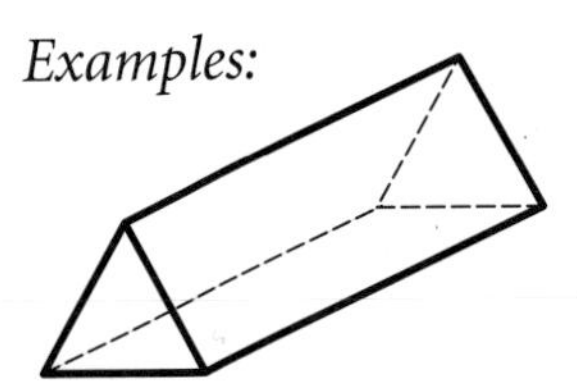

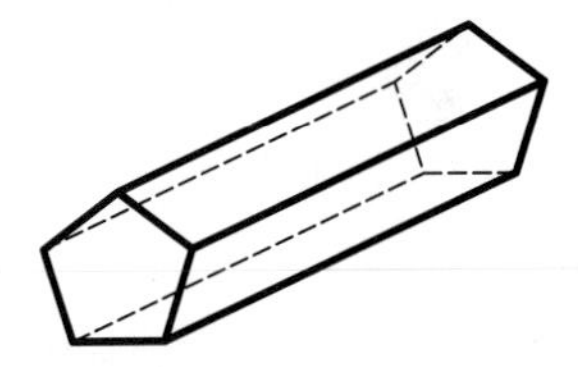

prisms

HOT WORDS

probability the study of likelihood or chance that describes the chances of an event occurring *see 4•6 Probability*

probability line a line used to order events from least likely to most likely to happen *see 4•6 Probability*

probability of events the likelihood or chance that events will occur

product the result obtained by multiplying two numbers or variables

profit the gain from a business; what is left when the cost of goods and of carrying on the business is subtracted from the amount of money taken in

project (v.) to extend a numerical model, to either greater or lesser values, in order to guess likely quantities in an unknown situation

proportion a statement that two ratios are equal *see 6•5 Ratio and Proportion*

pyramid a solid geometrical figure that has a polygonal base and triangular faces that meet at a common vertex *see 7•2 Naming and Classifying Polygons and Polyhedrons*

Examples:

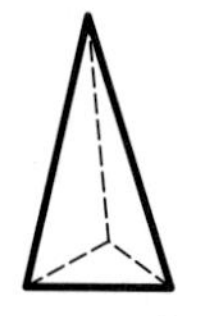

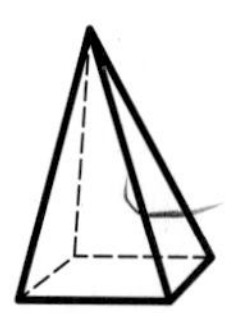

pyramids

Pythagorean Theorem a mathematical idea stating that the sum of the squared lengths of the two shorter sides of a right triangle is equal to the squared length of the hypotenuse *see 7•9 Pythagorean Theorem*

Example:

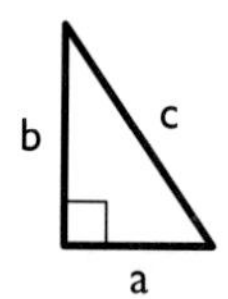

for a right triangle, $a^2 + b^2 = c^2$

Pythagorean triple a set of three positive integers a, b, and c, such that $a^2 + b^2 = c^2$

Example: for the *Pythagorean triple* {3, 4, 5}
$3^2 + 4^2 = 5^2$
$9 + 16 = 25$

quadrant [1] one quarter of the circumference of a circle; [2] on a coordinate graph, one of the four regions created by the intersection of the x-axis and the y-axis *see 6•7 Graphing on the Coordinate Plane*

quadratic equation a polynomial equation of the second degree, generally expressed as $ax^2 + bx + c = 0$, where a, b, and c are real numbers and a is not equal to zero *see degree*

quadrilateral a polygon that has four sides *see 7•2 Naming and Classifying Polygons and Polyhedrons*

Examples:

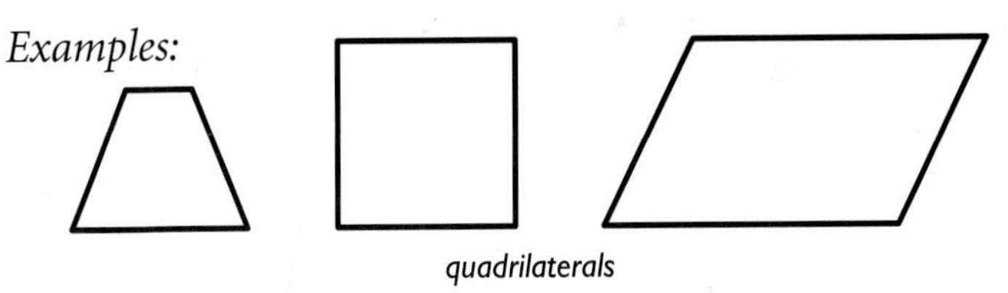

quadrilaterals

qualitative graphs a graph with words that describes such things as a general trend of profits, income, and expenses over time. It has no specific numbers.

quantitative graphs a graph that, in contrast to a qualitative graph, has specific numbers

quotient the result obtained from dividing one number or variable (the divisor) into another number or variable (the dividend)

Example: $24 \div 4 = 6$
dividend (24), divisor (4), quotient (6)

radical the indicated root of a quantity *see 3•2 Square and Cube Roots*

Examples: $\sqrt{3}$, $\sqrt[4]{14}$, $\sqrt[12]{-23}$

radical sign the root symbol $\sqrt{\ }$

radius a line segment from the center of a circle to any point on its circumference *see 7•8 Circles*

random sample a population sample chosen so that each member has the same probability of being selected *see 4•1 Collecting Data*

range in statistics, the difference between the largest and smallest values in a sample *see 4•4 Statistics*

rank to order the data from a statistical sample on the basis of some criterion—for example, in ascending or descending numerical order *see 4•4 Statistics*

ranking the position on a list of data from a statistical sample based on some criterion

rate [1] fixed ratio between two things; [2] a comparison of two different kinds of units, for example, miles per hour or dollars per hour *see 6•5 Ratio and Proportion*

ratio a comparison of two numbers *see 6•5 Ratio and Proportion*

Example: the *ratio* of consonants to vowels in the alphabet is 21:5

rational numbers the set of numbers that can be written in the form $\frac{a}{b}$, where *a* and *b* are integers and *b* does not equal zero

Examples: $1 = \frac{1}{1}$, $\frac{2}{9}$, $3\frac{2}{7} = \frac{23}{7}$, $-.333 = -\frac{1}{3}$

ray the part of a straight line that extends infinitely in one direction from a fixed point
see 7•1 Naming and Classifying Angles and Triangles

Example:

a *ray*

real numbers the set consisting of zero, all positive numbers, and all negative numbers. *Real numbers* include all rational and irrational numbers.

real-world data information processed by real people in everyday situations

reciprocal the result of dividing a given quantity into 1 *see 2•4 Multiplication and Division of Fractions*

Examples: the *reciprocal* of 2 is $\frac{1}{2}$; of $\frac{3}{4}$ is $\frac{4}{3}$; of x is $\frac{1}{x}$

rectangle a parallelogram with four right angles *see 7•2 Naming and Classifying Polygons and Polyhedrons*

Example:

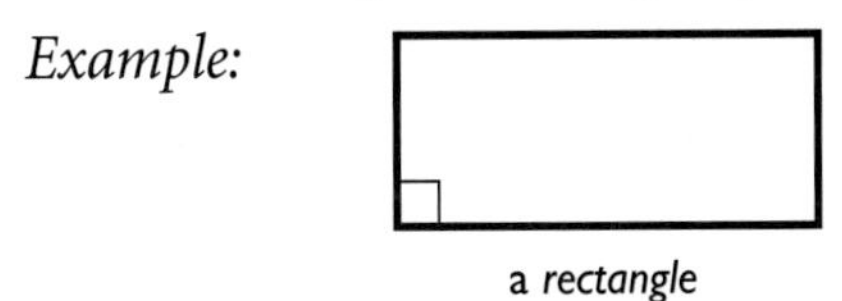

a *rectangle*

rectangular prism a prism that has rectangular bases and four rectangular faces *see 7•2 Naming and Classifying Polygons and Polyhedrons*

reflection *see flip, 7•3 Symmetry and Transformations*

Example:

the *reflection* of a trapezoid

reflex angle any angle whose measure is greater than 180° but less than 360°

Example:

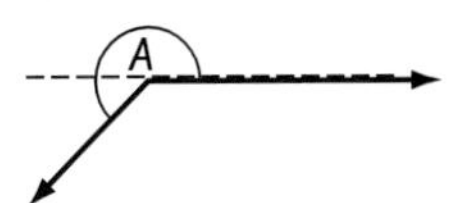

A is a *reflex angle*

regular polygon a polygon in which all sides are equal and all angles are equal

regular shape a figure in which all sides are equal and all angles are equal

relationship a connection between two or more objects, numbers, or sets. A mathematical *relationship* can be expressed in words or with numbers and letters.

repeating decimal a decimal in which a digit or a set of digits repeat infinitely

Example: 0.121212 . . .

rhombus a parallelogram with all sides of equal length
see 7•2 Naming and Classifying Polygons and Polyhedrons

Example:

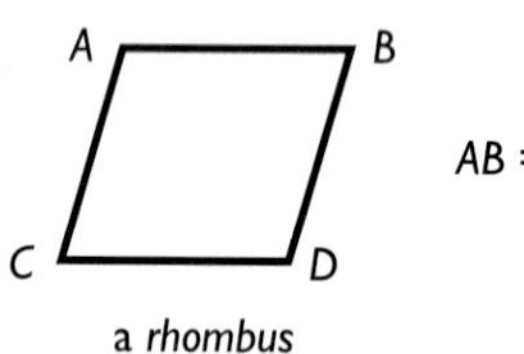

a *rhombus*

right angle an angle that measures 90°

Example:

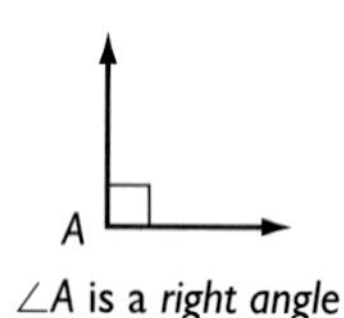

∠A is a *right angle*

right triangle a triangle with one right angle
see 7•1 Naming and Classifying Angles and Triangles

Example:

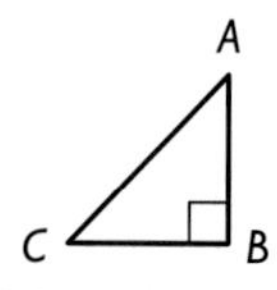

△ABC is a *right triangle*

rise the amount of vertical increase between two points
see 6•8 Slope and Intercept

Roman numerals the numeral system consisting of the symbols I (1), V (5), X (10), L (50), C (100), D (500), and

M (1,000). When a Roman symbol is preceded by a symbol of equal or greater value, the values of a symbol are added (XVI = 16). When a symbol is preceded by a symbol of lesser value, the values are subtracted (IV = 4).

root [1] the inverse of an exponent; [2] the radical sign $\sqrt{}$ indicates square root
see 3•2 Square and Cube Roots

rotation a transformation in which a figure is turned a certain number of degrees around a fixed point or line
see turn, 7•3 Symmetry and Transformations

Example:

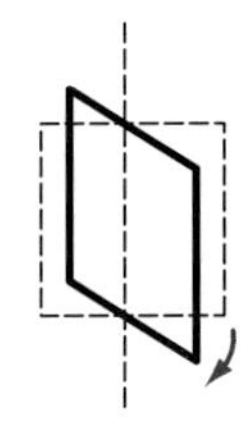

rotation of a square

round to approximate the value of a number to a given decimal place

Examples: 2.56 rounded to the nearest tenth is 2.6;
2.54 rounded to the nearest tenth is 2.5;
365 rounded to the nearest hundred is 400

row a horizontal list of numbers or terms. In spreadsheets, the labels of cells in a *row* all end with the same number (A3, B3, C3, D3 . . .) *see 9•4 Spreadsheets*

rule a statement that describes a relationship between numbers or objects

run the horizontal distance between two points
see 6•8 Slope and Intercept

sample a finite subset of a population, used for statistical analysis *see 4•6 Probability*

sampling with replacement a sample chosen so that each element has the chance of being selected more than once *see 4•6 Probability*

Example: A card is drawn from a deck, placed back into the deck, and a second card is drawn. Since the first card is replaced, the number of cards remains constant.

scale the ratio between the actual size of an object and a proportional representation *see 8•6 Size and Scale*

scale drawing a proportionally correct drawing of an object or area at actual, enlarged, or reduced size *see 8•6 Size and Scale*

scale factor the factor by which all the components of an object are multiplied in order to create a proportional enlargement or reduction *see 8•6 Size and Scale*

scale size the proportional size of an enlarged or reduced representation of an object or area *see 8•6 Size and Scale*

scalene triangle a triangle with no sides of equal length

Example:

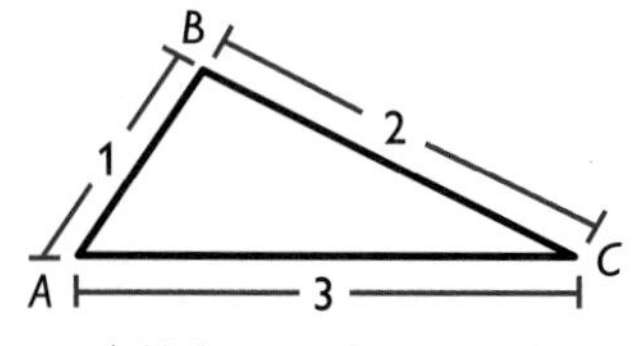

$\triangle ABC$ is a *scalene triangle*

scatter plot (or scatter diagram) a two-dimensional graph in which the points corresponding to two related factors (for example, smoking and life expectancy) are graphed and observed for correlation *see 4•3 Analyzing Data*

Example:

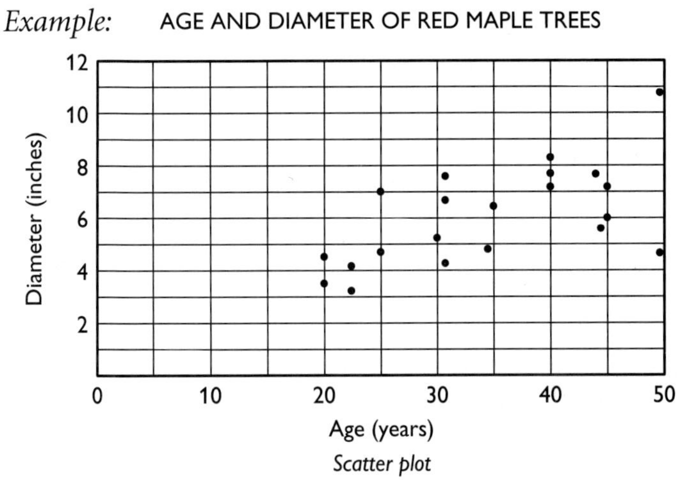

Scatter plot

scientific notation a system of writing numbers using exponents and powers of ten. A number in scientific notation is written as a number between 1 and 10 multiplied by a power of ten. *see 3•3 Scientific Notation*

Examples: $9{,}572 = 9.572 \times 10^3$ and $0.00042 = 4.2 \times 10^{-4}$

segment two points and all the points on the line between them *see 7•1 Naming and Classifying Angles and Triangles*

sequence *see page 66*

series *see page 66*

side a line segment that forms an angle or joins the vertices of a polygon *see 7•4 Perimeter*

sighting measuring a length or angle of an inaccessible object by lining up a measuring tool with one's line of vision

signed number a number preceded by a positive or negative sign *see 1•5 Integer Operations*

significant digit the digit in a number that indicates its precise magnitude

Example: 297,624 rounded to three significant digits is 298,000; 2.97624 rounded to three significant digits is 2.98

similar figures have the same shape but are not necessarily the same size *see 8•6 Size and Scale*

Example:

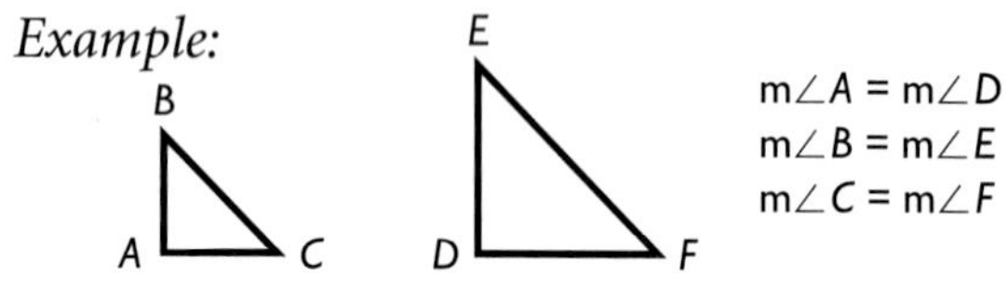

triangles ABC and DEF are *similar figures*

similarity *see similar figures*

simulation a mathematical experiment that approximates real-world process

single-bar graph a way of displaying related data using one horizontal or vertical bar to represent each data item *see 4•2 Displaying Data*

skewed distribution an asymmetrical distribution curve representing statistical data that is not balanced around the mean *see 4•3 Analyzing Data*

Example:

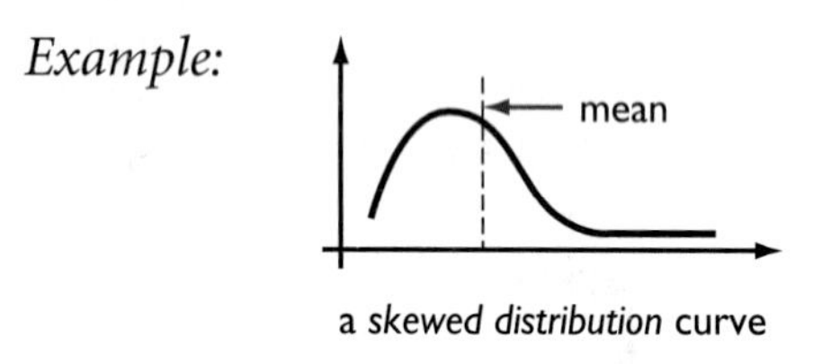

a *skewed distribution* curve

slide to move a shape to another position without rotating or reflecting it
see translation, 7•3 Symmetry and Transformations

Example:

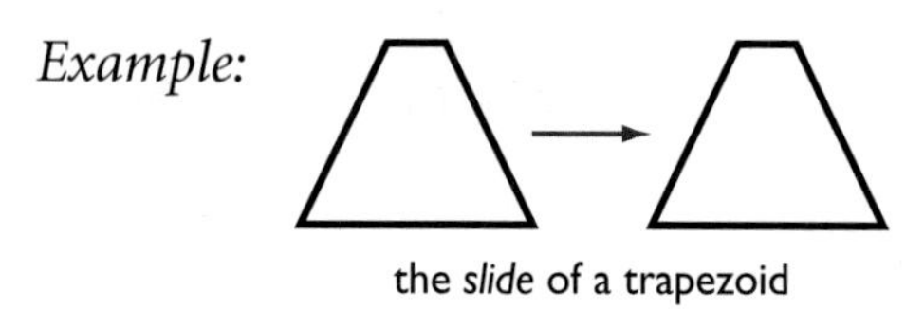

the *slide* of a trapezoid

slope [1] a way of describing the steepness of a line, ramp, hill, and so on; [2] the ratio of the rise to the run
see 6•8 Slope and Intercept

slope angle the angle that a line forms with the x-axis or other horizontal

slope ratio the slope of a line as a ratio of the rise to the run

solid a three-dimensional shape

solution the answer to a mathematical problem. In algebra, a *solution* usually consists of a value or set of values for a variable.

special cases a number or set of numbers, such as 0, 1, fractions and negative numbers, that is considered when determining whether or not a rule is always true

speed the rate at which an object moves

speed-time graph a graph used to chart how the speed of an object changes over time

sphere a perfectly round geometric solid, consisting of a set of points equidistant from a center point

Example:

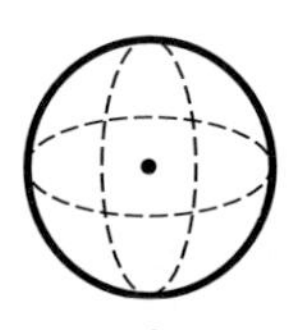

a *sphere*

spinner a device for determining outcomes in a probability experiment

Example:

a *spinner*

spiral *see page 67*

spreadsheet a computer tool where information is arranged into cells within a grid and calculations are performed within the cells. When one cell is changed, all other cells that depend on it automatically change.
see 9•4 Spreadsheets

square a rectangle with congruent sides *see 7•2 Naming and Classifying Polygons and Polyhedrons*

Example:

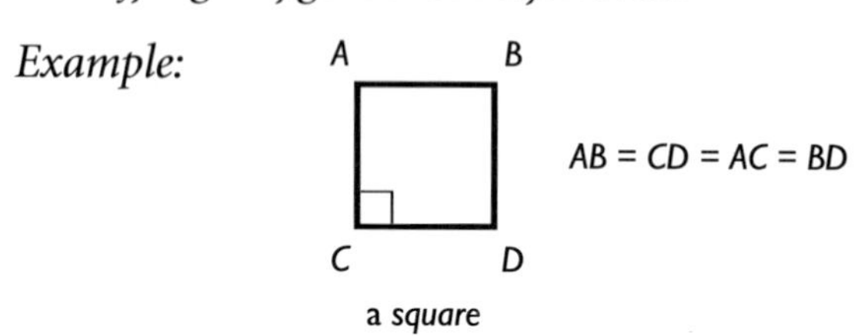

a square

square to multiply a number by itself; shown by the exponent 2 *see exponent, see 3•1 Powers and Exponents*

Example: $4^2 = 4 \times 4 = 16$

square centimeter a unit used to measure the size of a surface; the equivalent of a square measuring one centimeter on each side *see 8•3 Area, Volume, and Capacity*

square foot a unit used to measure the size of a surface; the equivalent of a square measuring one foot on each side *see 8•3 Area, Volume, and Capacity*

square inch a unit used to measure the size of a surface; the equivalent of a square measuring one inch on each side *see 8•3 Area, Volume, and Capacity*

square meter a unit used to measure the size of a surface; the equivalent of a square measuring one meter on each side *see 8•3 Area, Volume, and Capacity*

square number *see page 67*

Examples: 1, 4, 9, 16, 25, 36

square pyramid a pyramid with a square base

square root a number that when multiplied by itself produces a given number. For example, 3 is the *square root* of 9. *see 3•2 Square and Cube Roots*

Example: $3 \times 3 = 9$; $\sqrt{9} = 3$

square root sign the mathematical sign $\sqrt{\ }$; indicates that the square root of a given number is to be calculated *see 3•2 Square and Cube Roots*

standard measurement commonly used measurements, such as the meter used to measure length, the kilogram used to measure mass, and the second used to measure time *see Chapter 8 Measurement*

statistics the branch of mathematics concerning the collection and analysis of data *see 4•4 Statistics*

steepness a way of describing the amount of incline (or slope) of a ramp, hill, line, and so on

stem the ten-digit of an item of numerical data between 1 and 99 *see 4•2 Displaying Data*

stem-and-leaf plot a method of presenting numerical data between 1 and 99 by separating each number into its ten-digit (stem) and its unit-digit (leaf) and then arranging the data in ascending order of the ten-digits *see 4•2 Displaying Data*

Example:

stem	leaf
0	6
1	1 8 2 2 5
2	6 1
3	7
4	3
5	8

a *stem-and-leaf plot* for the data set 11, 26, 18, 12, 12, 15, 43, 37, 58, 6, and 21

straight angle an angle that measures 180°; a straight line

stratified random sampling a series of random samplings, each of which is taken from a specific part of the population. For example, a two-part sampling might involve taking separate samples of men and women.

strip graph a graph indicating the sequence of outcomes. A *strip graph* helps to highlight the differences among individual results and provides a strong visual representation of the concept of randomness.

Example: Outcomes of a coin toss
H = heads
T = tails

a *strip graph*

HOT WORDS

subtraction one of the four basic arithmetical operations, taking one number or quantity away from another

sum the result of adding two numbers or quantities

Example: $6 + 4 = 10$
10 is the *sum* of the two addends, 6 and 4

surface area the sum of the areas of all the faces of a geometric solid, measured in square units
see 7•6 Surface Area

Example:

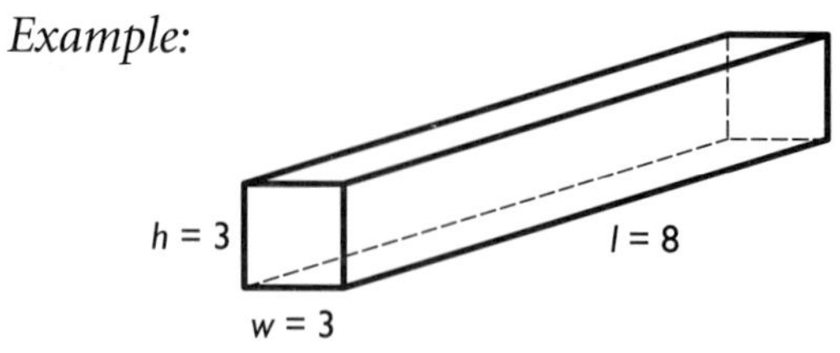

the *surface area* of this rectangular prism is
$2(3 \times 3) + 4(3 \times 8) = 114$ square units

survey a method of collecting statistical data in which people are asked to answer questions *see 4•1 Collecting Data*

symmetry *see line of symmetry*

Example:

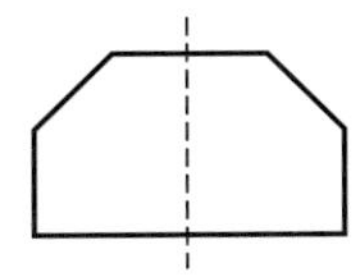

this hexagon has *symmetry* around the dotted line

table a collection of data arranged so that information can be easily seen *see 4•2 Displaying Data*

tally marks marks made for certain numbers of objects in keeping account. For example, 𝍸 /// = 8

tangent [1] a line that intersects a circle in exactly one point; [2] The *tangent* of an acute angle in a right triangle is the ratio of the length of the opposite side to the length of the adjacent side *see tangent ratio*

Example:

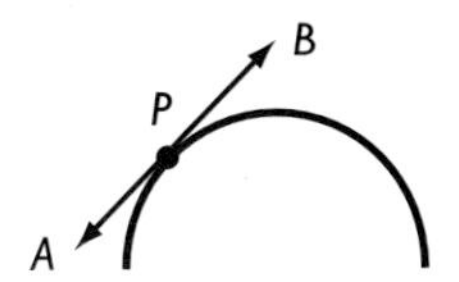

$\overleftrightarrow{AB}$ is *tangent* to the curve at point *P*

tangent ratio the ratio of the length of the side opposite a right triangle's acute angle to the length of the side adjacent to it *see 7•10 Tangent Ratio*

Example:

$$\tan S = \frac{\text{length of the side opposite to } \angle S}{\text{length of the side adjacent to } \angle S}$$

$\tan S = \frac{3}{4}$ or 0.75

tangent ratio of *S* is $\frac{3}{4}$ or 0.75

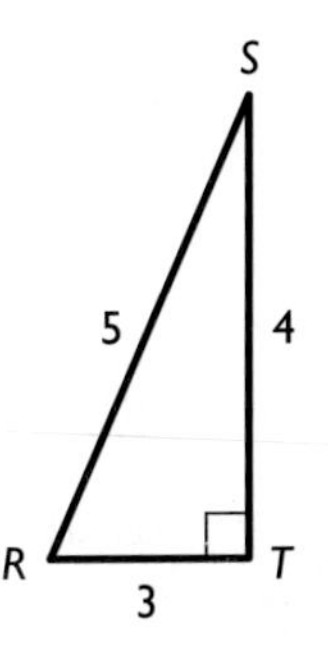

HOT WORDS

term product of numbers and variables;
x, ax^2, $2x^4y^2$, and $-4ab$ are four examples of a *term*

terminating decimal a decimal with a finite number of digits

tessellation *see page 67*

Examples:

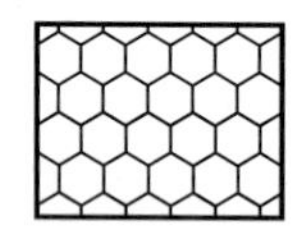

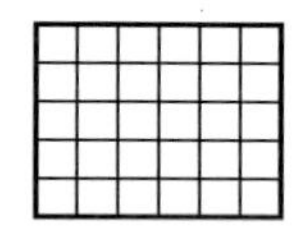

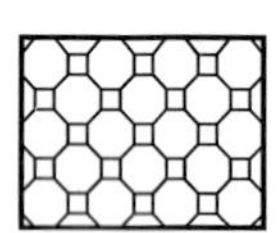

tessellations

tetrahedron a geometrical solid that has four triangular faces
see 7•2 Naming and Classifying Polygons and Polyhedrons

Example:

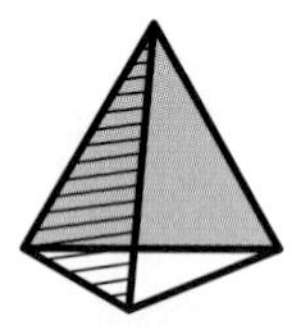

a tetrahedron

theoretical probability the ratio of the number of favorable outcomes to the total number of possible outcomes
see 4•6 Probability

three-dimensional having three measurable qualities: length, height, and width

tiling completely covering a plane with geometric shapes
see tessellation page 67

time in mathematics, the element of duration, usually represented by the variable t *see 8•5 Time*

total distance the amount of space between a starting point and an endpoint, represented by d in the equation $d = s$ (speed) $\times$ t (time)

total distance graph a coordinate graph that shows cumulative distance traveled as a function of time

total time the duration of an event, represented by t in the equation $t = d$ (distance) / s (speed)

HOT WORDS

transformation a mathematical process that changes the shape or position of a geometric figure *see reflection, rotation, translation, 7•3 Symmetry and Transformations*

translation a transformation in which a geometric figure is slid to another position without rotation or reflection *see slide, 7•3 Symmetry and Transformations*

trapezoid a quadrilateral with only one pair of parallel sides *see 7•2 Naming and Classifying Polygons and Polyhedrons*

Example:

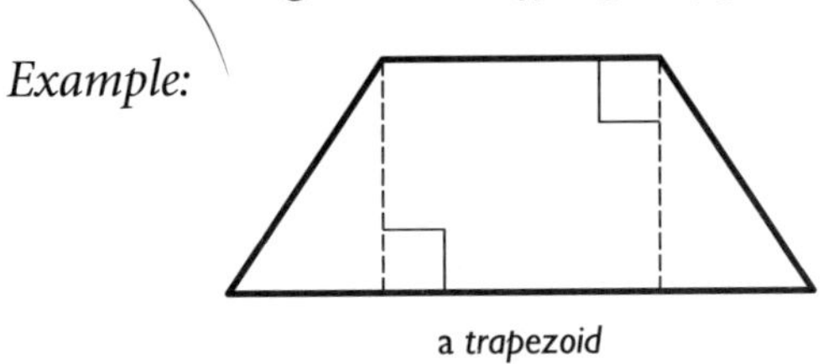

a *trapezoid*

tread the horizontal depth of one step on a stairway

tree diagram a connected, branching graph used to diagram probabilities or factors *see 1•4 Factors and Multiples, 4•5 Combinations and Permutations*

Example:

a *tree diagram*

trend a consistent change over time in the statistical data representing a particular population

triangle a polygon that has three sides *see 7•1 Naming and Classifying Angles and Triangles*

triangular numbers *see page 67*

triangular prism a prism with two triangular bases and three rectangular sides *see prism*

turn to move a geometric figure by rotating it around a point *see rotation, 7•3 Symmetry and Transformations*

Example:

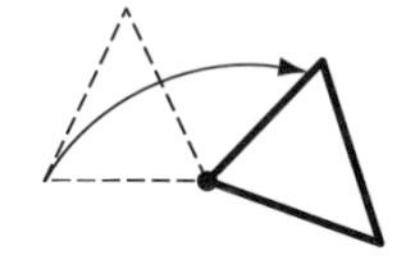

the *turning* of a triangle

two-dimensional having two measurable qualities: length and width

hot words U

unequal probabilities different likelihoods of occurrence. Two events have *unequal probabilities* if one is more likely to occur than the other.

unfair where the probability of each outcome is not equal

union a set that is formed by combining the members of two or more sets, as represented by the symbol ∪. The *union* contains all members previously contained in either set *see 5•3 Sets*

Example:

Set *A*: 3 6 9 12

Set *B*: 4 8 12 16

Set *A*∪B: 3 4 6 8 9 12 16

the *union* of sets *A* and *B*

unit price the price of an item expressed in a standard measure, such as *per ounce* or *per pint* or *each*

unit rate the rate in lowest terms

Example: 120 miles in two hours is equivalent to a *unit rate* of 60 miles per hour

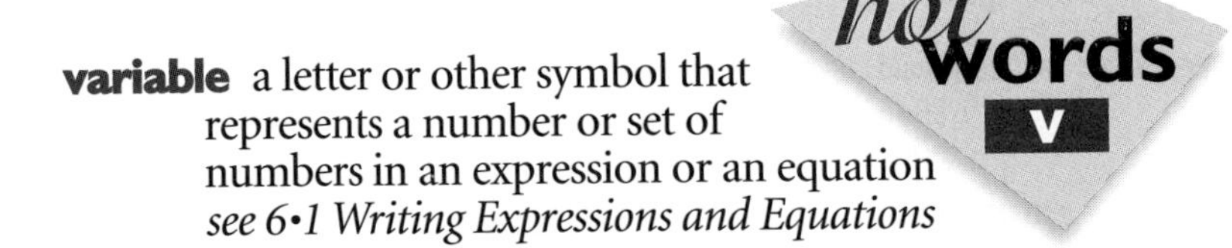

variable a letter or other symbol that represents a number or set of numbers in an expression or an equation *see 6•1 Writing Expressions and Equations*

Example: in the equation $x + 2 = 7$, the variable is x

variation a relationship between two variables. Direct variation, represented by the equation $y = kx$, exists when the increase in the value of one variable results in an increase in the value of the other. Inverse variation, represented by the equation $y = \frac{k}{x}$, exists when an increase in the value of one variable results in a decrease in the value of the other.

Venn diagram a pictorial means of representing the relationships between sets

Example:

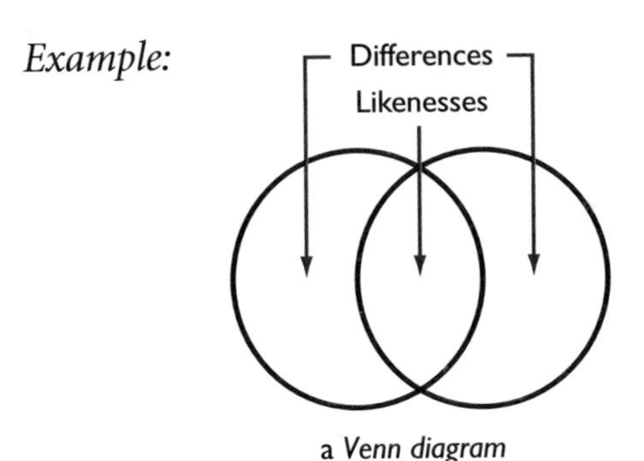

a *Venn diagram*

vertex (pl. *vertices*) the common point of two rays of an angle, two sides of a polygon, or three or more faces of a polyhedron

Examples:

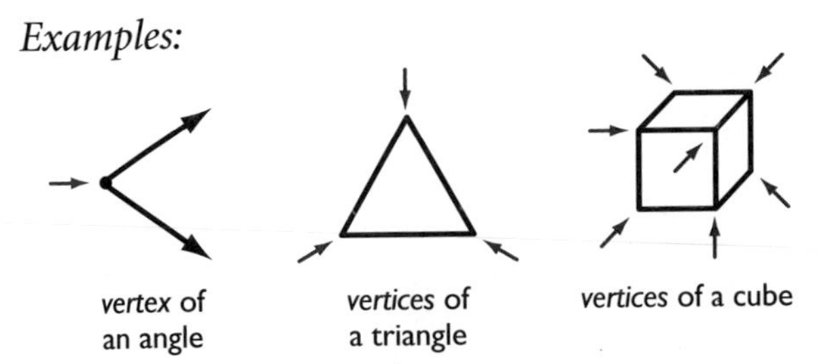

vertex of an angle

vertices of a triangle

vertices of a cube

vertex of tessellation the point where three or more tessellating figures come together

Example:

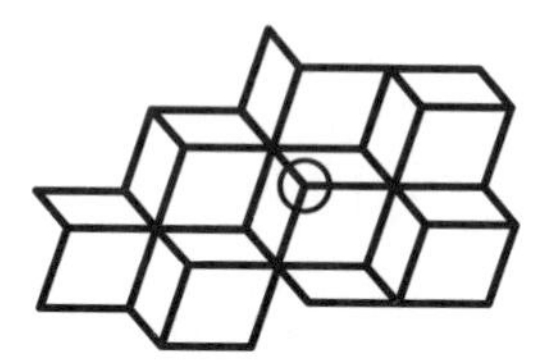

vertex of tessellation
(in the circle)

HOT WORDS

vertical a line that is perpendicular to a horizontal base line

Example:

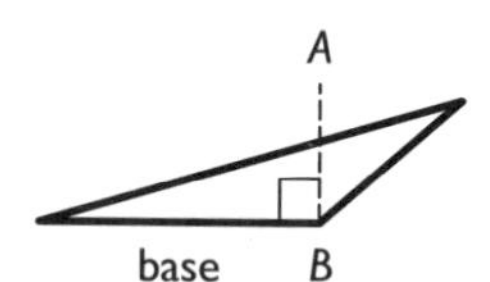

$\overline{AB}$ is *vertical* to the base of this triangle

volume the space occupied by a solid, measured in cubic units
see 7•7 Volume

Example:

$h = 2$
$l = 5$
$w = 3$

the *volume* of this rectangular prism is 30 cubic units
$2 \times 3 \times 5 = 30$

hot words
W

weighted average a statistical average in which each element in the sample is given a certain relative importance, or weight. For example, to find the accurate average percentage of people who own cars in three towns with different-sized populations, the largest town's percentage would have to be *weighted.*
see 4•4 Statistics

what-if questions a question posed to frame, guide, and extend a problem

whole numbers the set of all counting numbers plus zero

Examples: 0, 1, 2, 3, 4, 5

width a measure of the distance of an object from side to side

x-axis the horizontal reference line in the coordinate graph *see 6•7 Graphing on the Coordinate Plane*

x-intercept the point at which a line or curve cuts across the x-axis

y-axis the vertical reference line in the coordinate graph

y-intercept the point at which a line or curve cuts across the y-axis *see 6•8 Slope and Intercept*

zero-pair one positive cube and one negative cube used to model signed number arithmetic

Formulas

Area (see 7•5)

circle $A = \pi r^2$ (pi × square of the radius)

parallelogram $A = bh$ (base × height)

rectangle $A = lw$ (length × width)

square $A = s^2$ (side squared)

trapezoid $A = \frac{1}{2}h(b_1 + b_2)$
($\frac{1}{2}$ × height × sum of the bases)

triangle $A = \frac{1}{2}bh$ ($\frac{1}{2}$ × base × height)

Volume (see 7•7)

cone $V = \frac{1}{3}\pi r^2 h$
($\frac{1}{3}$ × pi × square of the radius × height)

cylinder $V = \pi r^2 h$
(pi × square of the radius × height)

prism $V = Bh$ (area of the base × height)

pyramid $V = \frac{1}{3}Bh$ ($\frac{1}{3}$ × area of the base × height)

rectangular prism $V = lwh$ (length × width × height)

sphere $V = \frac{4}{3}\pi r^3$ ($\frac{4}{3}$ × pi × cube of the radius)

Perimeter (see 7•4)

parallelogram $P = 2a + 2b$ (2 × side a + 2 × side b)

rectangle $P = 2l + 2w$ (twice length + twice width)

square $P = 4s$ (4 × side)

triangle $P = a + b + c$ (side a + side b + side c)

Circumference (see 7•8)

circle $C = \pi d$ (pi × diameter)
or
$C = 2\pi r$ (2 × pi × radius)

Formulas

Probability (see 4•6)

The *Experimental Probability* of an event is equal to the total number of times a favorable outcome occurred, divided by the total number of times the experiment was done.

$$\text{Experimental Probability} = \frac{\textit{favorable outcomes that occurred}}{\textit{total number of experiments}}$$

The *Theoretical Probability* of an event is equal to the number of favorable outcomes, divided by the total number of possible outcomes.

$$\text{Theoretical Probability} = \frac{\textit{favorable outcomes}}{\textit{possible outcomes}}$$

Other

Distance	$d = rt$ (rate $\times$ time)
Interest	$i = prt$ (principle $\times$ rate $\times$ time)
PIE	Profit = Income − Expenses

Symbols

$\{\ \}$	set
$\emptyset$	the empty set
$\subseteq$	is a subset of
$\cup$	union
$\cap$	intersection
$>$	is greater than
$<$	is less than
$\geq$	is greater than or equal to
$\leq$	is less than or equal to
$=$	is equal to
$\neq$	is not equal to
$^\circ$	degree
%	percent
$f(n)$	function, f of n
$a{:}b$	ratio of a to b, $\frac{a}{b}$
$\lvert a \rvert$	absolute value of a
$P(E)$	probability of an event E
π	pi
$\perp$	is perpendicular to
$\parallel$	is parallel to
$\cong$	is congruent to
$\sim$	is similar to
$\approx$	is approximately equal to
$\angle$	angle
$\llcorner$	right angle
$\triangle$	triangle
$\overline{AB}$	segment AB
$\overrightarrow{AB}$	ray AB
$\overleftrightarrow{AB}$	line AB
$\triangle ABC$	triangle ABC
$\angle ABC$	angle ABC
$m\angle ABC$	measure of angle ABC
AB or $m\overline{AB}$	length of segment AB
$\overset{\frown}{AB}$	arc AB
$!$	factorial
${}_nP_r$	permutations of n things taken r at a time
${}_nC_r$	combinations of n things taken r at a time
$\sqrt{\ }$	square root
$\sqrt[3]{\ }$	cube root
$'$	foot
$''$	inch
$\div$	divide
$/$	divide
$*$	multiply
$\times$	multiply
$\cdot$	multiply
$+$	add
$-$	subtract

Patterns

arithmetic sequence a sequence of numbers or terms that have a common difference between any one term and the next in the sequence. In the following sequence, the common difference is seven, so $8 - 1 = 7$; $15 - 8 = 7$; $22 - 15 = 7$, and so forth.

Example: 1, 8, 15, 22, 29, 36, 43, . . .

Fibonacci numbers a sequence in which each number is the sum of its two predecessors. Can be expressed as $x_n = x_{n-2} + x_{n-1}$. The sequence begins: 1, 1, 2, 3, 5, 8, 13, 21, 34, 55, . . .

Example: 1, 1, 2, 3, 5, 8, 13, 21, 34, 55 . . .

1+1=2

1+2=3

2+3=5

3+5=8

geometric sequence a sequence of terms in which each term is a constant multiple, called the *common ratio,* of the one preceding it. For instance, in nature, the reproduction of many single-celled organisms is represented by a progression of cells splitting in two in a growth progression of 1, 2, 4, 8, 16, 32, . . ., which is a geometric sequence in which the common ratio is 2.

harmonic sequence a progression $a_1, a_2, a_3, \ldots$ for which the reciprocals of the terms, $\frac{1}{a_1}, \frac{1}{a_2}, \frac{1}{a_3}, \ldots$ form an arithmetic sequence. For instance, in most musical tones, the frequencies of the sound waves are integer multiples of the fundamental frequency.

Lucas numbers a sequence in which each number is the sum of its two predecessors. Can be expressed as $x_n = x_{n-2} + x_{n-1}$

The sequence begins: 1, 3, 4, 7, 11, 18, 29, 47, . . .

magic square a square array of different numbers in which rows, columns, and diagonals add up to the same total

Example:

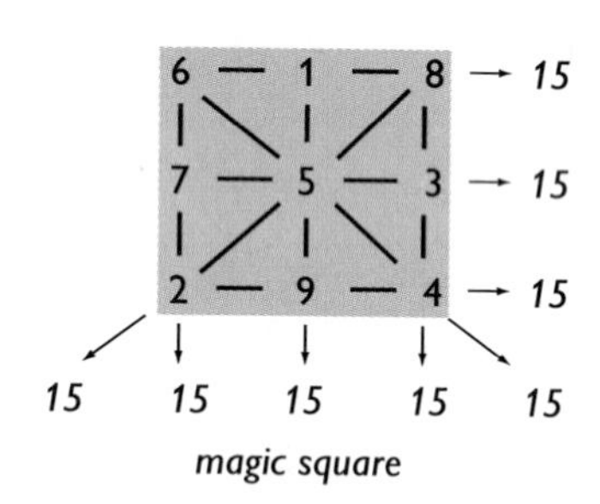

magic square

PATTERNS

Pascal's triangle a triangular arrangement of numbers. Blaise Pascal (1623–1662) developed techniques for applying this arithmetic triangle to probability patterns.

Example:

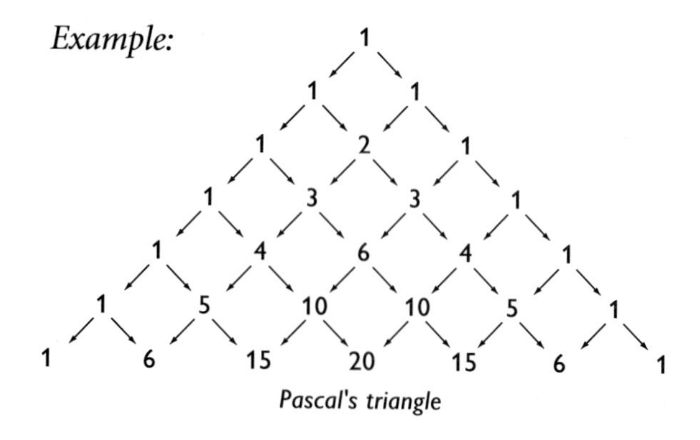

Pascal's triangle

sequence a set of elements, especially numbers, arranged in order according to some rule

series the sum of the terms of a sequence

spiral a plane curve traced by a point moving around a fixed point while continuously increasing or decreasing its distance from it

Example:

the shape of a chambered nautilus shell is a *spiral*

square numbers a sequence of numbers that can be shown by dots arranged in the shape of a square. Can be expressed as x^2. The sequence begins 1, 4, 9, 16, 25, 36, 49, . . .

Example:

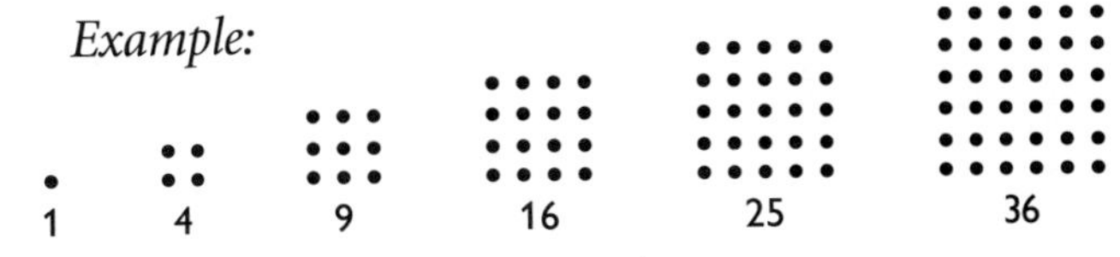

square numbers

tessellation a tiling pattern made of repeating polygons that fills a plane completely, leaving no gaps

Example:

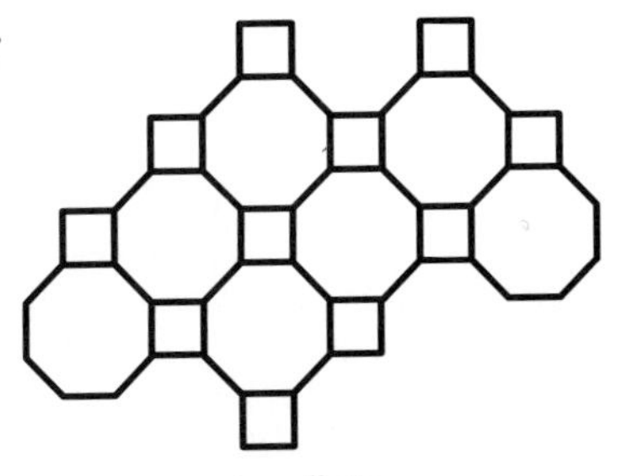

tessellation

triangular numbers a sequence of numbers that can be shown by dots arranged in the shape of a triangle. Any number in the sequence can be expressed as $x_n = x_{n-1} + n$. The sequence begins 1, 3, 6, 10, 15, 21, . . .

Example:

triangular numbers

PART TWO

hot topics

1	Numbers and Computation	70
2	Fractions, Decimals, and Percents	98
3	Powers and Roots	164
4	Data, Statistics, and Probability	192
5	Logic	254
6	Algebra	272
7	Geometry	338
8	Measurement	404
9	Tools	430

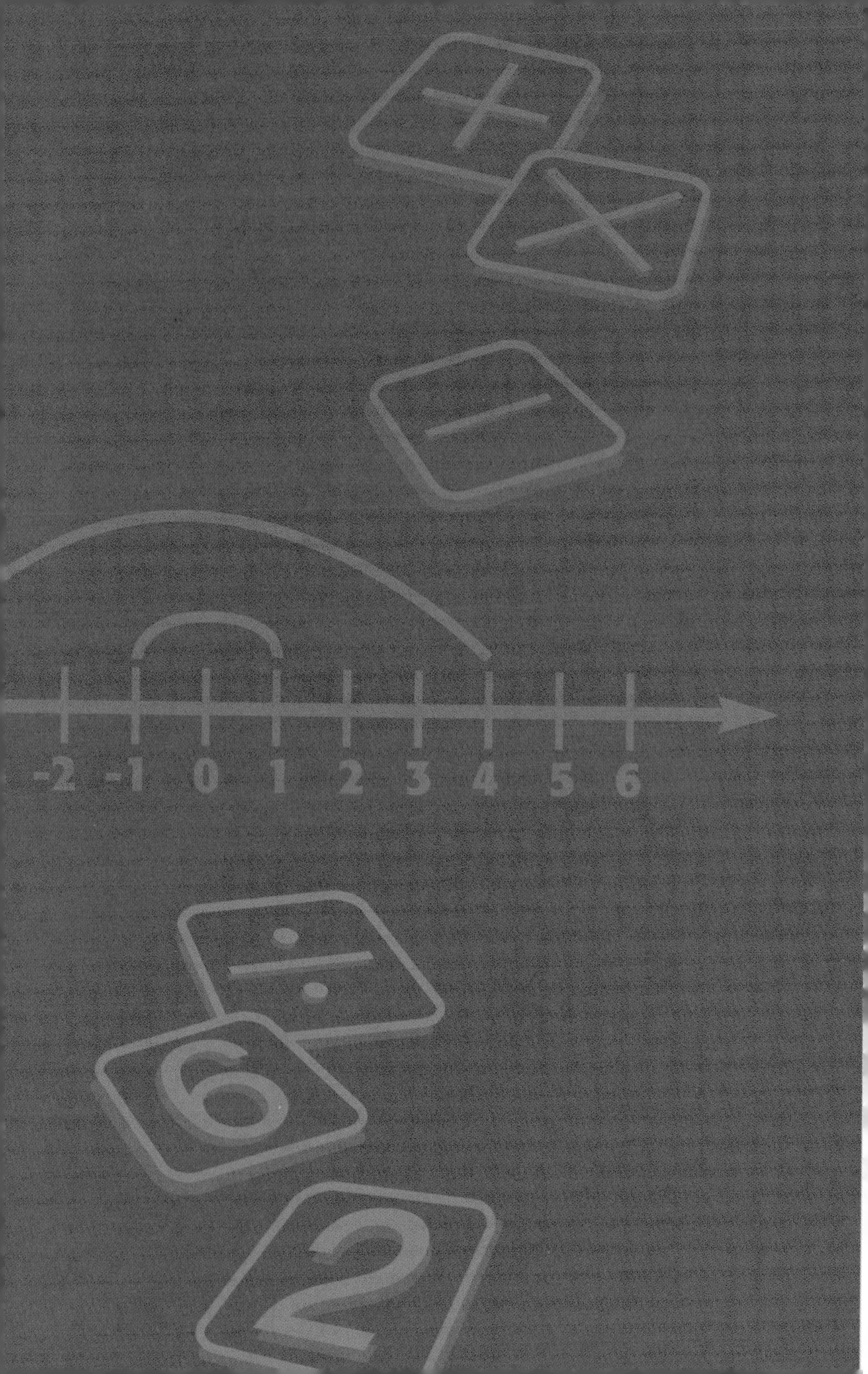
-2 -1 0 1 2 3 4 5 6
6
2

Numbers and Computation

1•1	Place Value of Whole Numbers	74
1•2	Properties	78
1•3	Order of Operations	82
1•4	Factors and Multiples	84
1•5	Integer Operations	92

What do you already know?

You can use the problems and the list of words that follow to see what you already know about this chapter. The answers to the problems are in Hot Solutions at the back of the book, and the definitions of the words are in Hot Words at the front of the book. You can find out more about a particular problem or word by referring to the boldfaced topic number (for example, **1•2**).

Problem Set

Give the value of the 6 in each number. **1•1**

1. 237,614
2. 765,134,987
3. Write 24,735, using expanded notation.
4. Write in order from greatest to least: 46,758; 406,758; 4,678; 396,758
5. Round 52,534,883 to the nearest ten, thousand, and million.

Solve. **1•2**

6. 236×0
7. $(4 \times 3) \times 1$
8. $5{,}889 + 0$
9. 1×0

Solve. Use mental math if you can. **1•2**

10. $6 \times (32 + 68)$
11. $25 \times 17 \times 4$

Use parentheses to make each expression true. **1•3**

12. $4 + 7 \times 3 = 33$
13. $30 + 15 \div 5 + 5 = 14$

Is it a prime number? Write Yes or No. **1•4**

14. 77
15. 111
16. 131
17. 301

Write the prime factorization for each. **1•4**

18. 40
19. 110
20. 230

Find the GCF for each pair. **1•4**

21. 12 and 40
22. 15 and 50
23. 18 and 171

Find the LCM for each pair. **1•4**

24. 5 and 12
25. 15 and 8
26. 18 and 30

27. A mystery number is a common multiple of 2, 4, and 15. It is also a factor of 120 but does not equal 120. What is the number? **1•4**

Give the absolute value of the integer. Then write its opposite. **1•5**

28. −7
29. 15
30. −12
31. 10

Add or subtract. **1•5**

32. $9 + (-7)$
33. $4 - 8$
34. $-5 + (-6)$
35. $8 - (-8)$
36. $-6 - (-6)$
37. $-3 + 9$

Compute. **1•5**

38. $-6 \times (-7)$
39. $48 \div (-12)$
40. $-56 \div (-8)$
41. $(-4 \times 3) \times (-2)$
42. $3 \times [-8 + (-4)]$
43. $-5\,[4 - (-6)]$

44. What can you say about the product of a negative integer and a positive integer? **1•5**
45. What can you say about the sum of two positive integers? **1•5**

CHAPTER 1

hot words

absolute value **1•5**
approximation **1•1**
associative property **1•2**
common factor **1•4**
commutative property **1•2**
composite number **1•4**
distributive property **1•2**
expanded notation **1•1**
exponent **1•4**
factor **1•4**
greatest common factor **1•4**
least common multiple **1•4**
multiple **1•4**
negative integer **1•5**
negative number **1•5**
number system **1•1**
operation **1•3**
PEMDAS **1•3**
place value **1•1**
positive integer **1•5**
prime factorization **1•4**
prime number **1•4**
round **1•1**

1·1 Place Value of Whole Numbers

Understanding Our Number System

Our **number system** is based on 10 and the value of each place is 10 times the value of the place to its right. The value of a digit is the product of that digit and its **place value.** For instance, in the number 6,400, the 6 has a value of six thousands and the 4 has a value of four hundreds.

A *place-value* chart can help you read numbers. In the chart, each group of three digits is called a *period.* Commas separate the periods. The chart below shows the speed of light, calculated at about 186,282 miles per second.

TRILLIONS PERIOD			BILLIONS PERIOD			MILLIONS PERIOD			THOUSANDS PERIOD			ONES PERIOD		
Hundred Trillions	Ten Trillions	One Trillions	Hundred Billions	Ten Billions	One Billions	Hundred Millions	Ten Millions	One Millions	Hundred Thousands	Ten Thousands	One Thousands	Hundreds	Tens	Ones
									1	8	6	2	8	2

To read a large number, think of the periods. At each comma, say the name of the period.

186,282 reads: one hundred eighty-six thousand, two hundred eighty-two.

Check It Out

Give the value of the 4 in each number.

1. 41,083
2. 824,000,297

Write each number in words.

3. 40,376,500
4. 57,320,100,000,000

Using Expanded Notation

To show the place values of the digits in a number, you can write the number using **expanded notation.**

Write 86,082 using expanded notation.

$86{,}082 = 80{,}000 + 6{,}000 + 80 + 2$

- Write the ten thousands. $(8 \times 10{,}000)$
- Write the thousands. $(6 \times 1{,}000)$
- Write the hundreds. (0×100)
- Write the tens. (8×10)
- Write the ones. (2×1)

So $86{,}082 = (8 \times 10{,}000) + (6 \times 1{,}000) + (8 \times 10) + (2 \times 1)$

Check It Out

Use expanded notation to write each number.

5. 98,025
6. 400,637

Comparing and Ordering Numbers

When you compare numbers, there are exactly three possibilities: the first number is greater than the second ($3 > 2$); the second is greater than the first ($3 < 4$); or the two numbers are equal ($5 = 5$).

When ordering several numbers, compare the numbers two at a time.

COMPARING AND ORDERING NUMBERS

Compare 54,186 and 52,998.

- Line up the digits, starting with the ones.

  54,186
  52,998

- Look at the digits in order, beginning at the left. Find the first place where they differ.

  The digits in the thousands place differ.

- The number with the greater digit is the greater.

Since $4 > 2$, 54,186 is greater than 52,998.

Check It Out

Write > , < , or =.

7. 438,297 ___ 439,366
8. 51,006 ___ 50,772

Write in order from least to greatest.

9. 77,302; 72,617; 7,520; 740,009

Using Approximations

For many situations, using an **approximation** makes sense. For instance, it is reasonable to use a rounded number to express population. You might say that the population of a place is "about 50,000" rather than saying it's "49,889."

You can use a rule to **round** numbers. Look at the digit to the right of the place you are rounding to. If the digit is 5 or greater, round up. If it is less than 5, round down.

Round 618,762 to the nearest thousand.

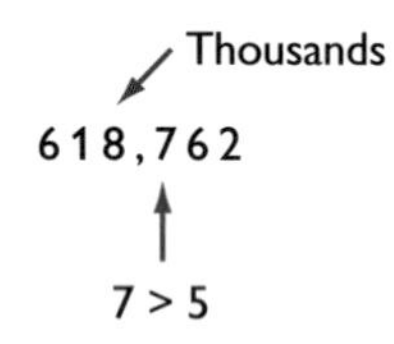

So 618,762 rounds to 619,000.

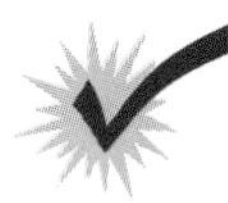

Check It Out

Round the following numbers.

10. Round 37,318 to the nearest hundred.
11. Round 488,225 to the nearest ten thousand.
12. Round 2,467,000 to the nearest million.
13. Round 764,335 to the nearest hundred thousand.

1•1 EXERCISES

Give the value of the 7 in each number.

1. 741,066
2. 624,010,297

Write each number in words.

3. 31,306,700
4. 5,020,460,000,000

Use expanded notation to write each number.

5. 37,024
6. 2,800,134

Write $>$, $<$, or $=$.

7. 638,297 ___ 638,366
8. 82,106 ___ 28,775

Write in order from least to greatest.

9. 57,388; 52,725; 15,752; 570,019

Round 68,253,522 to each place indicated.

10. nearest ten
11. nearest thousand
12. nearest hundred thousand
13. nearest ten million

Solve.

14. The first time around, a hit movie took in \$238,560,000. Ten years later it was re-released and it took in \$281,895,900. Did the movie earn more money or less money the second time around? How do you know?
15. When they rounded their sales to the nearest million, a music group found that they sold about 4,000,000 CDs. What is the greatest number of CDs they might have sold? What is the least number?

1·2 Properties

Commutative and Associative Properties

The operations of addition and multiplication share special properties because multiplication is repeated addition.

Both addition and multiplication are **commutative.** This means that order doesn't change the sum or the product. If we let a, b, and c be any whole numbers, then

$7 + 4 = 4 + 7$ and $7 \times 4 = 4 \times 7$
$a + b = b + a$ and $a \times b = b \times a$

Both addition and multiplication are **associative.** This means that grouping addends or factors will not change the sum or the product.

$(5 + 8) + 6 = 5 + (8 + 6)$ and $(5 \times 2) \times 4 = 5 \times (2 \times 4)$
$(a + b) + c = a + (b + c)$ and $(a \times b) \times c = a \times (b \times c)$

Subtraction and division do not share these properties. For example:

$7 - 4 = 3$, but $4 - 7 = -3$; therefore $7 - 4 \neq 4 - 7$
$7 \div 4 = 1.75$, but $4 \div 7$ is about 0.57;
therefore $7 \div 4 \neq 4 \div 7$

$(5 - 8) - 6 = -9$, but $5 - (8 - 6) = 3$;
therefore $(5 - 8) - 6 \neq 5 - (8 - 6)$
$(5 \div 2) \div 4 = 0.625$, but $5 \div (2 \div 4) = 10$;
therefore $(5 \div 2) \div 4 \neq 5 \div (2 \div 4)$

Check It Out

Write Yes or No.

1. $6 \times 4 = 4 \times 6$
2. $15 - 5 = 5 - 15$
3. $(12 \div 2) \div 4 = 12 \div (2 \div 4)$
4. $7 + (8 + 3) = (7 + 8) + 3$

Properties of One and Zero

When you add 0 to any number, the sum is that number. This is called the *zero (or identity) property of addition.* When you multiply any number by 1, the product is that number. This is called the *one (or identity) property of multiplication.* But the product of any number and 0 is 0. This is called the *zero property of multiplication.*

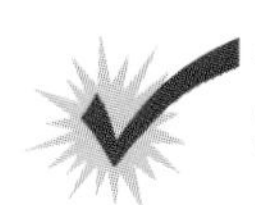

Check It Out

Solve.

5. $28{,}407 \times 1$
6. $299 + 0$
7. $8 \times (9 \times 0)$
8. $(6 \times 0.8) \times 1$

Distributive Property

The **distributive property** is important because it combines both addition and multiplication. It states that multiplying a sum by a number is the same as multiplying each addend by that number and then adding the two products.

$$3(8 + 2) = (3 \times 8) + (3 \times 2)$$

If we let *a, b,* and *c* be any whole numbers, then:

$$a(b + c) = ab + ac$$

Check It Out

Rewrite each expression, using the distributive property.

9. $3 \times (2 + 5)$
10. $(6 \times 8) + (6 \times 4)$

Shortcuts for Adding and Multiplying

Use the properties to help you perform some computations mentally.

$$77 + 56 + 23 = (77 + 23) + 56 = 100 + 56 = 156$$

Use commutative and associative properties.

$$4 \times 9 \times 25 = (4 \times 25) \times 9 = 100 \times 9 = 900$$

$$8 \times 340 = (8 \times 300) + (8 \times 40) = 2,400 + 320 = 2,720$$

Use distributive property.

Number Palindromes

Do you notice anything unusual about this word, name, or sentence?

noon Otto

Was it a can on a cat I saw?

Each one is a *palindrome*—a word, name, or sentence that reads the same forwards and backwards. It is easy to make up number palindromes using three or more digits (like 323 or 7227). But it is harder to make up a number sentence, such as 10989 x 9 = 98901, that is the same when you read its digits from either direction. Try it and see!

1·2 EXERCISES

Write Yes or No.

1. $7 \times 41 = 41 \times 7$
2. $3 \times 4 \times 5 = 3 \times 5 \times 4$
3. $6 \times 120 = (6 \times 100) + (6 \times 20)$
4. $m \times (n + p) = mn + mp$
5. $(4 \times 6 \times 5) = (4 \times 6) + (4 \times 5)$
6. $a \times (b + c + d) = ab + ac + ad$
7. $15 - 8 = 8 - 15$
8. $12 \div 3 = 3 \div 12$

Solve.

9. $32{,}450 \times 1$
10. $688 + 0$
11. $7 \times (0 \times 6)$
12. $0 \times 5 \times 12$
13. 0×1
14. $2.7 + 0$
15. 8.22×1
16. $(3 + 6 + 5) \times 1$

Rewrite each expression using the distributive property.

17. $4 \times (7 + 5)$
18. $(8 \times 13) + (8 \times 4)$
19. 6×250

Solve. Use mental math if you can.

20. $5 \times (54 + 6)$
21. $8 \times (26 + 74)$
22. 7×520
23. 15×12
24. $17 + 87 + 83$
25. $150 + 350 + 250$
26. 130×8
27. $12 \times 25 \times 4$
28. Give an example to show that subtraction is not commutative.
29. Give an example to show that division is not associative.
30. Think about the number properties. How would you describe the zero (or identity) property of subtraction?

1·3 Order of Operations

Understanding the Order of Operations

Solving a problem may involve using more than one **operation.** Your answer can depend on the order in which you do those operations.

For instance, take the expression $3^2 + 5 \times 7$.

$3^2 + 5 \times 7$ or $3^2 + 5 \times 7$

$9 + 5 \times 7$ $\quad$ $9 + 35 = 44$

$14 \times 7 = 98$

The order in which you perform operations really makes a difference.

To make sure that there is just one answer to a series of computations, mathematicians have agreed upon an order in which to do the operations.

USING THE ORDER OF OPERATIONS

How can you simplify $(4 + 5) \times 3^2 - 5$?

- Simplify within the parentheses. $= (9) \times 3^2 - 5$
- Evaluate the exponent. $= 9 \times (9) - 5$
- Multiply and divide from left to right. $= (81) - 5$
- Add and subtract from left to right. $= (76)$

So $(4 + 5) \times 3^2 - 5 = 76$

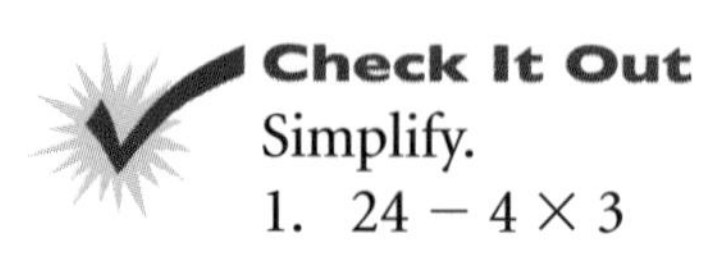

Check It Out

Simplify.

1. $24 - 4 \times 3$
2. $3 \times (4 + 5^2)$

1·3 EXERCISES

Is each expression true? Write Yes or No.

1. $7 \times 4 + 5 = 33$
2. $3 + 4 \times 8 = 56$
3. $6 \times (4 + 6 \div 2) = 30$
4. $4^2 - 1 = 9$
5. $(3 + 5)^2 = 64$
6. $(2^3 + 3 \times 4) + 5 = 49$
7. $25 - 4^2 = 9$
8. $(4^2 \div 2)^2 = 64$

Simplify.

9. $24 - (4 \times 5)$
10. $2 \times (6 + 5^2)$
11. $2^4 \times (12 - 8)$
12. $5^2 + (5 - 3)^2$
13. $(16 - 10)^2 \times 5$
14. $12 + 4 \times 3^2$
15. $(4^2 + 4)^2$
16. $60 \div (12 + 3)$
17. $30 - (10 - 7)^2$
18. $44 + 5 \times (4^2 \div 8)$

Use parentheses to make each expression true.

19. $5 + 5 \times 6 = 60$
20. $4 \times 25 + 75 = 400$
21. $36 \div 6 + 3 = 4$
22. $20 + 20 \div 4 - 4 = 21$
23. $12 \times 3^2 + 7 = 192$
24. $6^2 - 15 \div 3 \times 2^2 = 124$

25. Use five 2's, a set of parentheses (as needed), and any of the operations to make the numbers 1 through 5.

Parentheses
Exponents
Multiplication &
Division
Addition &
Subtraction

1•4 Factors and Multiples

Factors

Suppose that you want to arrange 12 small squares into a rectangular pattern.

$1 \times 12 = 12$

$2 \times 6 = 12$

$3 \times 4 = 12$

Two numbers multiplied by each other to produce 12 are considered **factors** of 12. So the factors of 12 are 1, 2, 3, 4, 6, and 12.

To decide whether one number is a factor of another, divide. If there is a remainder of 0, the number is a factor.

FINDING THE FACTORS OF A NUMBER

What are the factors of 18?

- Find all pairs of numbers that multiply to give the product.

  $1 \times 18 = 18$ $2 \times 9 = 18$ $3 \times 6 = 18$

- List the factors in order, starting with 1.

The factors of 18 are 1, 2, 3, 6, 9, and 18.

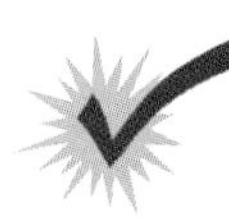

Check It Out

Find the factors of each number.

1. 8
2. 48

Common Factors

Factors that are the same for two or more numbers are **common factors.**

FINDING COMMON FACTORS

What numbers are factors of both 12 and 40?

- List the factors of the first number.

  1, 2, 3, 4, 6, 12
- List the factors of the second number.

  1, 2, 4, 5, 8, 10, 20, 40
- Common factors are the numbers that are in both lists.

  1, 2, 4

The common factors of 12 and 40 are 1, 2, and 4.

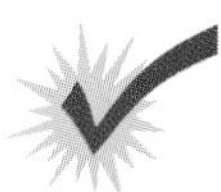

Check It Out

List the common factors of each set of numbers.

3. 8 and 18
4. 10, 30, and 45

Greatest Common Factor

The **greatest common factor** (GCF) of two whole numbers is the greatest number that is a factor of both the numbers.

One way to find the GCF is to follow these steps:

- Find the common factors.
- Choose the greatest common factor.

What is the GCF of 24 and 60?

- Factors of 24 are 1, 2, 3, 4, 6, 8, 12, 24.
- Factors of 60 are 1, 2, 3, 4, 5, 6, 10, 12, 15, 20, 30, 60.
- Common factors that are in both lists are 1, 2, 3, 4, 6, 12.

The greatest common factor of 24 and 60 is 12.

Check It Out

Find the GCF for each pair.

5. 8 and 18
6. 12 and 30

Divisibility Rules

Sometimes you may wish to know if a number is a factor of a much larger number. For instance, if you want to form teams of 3 from a group of 246 basketball players entered in a tournament, you will need to know whether 3 is a factor of 246.

You can quickly figure out whether 246 is divisible by 3 if you know the divisibility rule for 3. A number is divisible by 3 if the sum of the digits is divisible by 3. For example, 246 is divisible by 3 because $2 + 4 + 6 = 12$, and 12 is divisible by 3.

It can be helpful to know other divisibility rules. A number is divisible by:

2, if the ones digit is an even number.
3, if the sum of the digits is divisible by 3.
4, if the number formed by the last two digits is divisible by 4.
5, if the ones digit is 0 or 5.
6, if the number is divisible by 2 and 3.
8, if the number formed by the last three digits is divisible by 8.
9, if the sum of the digits is divisible by 9.

And...

Any number is divisible by **10,** if the ones digit is 0.

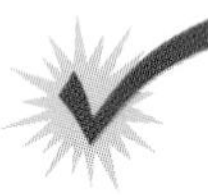

Check It Out

Check by using divisibility rules.

7. Is 424 divisible by 4?
8. Is 199 divisible by 9?
9. Is 534 divisible by 6?
10. Is 1,790 divisible by 5?

Prime and Composite Numbers

A **prime number** is a whole number greater than one with exactly two factors, itself and 1. Here are the first 10 prime numbers:

2, 3, 5, 7, 11, 13, 17, 19, 23, 29

Twin primes are pairs of primes whose difference is 2. (3, 5), (5, 7), and (11, 13) are examples of twin primes.

A number with more than two factors is called a **composite number.** When two composite numbers have no common factors, they are said to be *relatively prime.* The numbers 12 and 25 are relatively prime.

One way to find out whether a number is prime or composite is to use the "sieve of Eratosthenes." Here is how it works.

- Use a chart of numbers listed in order. First skip the number 1 because it is neither prime nor composite.
- Circle the number 2 and cross out every multiple of 2.
- Next circle the number 3 and cross out every multiple of 3.
- Then continue this procedure with 5, 7, 11, and with each succeeding number that has not been crossed out.
- The prime numbers are all the circled ones. The crossed-out numbers are the composite numbers.

1	(2)	(3)	~~4~~	(5)	~~6~~	(7)	~~8~~	~~9~~	~~10~~
(11)	~~12~~	(13)	~~14~~	~~15~~	~~16~~	(17)	~~18~~	(19)	~~20~~
~~21~~	~~22~~	(23)	~~24~~	~~25~~	~~26~~	~~27~~	~~28~~	(29)	~~30~~
(31)	~~32~~	~~33~~	~~34~~	~~35~~	~~36~~	(37)	~~38~~	~~39~~	~~40~~
(41)	~~42~~	(43)	~~44~~	~~45~~	~~46~~	(47)	~~48~~	~~49~~	~~50~~
~~51~~	~~52~~	(53)	~~54~~	~~55~~	~~56~~	~~57~~	~~58~~	(59)	~~60~~ ...

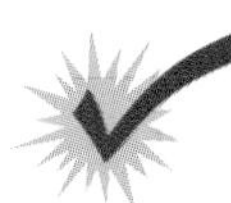

Check It Out

Is it a prime number? You can use the sieve of Eratosthenes method to decide.

11. 61
12. 77
13. 83
14. 91
15. List a pair of twin primes greater than 13.

Prime Factorization

Every composite number can be expressed as a product of prime factors.

You can use a factor tree to find the prime factors. The one below shows the **prime factorization** of 60.

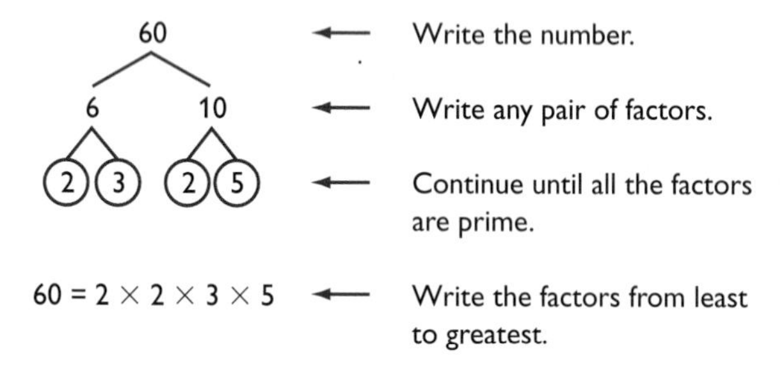

Although the order of the factors may be different because you can start with different pairs of factors, every factor tree for 60 has the same prime factorization. You also can write the prime factorization using **exponents.**

$$60 = 2^2 \times 3 \times 5$$

Check It Out

What is the prime factorization of each?

16. 80
17. 120

Shortcut to Finding GCF

Here is how you can use prime factorization to find the greatest common factor.

USING PRIME FACTORIZATION TO FIND THE GCF

Find the greatest common factor of 12 and 20.

- Find the prime factors of each number. Use a factor tree if it helps you.

  $12 = 2 \times 2 \times 3$

  $20 = 2 \times 2 \times 5$

- Find the prime factors common to both numbers.

  2 and 2

- Find their product.

  $2 \times 2 = 4$

The GCF of 12 and 20 is 2^2, or 4.

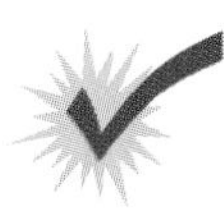

Check It Out

Use prime factorization to find the GCF of each pair of numbers.

18. 6 and 24
19. 24 and 56

Multiples and Least Common Multiples

The **multiples** of a number are the whole-number products when that number is a factor. In other words, you can find a multiple of a number by multiplying it by −3, −2, −1, 0, 1, 2, 3, and so on.

The **least common multiple** (LCM) of two numbers is the smallest positive number that is a multiple of both. One way to find the LCM of a pair of numbers is to first list positive multiples of each and then identify the smallest one common to both. For instance, to find the LCM of 6 and 8:

- List multiples of 6: 6, 12, 18, 24, 30, …
- List multiples of 8: 8, 16, 24, 32, …
- LCM of 6 and 8 is 24.

Another way to find the LCM is to use prime factorization.

USING PRIME FACTORIZATION TO FIND THE LCM

Here is how you can use prime factorization to find the least common multiple of 6 and 8.

- Find the prime factors of each number.

  $6 = 2 \times 3$

  $8 = 2 \times 2 \times 2$

- Multiply the prime factors of the least number by the prime factors of the greater number that are not factors of the least number.

  $2 \times 3 \times 2 \times 2 = 24$

Check It Out

Use either method to find the LCM.

20. 6 and 9

21. 20 and 35

1•4 EXERCISES

Find the factors of each number.

1. 16
2. 21
3. 36
4. 54

Is it a prime number? Write Yes or No.

5. 71
6. 87
7. 103
8. 291

Write the prime factorization for each.

9. 50
10. 130
11. 180
12. 320

Find the GCF for each pair.

13. 12 and 30
14. 8 and 40
15. 18 and 60
16. 20 and 25
17. 16 and 50
18. 15 and 32

Find the LCM for each pair.

19. 6 and 9
20. 12 and 60
21. 18 and 24
22. 20 and 35
23. What is the divisibility rule for 9? Is 118 divisible by 9?
24. Describe how to use prime factorization to find the GCF of two numbers.
25. A mystery number is a factor of 100 and a common multiple of 2 and 5. The sum of its digits is 5. What is the number?

1·5 Integer Operations

Positive and Negative Integers

A glance through any newspaper will show that many quantities are expressed using **negative numbers.** For example, negative numbers show below-zero temperatures, drops in the value of stocks, or business losses.

Whole numbers less than zero are called **negative integers.** Whole numbers greater than zero are called **positive integers.**

Here is the set of all integers:

$$..., -5, -4, -3, -2 -1, 0, 1, 2, 3, 4, 5, ...$$

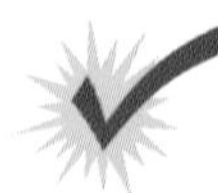

Check It Out

Write an integer to describe the situation.

1. 6° below zero
2. a gain of $200

Opposites of Integers and Absolute Value

Integers can describe opposite ideas. Each integer has an opposite.

The opposite of a gain of 4 inches is a loss of 4 inches.
The opposite of $+4$ is -4.
The opposite of spending $5 is earning $5.
The opposite of -5 is $+5$.

The **absolute value** of an integer is its distance from 0 on the number line. You write the absolute value of -7 as $|-7|$.

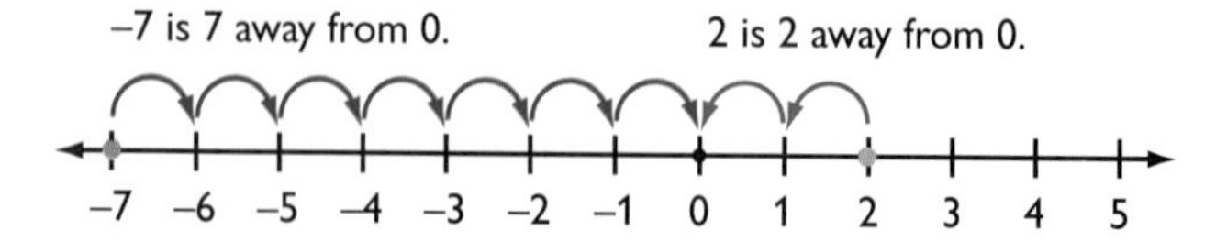

The absolute value of 2 is 2. You write $|2| = 2$.
The absolute value of -7 is 7. You write $|-7| = 7$.

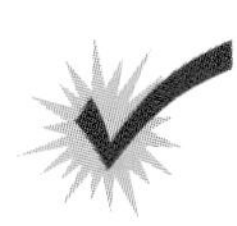

Check It Out

Give the absolute value of the integer. Then write the opposite of the original integer.

3. -12
4. 5
5. -9
6. 0

Adding and Subtracting Integers

Use a number line to model adding and subtracting integers.

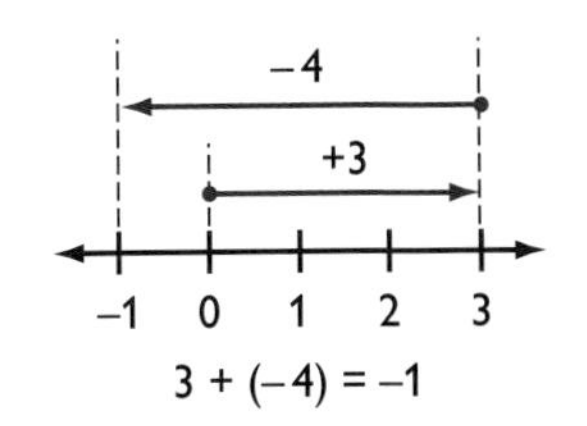

$3 + (-4) = -1$

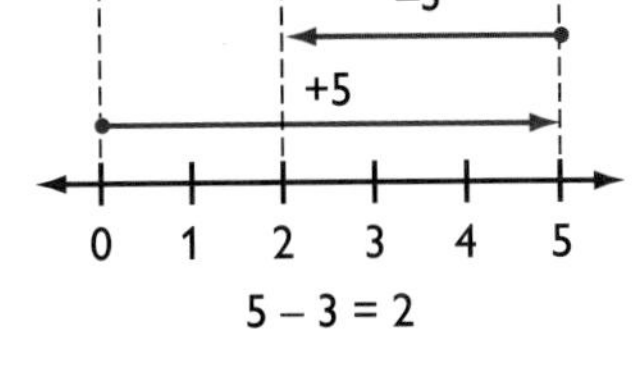

$5 - 3 = 2$

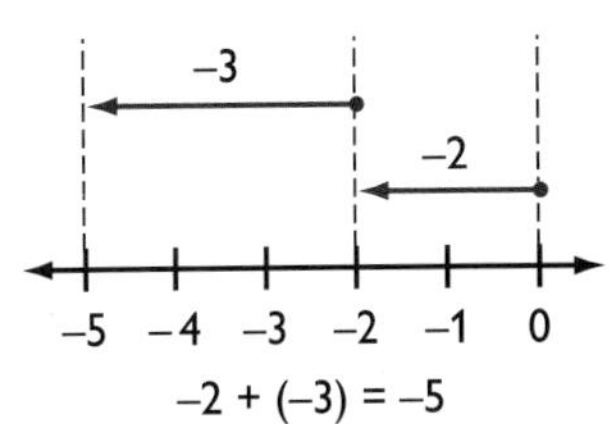

$-2 + (-3) = -5$

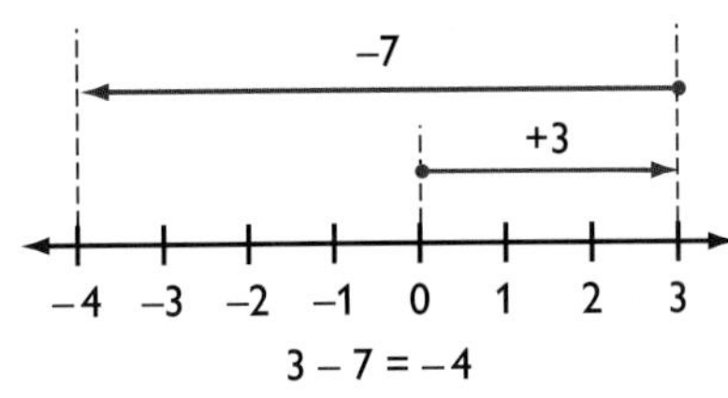

$3 - 7 = -4$

Rules for Adding or Subtracting Integers		
To	Solve	Example
Add same sign	Add absolute values. Use original sign in answer.	$-3 + (-3)$: $\lvert -3\rvert + \lvert -3\rvert = 3 + 3 = 6$ So $-3 + (-3) = -6$.
Add different signs	Subtract absolute values. Use sign of addend with greatest absolute value in answer.	$-6 + 4$: $\lvert -6\rvert - \lvert 4\rvert = 6 - 4 = 2$ $\lvert -6\rvert > \lvert 4\rvert$ So $-6 + 4 = -2$.
Subtract	Add opposite.	$-4 - 2 = -4 + (-2) = -6$

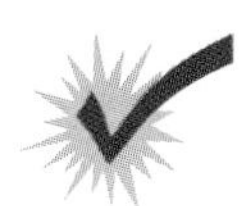

Check It Out

Solve.

7. $5 - 7$
8. $4 + (-4)$
9. $-9 - (-4)$
10. $0 + (-3)$

Multiplying and Dividing Integers

Multiply and divide integers as you would whole numbers. Then use these rules for writing the sign of the answer.

The product of two integers with like signs is positive. So is the quotient.

$$-4 \times (-3) = 12 \quad \text{or} \quad -24 \div (-4) = 6$$

When the signs of the two integers are different, the product is negative. So is the quotient.

$$-8 \div 2 = -4 \quad \text{or} \quad -3 \times 6 = -18$$

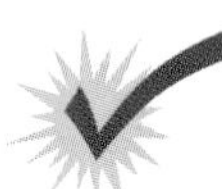

Check It Out

Find the product or quotient.

11. $-2 \times (-5)$
12. $9 \div (-3)$
13. $-15 \div (-5)$
14. -6×8

Oops!

Seventeen-year-old Colin Rizzio took the SAT test and found a mistake in its math portion. One of the questions used the letter *a* to represent a number. The test makers assumed *a* was a positive number. But Colin Rizzio thought it could stand for any integer. Rizzio was right!

He notified the test makers by email. They had to change the test scores of 45,000 students.

Explain how $2 + a > 2$ changes if *a* can be positive, zero, or negative. See Hot Solutions for answer.

1·5 EXERCISES

Give the absolute value of the integer. Then write its opposite.

1. -14
2. 6
3. -8
4. 1

Add or subtract.

5. $5 - 3$
6. $4 + (-6)$
7. $-7 - (-4)$
8. $0 + (-5)$
9. $-2 + 6$
10. $0 - 8$
11. $0 - (-8)$
12. $-2 - 8$
13. $4 + (-4)$
14. $-9 - (-5)$
15. $-5 - (-5)$
16. $-7 + (-8)$

Find the product or quotient.

17. $-2 \times (-6)$
18. $8 \div (-4)$
19. $-15 \div 5$
20. -6×7
21. $4 \times (-9)$
22. $-24 \div 8$
23. $-18 \div (-3)$
24. $3 \times (-7)$

Compute.

25. $[-3 \times (-2)] \times 4$
26. $6 \times [3 \times (-4)]$
27. $[-2 \times (-5)] \times -3$
28. $-4 \times [3 + (-5)]$
29. $(-8 - 2) \times 3$
30. $-4 \times [6 - (-3)]$
31. Is the absolute value of a negative integer positive or negative?
32. If you know that the absolute value of an integer is 4, what are the possible values for that integer?
33. What can you say about the sum of two negative integers?
34. What can you say about the product of two negative integers?
35. The temperature at noon was 18°F. For the next 4 hours it dropped at a rate of 3 degrees an hour. First express this change as an integer. Then give the temperature at 4 P.M.

What have you learned?

You can use the problems and the list of words that follow to see what you have learned in this chapter. You can find out more about a particular problem or word by referring to the boldfaced topic number (for example, **1•2**).

Problem Set

Give the value of the 4 in each number. **1•1**

1. 247,617
2. 784,122,907

3. Write 28,356, using expanded notation. **1•1**
4. Write in order from greatest to least: 346,258; 386,258; 3,258; 396,258. **1•1**
5. Round 65,434,486 to the nearest ten, thousand, and million. **1•1**

Solve. **1•2**

6. 516×0
7. $(6 \times 3) \times 1$
8. $7{,}243 + 0$
9. 0×1

Solve. Use mental math if you can. **1•2**

10. $4 \times (39 + 61)$
11. $50 \times 14 \times 2$

Use parentheses to make each expression true. **1•3**

12. $4 + 9 \times 2 = 26$
13. $25 + 10 \div 2 + 7 = 37$

Is it a prime number? Write Yes or No. **1•4**

14. 87
15. 102
16. 143
17. 401

Write the prime factorization for each. **1•4**

18. 35
19. 150
20. 320

Find the GCF for each pair. **1•4**

21. 16 and 30
22. 12 and 50
23. 10 and 160

Find the LCM for each pair. **1•4**

24. 5 and 12
25. 15 and 8
26. 18 and 30

27. What is the divisibility rule for 6? Is 246 a multiple of 6? **1•4**

Give the absolute value of the integer. Then write the opposite of the original integer. **1•5**

28. -9
29. 13
30. -10
31. 20

Add or subtract. **1•5**

32. $9 + (-8)$
33. $6 - 7$
34. $-8 + (-9)$
35. $5 - (-5)$
36. $-7 - (-7)$
37. $-4 + 12$

Compute. **1•5**

38. $-8 \times (-9)$
39. $64 \div (-32)$
40. $-36 \div (-9)$
41. $(-4 \times 5) \times (-3)$
42. $4 \times [-3 + (-8)]$
43. $-6\,[5 - (-8)]$
44. What can you say about the product of two positive integers? **1•5**
45. What can you say about the difference of two negative integers? **1•5**

hot words

WRITE DEFINITIONS FOR THE FOLLOWING WORDS:

absolute value **1•5**
approximation **1•1**
associative property **1•2**
common factor **1•4**
commutative property **1•2**
composite number **1•4**
distributive property **1•2**
expanded notation **1•1**
exponent **1•4**
factor **1•4**
greatest common factor **1•4**
least common multiple **1•4**
multiple **1•4**
negative integer **1•5**
negative number **1•5**
number system **1•1**
operation **1•3**
place value **1•1**
positive integer **1•5**
prime factorization **1•4**
prime number **1•4**
round **1•1**

= 3/8 =
=.375=37.5%
7

hot topics 2

Fractions, Decimals, and Percents

2·1	Fractions and Equivalent Fractions	102
2·2	Comparing and Ordering Fractions	112
2·3	Addition and Subtraction of Fractions	116
2·4	Multiplication and Division of Fractions	122
2·5	Naming and Ordering Decimals	126
2·6	Decimal Operations	132
2·7	Meaning of Percent	140
2·8	Using and Finding Percents	144
2·9	Fraction, Decimal, and Percent Relationships	154

What do you already know?

You can use the problems and the list of words that follow to see what you already know about this chapter. The answers to the problems are in Hot Solutions at the back of the book, and the definitions are in Hot Words at the front of the book. You can find out more about a particular problem or word by referring to the boldfaced topic number (for example, **2•2**).

Problem Set

1. In one basketball game, Julian scored $\frac{3}{7}$ of his free throws. In a second basketball game, he scored $\frac{1}{2}$ of his free throws. In which game did he perform better? **2•2**
2. It takes Mr. Chen about $1\frac{1}{2}$ work days to install a tile floor in an average-size kitchen. How many days would it take him to install floors for 6 kitchens? **2•4**
3. Leslie has $7\frac{1}{2}$ cups of cooked pasta. She wants each serving to be $\frac{3}{4}$ cup. How many servings does she have? **2•4**
4. Nalani got 17 out of 20 questions correct on her science test. What percent did she get correct? **2•8**
5. Which fraction is not equivalent to $\frac{9}{12}$? **2•1**
   A. $\frac{3}{4}$ B. $\frac{6}{8}$ C. $\frac{8}{11}$ D. $\frac{75}{100}$

Add or subtract as indicated. Write your answers in lowest terms. **2•3**

6. $\frac{2}{3} + \frac{1}{2}$
7. $3\frac{3}{8} - 1\frac{5}{8}$
8. $6 - 2\frac{3}{4}$
9. $3\frac{1}{2} + 4\frac{4}{5}$

10. Find the improper fraction and write it as a mixed number. **2•1**
    A. $\frac{6}{12}$ B. $\frac{4}{3}$ C. $3\frac{5}{6}$

Multiply or divide. **2•4**

11. $\frac{4}{5} \times \frac{1}{2}$
12. $\frac{3}{4} \div 1\frac{1}{2}$
13. $3\frac{3}{8} \times \frac{2}{9}$
14. $7\frac{1}{2} \div 2\frac{1}{2}$
15. Give the place value of 6 in 35.063. **2•5**
16. Write 3.003 in expanded form. **2•5**

17. Write as a decimal: four hundred and four hundred four thousandths. **2•5**

18. Write the following numbers in order from least to greatest: 1.650; 1.605; 1.065; 0.165. **2•5**

Solve. **2•6**

19. 3.604 + 12.55
20. 11.4 − 10.08
21. 6.05 × 5.1
22. 67.392 ÷ 9.6

Use a calculator. Round answers to the nearest tenth. **2•8**

23. What percent of 80 is 24?
24. Find 23% of 121.
25. 44 is 80% of what number?

CHAPTER 2

hot words

benchmark **2•7**
common denominator **2•2**
cross product **2•1**
denominator **2•1**
discount **2•8**
equivalent fractions **2•1**
estimate **2•6**
factor **2•4**
fraction **2•1**
greatest common factor **2•1**
improper fraction **2•1**
mixed number **2•1**
numerator **2•1**
percent **2•7**
place value **2•5**
product **2•4**
reciprocal **2•4**
repeating decimal **2•9**
terminating decimal **2•9**
whole number **2•1**

2•1 Fractions and Equivalent Fractions

Naming Fractions

A **fraction** can be used to name a part of a whole. For example, the flag of France is divided into three equal parts: red, white, and blue. Each part, or color, of the French flag represents $\frac{1}{3}$ of the whole flag. $\frac{3}{3}$ or 1 represents the whole flag.

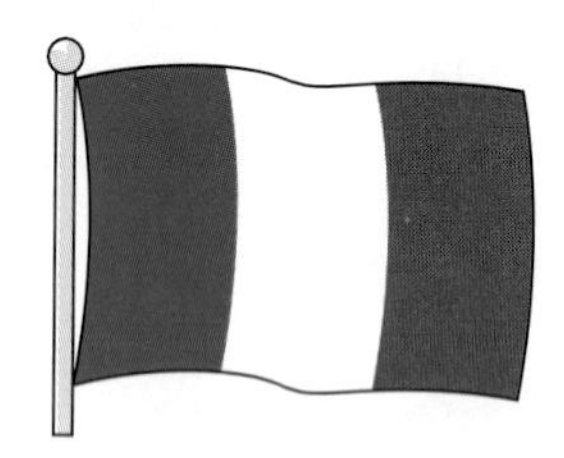

A fraction can also name part of a set.

There are five balls in the set of balls. Each ball is $\frac{1}{5}$ of the set. $\frac{5}{5}$ or 1 equals the whole set. Three of the balls are baseballs. The baseballs represent $\frac{3}{5}$ of the set. Two of the five balls are footballs. The footballs represent $\frac{2}{5}$ of the set.

You use **numerators** and **denominators** to name fractions.

NAMING FRACTIONS

Write a fraction for the number of shaded triangles.

- The denominator of the fraction tells the number of parts.

  There are 7 triangles.
- The numerator of the fraction tells the number of parts being considered.

  4 triangles are shaded.
- Write the fraction:

$$\frac{\text{parts under consideration}}{\text{parts that make a whole}} = \frac{\text{numerator}}{\text{denominator}}$$

$\frac{4}{7}$ is the fraction for the number of shaded triangles.

Check It Out

Write the fraction for each picture.

1. ___ of the circle is shaded.

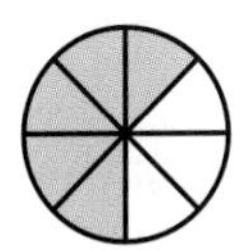

2. ___ of the shapes are shaded.

△▲▲△
△▲△△

3. Draw two pictures to represent the fraction $\frac{3}{5}$. Use regions and sets.

Methods for Finding Equivalent Fractions

Equivalent fractions are fractions that describe the same amount of a region. You can use fraction pieces to show equivalent fractions.

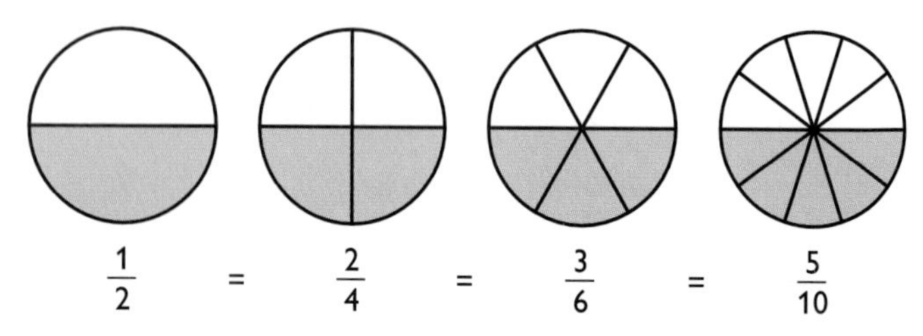

$$\frac{1}{2} = \frac{2}{4} = \frac{3}{6} = \frac{5}{10}$$

Fraction Names for One

There are an infinite number of fractions that are equal to one.

Names for One

$$\frac{4}{4} \quad \frac{5}{5} \quad \frac{123}{123} \quad \frac{8}{8}$$

Since any number multiplied by one is still equal to the original number, knowing different names for one can help you find equivalent fractions.

To find a fraction that is equivalent to another fraction, you can multiply the original fraction by a form of one. You can also divide the numerator and denominator by the same number to get an equivalent fraction.

METHODS FOR FINDING EQUIVALENT FRACTIONS

Find a fraction equal to $\frac{6}{12}$.

- Multiply by a form of one or divide the numerator and denominator by the same number.

Multiply	OR	Divide
$\frac{6}{12} \times \frac{5}{5} = \frac{30}{60}$		$\frac{6 \div 2}{12 \div 2} = \frac{3}{6}$
$\frac{6}{12} = \frac{30}{60}$		$\frac{6}{12} = \frac{3}{6}$

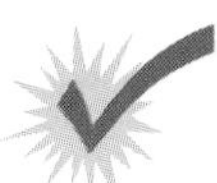

Check It Out

Write two fractions equivalent to each fraction.

4. $\frac{1}{4}$
5. $\frac{10}{20}$
6. $\frac{4}{5}$
7. Write three fractional names for one.

Deciding if Two Fractions Are Equivalent

Two fractions are equivalent if you can show that each fraction is just a different name for the same amount.

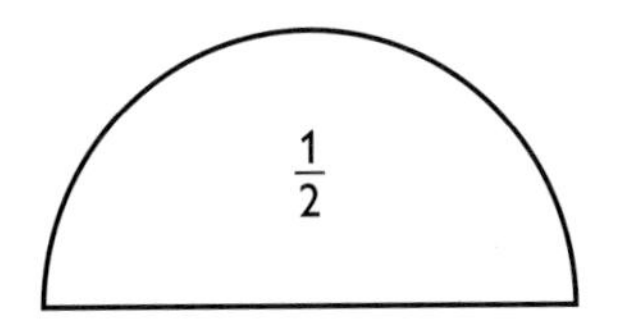

The fraction piece above shows $\frac{1}{2}$ of the whole circle.

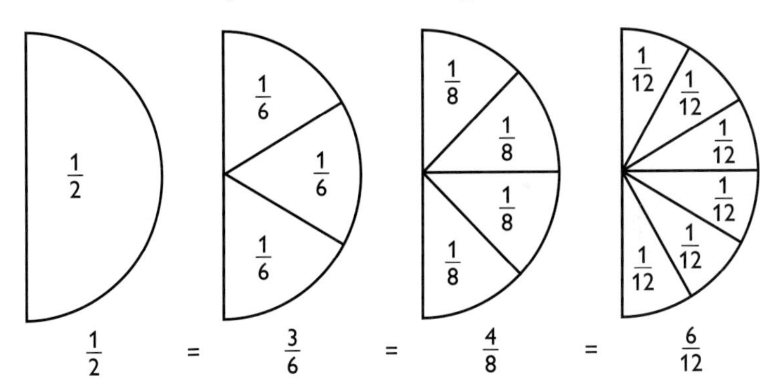

There are many fractional names for the same amount.

One method you can use to identify equivalent fractions is to find the **cross products** of the fractions.

DECIDING IF TWO FRACTIONS ARE EQUIVALENT

Find out whether or not $\frac{2}{4}$ is equivalent to $\frac{10}{20}$.

- Find the cross products of the fractions.

$$\frac{2}{4} \overset{?}{=} \frac{10}{20}$$

- Compare the cross products.

  $40 = 40$

- If the cross products are the same, then the fractions are equivalent.

  So $\frac{2}{4} = \frac{10}{20}$.

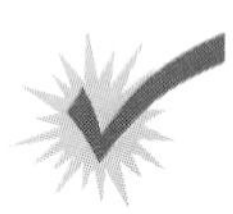

Check It Out

Use the cross-products method to determine whether or not each pair of fractions is equivalent.

8. $\frac{15}{20}, \frac{30}{20}$
9. $\frac{4}{5}, \frac{24}{30}$
10. $\frac{3}{4}, \frac{15}{24}$

Writing Fractions in Lowest Terms

When the numerator and the denominator of a fraction have no common factor other than 1, the fraction is in *lowest terms.*

You can use fraction pieces to show fractions in lowest terms.

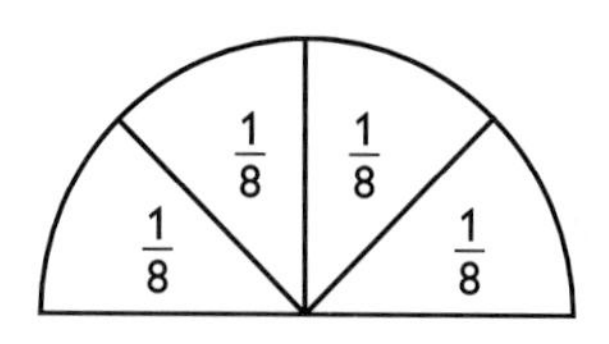

The fraction pieces show the fraction $\frac{4}{8}$.

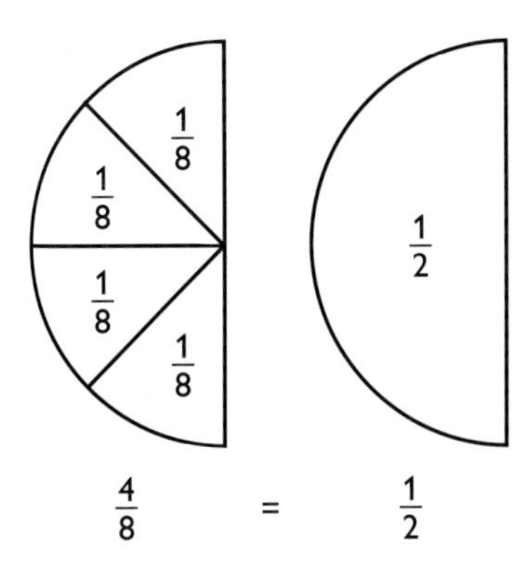

The fewest number of fraction pieces to show the equivalent of $\frac{4}{8}$ is $\frac{1}{2}$. Therefore the fraction $\frac{4}{8}$ is equal to $\frac{1}{2}$ in lowest terms.

To express fractions in lowest terms, you can divide numerators and denominators by **greatest common factors** (GCF).

FINDING LOWEST TERMS OF FRACTIONS

Express $\frac{18}{24}$ in lowest terms.

- List the factors of the numerator.

  The factors of 18 are:

  1, 2, 3, 6, 9, 18
- List the factors of the denominator.

  The factors of 24 are:

  1, 2, 3, 4, 6, 8, 12, 24
- Find the greatest common factor (GCF).

  The GCF is 6.
- Divide the numerator and the denominator of the fraction by the GCF.

  $18 \div 6 = 3 \qquad 24 \div 6 = 4$
- Write the fraction in lowest terms.

  $\frac{3}{4}$

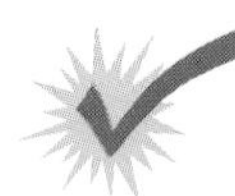

Check It Out

Express each fraction in lowest terms.

11. $\frac{8}{10}$ 12. $\frac{12}{16}$ 13. $\frac{24}{60}$

Go For It!

Calling all couch potatoes! To get in shape, you need to do an aerobic activity (such as walking, jogging, biking, or swimming) at least three times a week.

The goal is to get your heart beating at $\frac{1}{2}$ to $\frac{3}{4}$ of its maximum rate and to keep it there long enough to give it a good workout. You calculate your maximum heart rate by subtracting your age from 220.

Heart Rate (percent of maximum)	Number of Minutes You Need to Exercise
50%	45
55%	40
60%	35
65%	30
70%	25
75%	20

Writing Improper Fractions and Mixed Numbers

You can write fractions for amounts greater than 1. Fractions with a numerator greater than or equal to the denominator are called **improper fractions.**

$\frac{7}{2}$ cookies

$\frac{7}{2}$ is an improper fraction.

A whole number and a fraction make up a **mixed number.**

$3\frac{1}{2}$ cookies

$3\frac{1}{2}$ is a mixed number.

You can write any mixed number as an improper fraction and any improper fraction as a mixed number. You can use division to change an improper fraction to a mixed number.

CHANGING AN IMPROPER FRACTION TO A MIXED NUMBER

Change $\frac{17}{5}$ to a mixed number.

- Divide the numerator by the denominator.

$\frac{17}{5}$ divisor → $5\overline{)17}$ — quotient ← 3; $17 - 15 = 2$ ← remainder

- Write the mixed number.

quotient → $3\frac{2}{5}$ ← remainder (2), ← divisor (5)

You can use multiplication to change a mixed number to an improper fraction. Start by renaming the whole-number part. Rename it as an improper fraction with the same denominator as the fraction part. Then add the two parts.

CHANGING A MIXED NUMBER TO AN IMPROPER FRACTION

Change $3\frac{1}{4}$ to an improper fraction.

- Multiply the whole-number part by a version of 1 that has the same denominator as the fraction part.

  $3 \times \frac{4}{4} = \frac{12}{4}$

- Add the two parts.

  $3\frac{1}{4} = \frac{12}{4} + \frac{1}{4} = \frac{13}{4}$

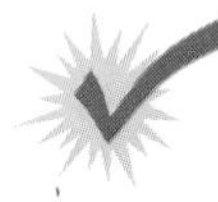

Check It Out

Write a mixed number for each improper fraction.

14. $\frac{43}{6}$
15. $\frac{34}{3}$
16. $\frac{32}{5}$
17. $\frac{37}{4}$

Write an improper fraction for each mixed number.

18. $4\frac{5}{8}$
19. $12\frac{5}{6}$
20. $24\frac{1}{2}$
21. $32\frac{2}{3}$

2·1 EXERCISES

Write the fraction for each picture.

1. ___ of the apples are green.

2. ___ of the circle is blue.

3. ___ of the stars are yellow.

4. ___ of the balls are basketballs.

Write the fraction.

5. three tenths
6. twelve seventeenths

Write one fraction equivalent to the given fraction.

7. $\frac{1}{2}$
8. $\frac{7}{8}$
9. $\frac{40}{60}$
10. $\frac{18}{48}$

Express each fraction in lowest terms.

11. $\frac{45}{90}$
12. $\frac{24}{32}$
13. $\frac{12}{34}$
14. $3\frac{12}{60}$
15. $\frac{38}{14}$
16. $\frac{82}{10}$

Find the GCF of each pair of numbers.

17. 16, 21
18. 81, 27
19. 18, 15

Write a mixed number for each improper fraction.

20. $\frac{25}{4}$
21. $\frac{12}{10}$
22. $\frac{11}{4}$

Write an improper fraction for each mixed number.

23. $5\frac{1}{6}$
24. $8\frac{3}{5}$
25. $13\frac{4}{9}$

2•2 Comparing and Ordering Fractions

Comparing Fractions

You can use fraction pieces to compare fractions.

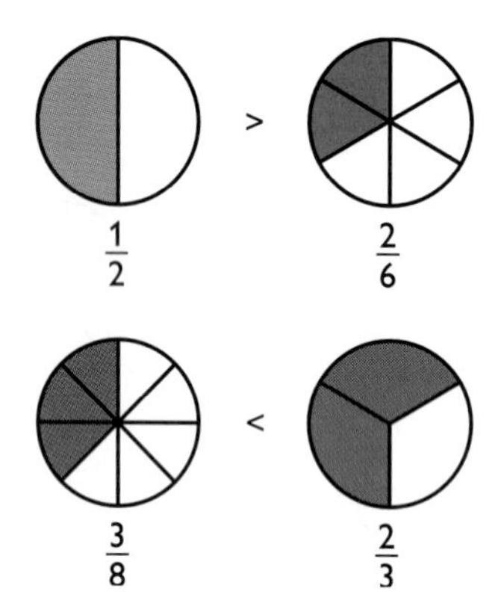

To compare fractions, you can find *equivalent fractions* (p. 104) and compare numerators.

COMPARING FRACTIONS

Compare the fractions $\frac{3}{4}$ and $\frac{2}{3}$.

- Look at the denominators.

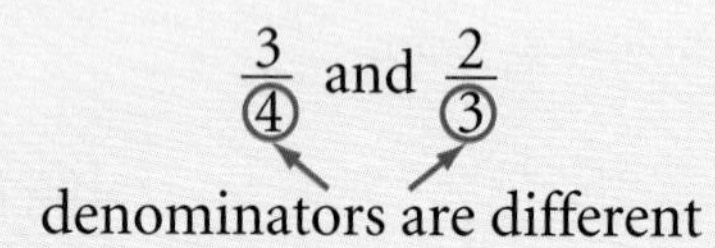

- Write equivalent fractions with a common denominator.

$$\frac{3}{4} \times \frac{3}{3} = \frac{9}{12} \qquad \frac{2}{3} \times \frac{4}{4} = \frac{8}{12}$$

- Compare the numerators.

  $9 > 8$

- The fractions compare as the numerators compare.

  $\frac{9}{12} > \frac{8}{12}$, so $\frac{3}{4} > \frac{2}{3}$.

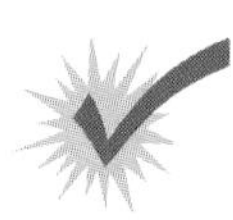

Check It Out

Compare the fractions. Use >, <, or =.

1. $\frac{3}{4} \square \frac{9}{16}$ 2. $\frac{1}{6} \square \frac{1}{10}$ 3. $\frac{6}{8} \square \frac{18}{24}$ 4. $\frac{1}{3} \square \frac{5}{9}$

Comparing Mixed Numbers

To compare *mixed numbers* (p. 109), first compare the whole numbers. Then compare the fractions if necessary.

COMPARING MIXED NUMBERS

Compare $2\frac{1}{5}$ and $2\frac{2}{3}$.

- Be sure fractions are not improper.
  $\frac{1}{5}$ and $\frac{2}{3}$ are not improper.
- Compare the whole-number parts. If they are different, the one that is greater is the greater mixed number. If they are equal, go on.
  $2 = 2$
- Compare the fraction parts by renaming them with a *common denominator.*

  15 is the least common multiple of 5 and 3.
  Use 15 for the common denominator.

  $$\frac{1}{5} = \frac{3}{15} \qquad \frac{2}{3} = \frac{10}{15}$$
- Compare the fractions.
  $\frac{3}{15} < \frac{10}{15}$, so $2\frac{1}{5} < 2\frac{2}{3}$.

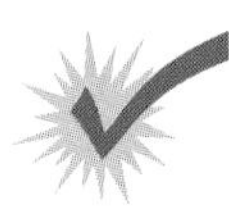

Check It Out

Compare the mixed numbers. Use <, >, or =.

5. $4\frac{1}{2} \square 3\frac{5}{8}$ 6. $5\frac{1}{3} \square 5\frac{2}{5}$ 7. $4\frac{7}{16} \square 4\frac{7}{9}$

Ordering Fractions

To compare and order fractions, you can find equivalent fractions and then compare the numerators of the fractions.

ORDERING FRACTIONS WITH UNLIKE DENOMINATORS

Order the fractions $\frac{3}{4}$, $\frac{5}{8}$, and $\frac{2}{3}$ from least to greatest.

- Find the *least common multiple* (LCM) (p. 90) of 4, 8, and 3.

Multiples of 4: 4, 8, 12, 16, 20, (24), 28, 32...
Multiples of 8: 8, 16, (24), 32...
Multiples of 3: 3, 6, 9, 12, 15, 18, 21, (24)

24 is the LCM of 4, 8, and 3.

- Write equivalent fractions with a common denominator.

$$\frac{3}{4} = \frac{3}{4} \times \frac{6}{6} = \frac{18}{24}$$
$$\frac{5}{8} = \frac{5}{8} \times \frac{3}{3} = \frac{15}{24}$$
$$\frac{2}{3} = \frac{2}{3} \times \frac{8}{8} = \frac{16}{24}$$

- The fractions compare as the numerators compare.

$\frac{15}{24} < \frac{16}{24} < \frac{18}{24}$ or $\frac{5}{8} < \frac{2}{3} < \frac{3}{4}$

Check It Out

Order the fractions from least to greatest.

8. $\frac{3}{5}; \frac{1}{2}; \frac{1}{4}; \frac{2}{5}$
9. $\frac{2}{3}; \frac{13}{18}; \frac{7}{9}; \frac{5}{6}$
10. $\frac{4}{7}; \frac{5}{8}; \frac{11}{12}; \frac{1}{2}; \frac{2}{3}$

2·2 EXERCISES

Compare each fraction. Use <, >, or =.

1. $\frac{3}{4} \square \frac{9}{16}$
2. $\frac{2}{3} \square \frac{12}{16}$
3. $\frac{9}{36} \square \frac{1}{4}$
4. $\frac{12}{20} \square \frac{16}{30}$
5. $\frac{18}{44} \square \frac{9}{34}$
6. $\frac{5}{10} \square \frac{9}{18}$

Compare each mixed number. Use <, >, or =.

7. $2\frac{1}{2} \square 2\frac{2}{3}$
8. $3\frac{3}{4} \square 3\frac{5}{16}$
9. $8\frac{1}{2} \square 9\frac{12}{16}$
10. $6\frac{9}{10} \square 6\frac{3}{4}$
11. $7\frac{11}{25} \square 7\frac{23}{50}$
12. $7\frac{1}{5} \square 6\frac{4}{5}$

Order the fractions and mixed numbers from least to greatest.

13. $\frac{1}{2}; \frac{3}{8}; \frac{5}{8}; \frac{1}{4}$
14. $\frac{2}{3}; \frac{9}{10}; \frac{7}{8}; \frac{3}{4}$
15. $\frac{5}{6}; \frac{5}{7}; \frac{3}{4}; \frac{2}{3}; \frac{9}{11}; \frac{3}{7}$
16. $2\frac{1}{2}; \frac{7}{2}; 2\frac{1}{3}; 2\frac{4}{5}; 2\frac{3}{8}$

Use the table below to answer items 17–20.

RECESS BASKETBALL FREE THROWS

Ryan $\frac{8}{12}$	Gwen $\frac{7}{9}$
Tomas $\frac{5}{6}$	Roberto $\frac{4}{7}$

numerator = baskets made
denominator = attempts

17. Who was more accurate in basketball free throws, Gwen or Tomas?
18. Order the players from most accurate to least accurate.
19. Who attempted the most shots?
20. Who made the most points?

2·3 Addition and Subtraction of Fractions

Adding and Subtracting Fractions with Like Denominators

When you add or subtract fractions that have the same or like denominators, you only add or subtract the numerators. The denominator stays the same.

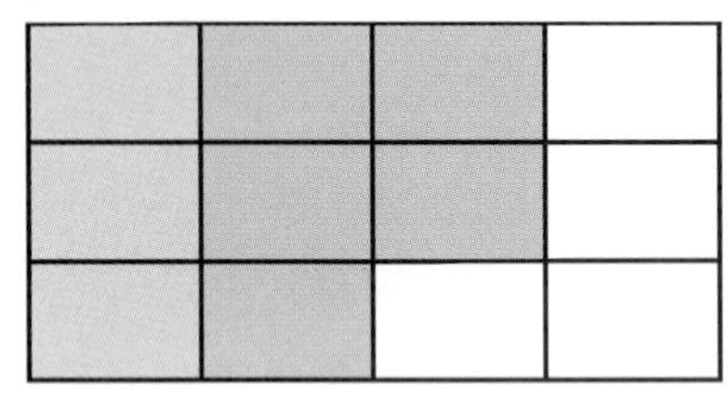

$\frac{3}{12} + \frac{5}{12} = \frac{8}{12}$

You can use fraction drawings to model the addition and subtraction of fractions with like denominators.

ADDING AND SUBTRACTING FRACTIONS WITH LIKE DENOMINATORS

Add $\frac{3}{4} + \frac{2}{4}$.

- Add or subtract the numerators.

  $3 + 2 = 5$
- Write the result over the like denominator.

  $\frac{3}{4} + \frac{2}{4} = \frac{5}{4}$
- Simplify if possible.

  $\frac{5}{4} = 1\frac{1}{4}$

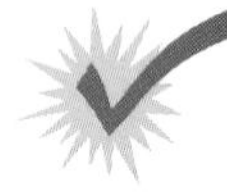

Check It Out

Add or subtract. Simplify if possible.

1. $\frac{12}{15} + \frac{6}{15}$
2. $\frac{24}{34} + \frac{13}{34}$
3. $\frac{11}{12} - \frac{5}{12}$
4. $\frac{7}{10} - \frac{2}{10}$

Adding and Subtracting Fractions with Unlike Denominators

You can use models to add fractions with unlike denominators.

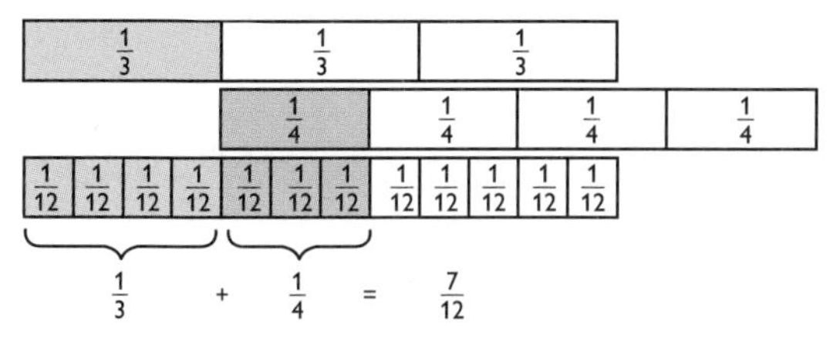

You can also use fraction pieces to model subtraction with unlike denominators.

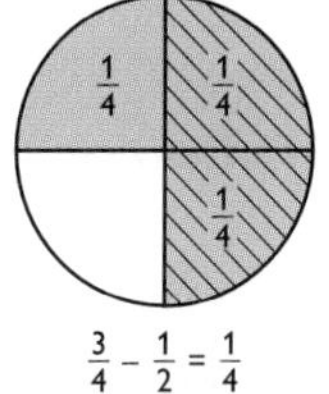

To add or subtract fractions with unlike denominators, you need to change the fractions to equivalent fractions with common, or like, denominators before you find the sum or difference.

ADDING AND SUBTRACTING FRACTIONS WITH UNLIKE DENOMINATORS

Add $\frac{4}{5} + \frac{3}{4}$.

- Find the least common denominator of the fractions.

  20 is the LCD of 4 and 5.

- Write equivalent fractions with the LCD.

  $\frac{4}{5} = \frac{4}{5} \times \frac{4}{4} = \frac{16}{20}$ and $\frac{3}{4} = \frac{3}{4} \times \frac{5}{5} = \frac{15}{20}$

- Add or subtract the numerators. Put the result over the common denominator.

  $\frac{16}{20} + \frac{15}{20} = \frac{31}{20}$

- Simplify if possible.

  $\frac{31}{20} = 1\frac{11}{20}$

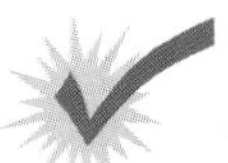

Check It Out

Add or subtract. Simplify if possible.

5. $\frac{9}{10} + \frac{1}{2}$
6. $\frac{1}{2} + \frac{5}{7}$
7. $\frac{4}{5} - \frac{3}{4}$
8. $\frac{5}{8} - \frac{1}{6}$

Adding and Subtracting Mixed Numbers

Adding and subtracting mixed numbers is similar to adding and subtracting fractions. Sometimes you have to rename your number before subtracting. Sometimes you will have an improper fraction to simplify in the answer.

Adding Mixed Numbers with Common Denominators

To add *mixed numbers* (p. 109) with common denominators, you just need to write the sum of the numerators over the common denominator. Then add the whole numbers.

ADDING MIXED NUMBERS WITH COMMON DENOMINATORS

Add $5\frac{3}{8} + 2\frac{3}{8}$.

Add the whole numbers. Add the fractions.

$$\begin{array}{r} 5\frac{3}{8} \\ +\ 2\frac{3}{8} \\ \hline 7\frac{6}{8} \end{array}$$

Simplify if possible. $7\frac{6}{8} = 7\frac{3}{4}$

Check It Out

Add. Simplify if possible.

9. $4\frac{2}{6} + 5\frac{3}{6}$
10. $21\frac{7}{8} + 12\frac{6}{8}$
11. $23\frac{7}{10} + 37\frac{3}{10}$

Adding Mixed Numbers with Unlike Denominators

You can use fraction pieces to model the addition of mixed numbers with unlike denominators.

	1	1	$\frac{1}{2}$ $\frac{1}{2}$	$2\frac{1}{2}$
		1	$\frac{1}{4}$ $\frac{1}{4}$ $\frac{1}{4}$ $\frac{1}{4}$	$+\ 1\frac{1}{4}$
1	1	1	$\frac{1}{4}$ $\frac{1}{4}$ $\frac{1}{4}$	$3\frac{3}{4}$

To add mixed numbers with unlike denominators, you need to write equivalent fractions with a common denominator.

ADDING MIXED NUMBERS WITH UNLIKE DENOMINATORS

Add $4\frac{2}{3} + 1\frac{3}{5}$.

- Write equivalent fractions with a common denominator.

  $\frac{2}{3} = \frac{10}{15}$ and $\frac{3}{5} = \frac{9}{15}$

- Add.

$$\begin{array}{r} 4\frac{10}{15} \\ +1\frac{9}{15} \\ \hline 5\frac{19}{15} \end{array}$$

Add the whole numbers. Add the fractions.

Simplify if possible. $5\frac{19}{15} = 6\frac{4}{15}$

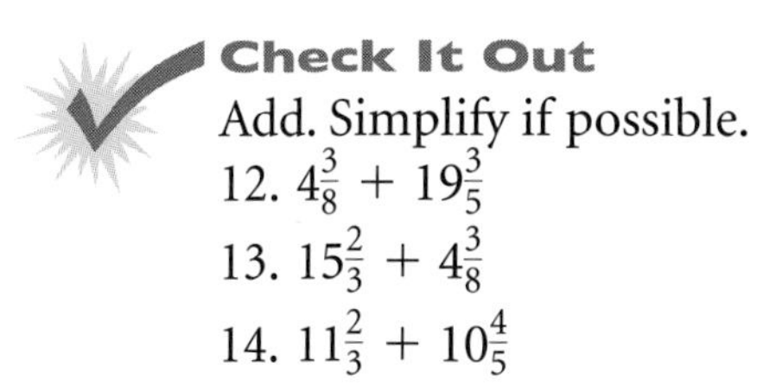

Check It Out

Add. Simplify if possible.

12. $4\frac{3}{8} + 19\frac{3}{5}$
13. $15\frac{2}{3} + 4\frac{3}{8}$
14. $11\frac{2}{3} + 10\frac{4}{5}$

Subtracting Mixed Numbers

You can model the subtraction of *mixed numbers* (p. 109) with unlike denominators.

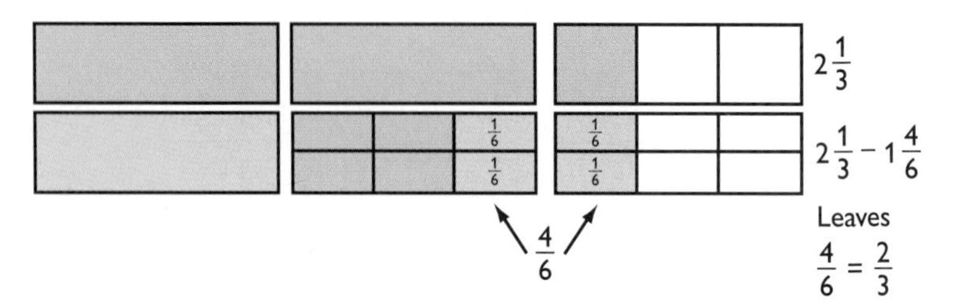

To subtract mixed numbers, you need to have or make *common denominators.*

SUBTRACTING MIXED NUMBERS

Subtract $6\frac{1}{2} - 1\frac{5}{6}$.

- If the denominators are unlike, write equivalent fractions with a common denominator.

  $6\frac{1}{2} = 6\frac{3}{6}$ and $1\frac{5}{6} = 1\frac{5}{6}$

- Then subtract.

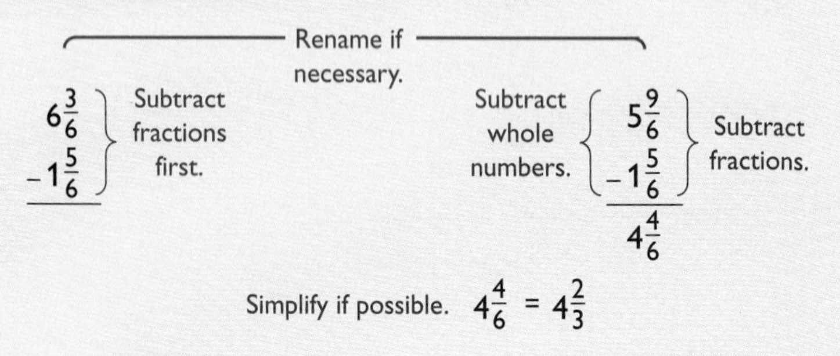

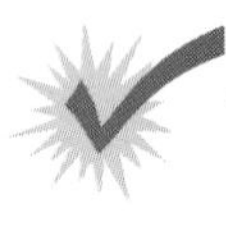

Check It Out

Subtract.

15. $12 - 4\frac{1}{2}$
16. $9\frac{1}{10} - 5\frac{4}{7}$
17. $14\frac{7}{8} - 3\frac{3}{4}$

2•3 ADDITION AND SUBTRACTION

2·3 EXERCISES

Add or subtract.

1. $\frac{2}{8} + \frac{3}{8}$
2. $\frac{7}{9} - \frac{4}{9}$
3. $\frac{3}{8} - \frac{1}{4}$
4. $\frac{1}{9} + \frac{1}{5}$
5. $\frac{5}{6} + \frac{3}{4}$
6. $\frac{2}{3} - \frac{5}{8}$
7. $\frac{7}{12} + \frac{9}{16}$
8. $\frac{9}{8} - \frac{1}{2}$
9. $1\frac{1}{2} + \frac{1}{6}$
10. $8\frac{3}{8} + 2\frac{1}{3}$
11. $12 - 11\frac{5}{9}$
12. $4\frac{1}{2} + 2\frac{1}{2}$
13. $4\frac{3}{4} - 2\frac{1}{4}$
14. $13\frac{7}{12} - 2\frac{5}{8}$
15. $7\frac{3}{8} - 2\frac{2}{3}$
16. The seventh-grade students sold $3\frac{3}{4}$ dozen chocolate chip cookies and $3\frac{1}{3}$ dozen oatmeal cookies at their bake sale last week. How many dozen cookies did they sell?
17. Rita's broad jumps during the middle-school track meet were 9′ $8\frac{1}{2}$″, 10′ $1\frac{1}{4}$″, and 9′ $11\frac{7}{8}$″. What was the difference between her longest and shortest jumps?
18. Last week Gabriel worked $9\frac{1}{4}$ hours baby-sitting and $6\frac{1}{2}$ hours giving gymnastic lessons. How many hours did he work in all?
19. Bill's Burger Palace had its grand opening on Tuesday. They had $164\frac{1}{2}$ lb of ground beef in stock. They had $18\frac{1}{4}$ lb left at the end of the day. Each burger requires $\frac{1}{4}$ lb of ground beef. How many hamburgers did they sell?
20. Rebecca is wrapping presents for people at a senior center. She had $8\frac{1}{2}$ yards of ribbon when she began. She used $2\frac{1}{3}$ yards of ribbon to wrap one present. She needs to wrap two more presents of the same size. Does she have enough ribbon?

2·4 Multiplication and Division of Fractions

Multiplying Fractions

You know that 3 × 2 means "3 groups of 2." Multiplying fractions involves the same concept: $3 \times \frac{1}{2}$ means "3 groups of $\frac{1}{2}$." You will find it helpful to think of *times* as meaning *of.*

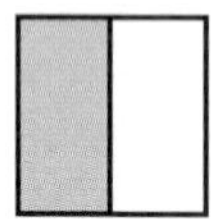

One group of $\frac{1}{2}$

$1 \times \frac{1}{2} = \frac{1}{2}$

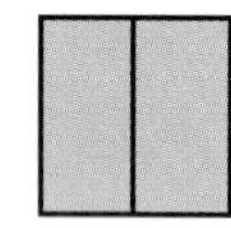

Two groups of $\frac{1}{2}$

$2 \times \frac{1}{2} = \frac{2}{2} = 1$

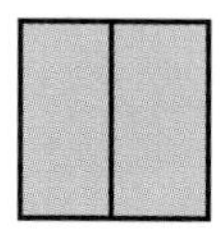
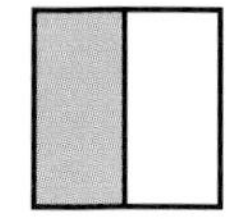

Three groups of $\frac{1}{2}$

$3 \times \frac{1}{2} = \frac{3}{2} = 1\frac{1}{2}$

The same is true when you are multiplying a fraction by a fraction. For example, $\frac{1}{2} \times \frac{1}{3}$ means you would actually be finding $\frac{1}{2}$ of $\frac{1}{3}$.

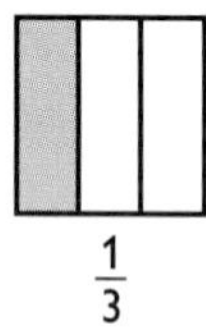

$\frac{1}{3}$

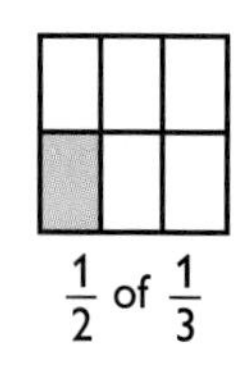

$\frac{1}{2}$ of $\frac{1}{3}$

When you are not using models to multiply fractions, you multiply the numerators and then the denominators. There's no need to find a common denominator.

$$\frac{1}{2} \times \frac{1}{4} = \frac{1}{8}$$

MULTIPLYING FRACTIONS

Multiply $\frac{3}{4}$ and $2\frac{2}{5}$.

- Convert mixed numbers, if any, to *improper fractions* (p. 109).

  $$\frac{3}{4} \times 2\frac{2}{5} = \frac{3}{4} \times \frac{12}{5}$$

- Multiply the numerators and the denominators.

  $$\frac{3}{4} \times \frac{12}{5} = \frac{3 \times 12}{4 \times 5} = \frac{36}{20}$$

- Write the **products** in *lowest terms* (p. 106) if necessary.

  $$\frac{36}{20} = 1\frac{16}{20} = 1\frac{4}{5}$$

Check It Out

1. $\frac{2}{5} \times \frac{5}{6}$
2. $\frac{3}{8} \times \frac{2}{9}$
3. $5\frac{1}{3} \times \frac{3}{8}$

Shortcut for Multiplying Fractions

You can use a shortcut when you multiply fractions. Instead of multiplying across and then writing the product in lowest terms, you can cancel **factors** first.

CANCELING FACTORS

Multiply $\frac{5}{8}$ and $1\frac{1}{5}$.

$\frac{5}{8} \times 1\frac{1}{5}$

$= \frac{5}{8} \times \frac{6}{5}$

$= \frac{\overset{1}{\cancel{5}}}{\underset{1}{\cancel{2}} \times 4} \times \frac{\overset{1}{\cancel{2}} \times 3}{\underset{1}{\cancel{5}}}$

$= \frac{3}{4}$

- Write the numbers as improper fractions.
- Cancel factors if you can.
- Multiply across.
- Write the product in lowest terms if necessary.

The answer is $\frac{3}{4}$.

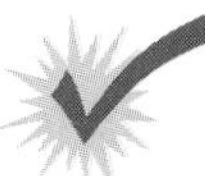

Check It Out

4. $\frac{3}{5} \times \frac{1}{6}$
5. $\frac{4}{7} \times \frac{14}{15}$
6. $1\frac{1}{2} \times 1\frac{1}{3}$

Finding the Reciprocal of a Number

To find the **reciprocal** of a number, you switch the numerator and the denominator.

Number	*Reciprocal*
$\frac{3}{5}$	$\frac{5}{3}$
$2 = \frac{2}{1}$	$\frac{1}{2}$
$3\frac{1}{2} = \frac{7}{2}$	$\frac{2}{7}$

When you multiply a number by its reciprocal, the product is 1.

$\frac{3}{8} \times \frac{8}{3} = \frac{24}{24} = 1$

The number 0 does not have a reciprocal.

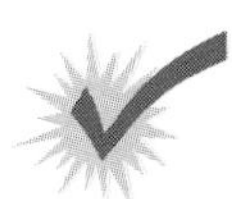

Check It Out

Find the reciprocal of each number.

7. $\frac{3}{7}$
8. 3
9. $4\frac{2}{5}$

Dividing Fractions

When you divide a fraction by a fraction, such as $\frac{1}{2} \div \frac{1}{4}$, you are really finding out how many $\frac{1}{4}$'s are in $\frac{1}{2}$. That's why the answer is 2. To divide fractions, you replace the divisor with its reciprocal and then multiply to get your answer.

$$\frac{1}{2} \div \frac{1}{4} = \frac{1}{2} \times \frac{4}{1} = 2$$

DIVIDING FRACTIONS

Divide $\frac{5}{8} \div 3\frac{3}{4}$.

- Write any mixed numbers as improper fractions.

  $\frac{5}{8} \div \frac{15}{4}$

- Replace the divisor with its reciprocal and cancel factors.

  $\frac{5}{8} \times \frac{4}{15} = \frac{\overset{1}{\cancel{5}}}{\underset{2}{\cancel{8}}} \times \frac{\overset{1}{\cancel{4}}}{\underset{3}{\cancel{15}}} = \frac{1}{2} \times \frac{1}{3}$

- Multiply.

  $\frac{1}{2} \times \frac{1}{3} = \frac{1}{6}$

The answer is $\frac{1}{6}$.

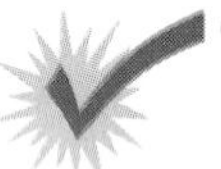

Check It Out

10. $\frac{3}{4} \div \frac{1}{2}$

11. $\frac{5}{7} \div 10$

12. $1\frac{1}{8} \div 4\frac{1}{2}$

Oseola McCarty

Miss Oseola McCarty had to leave school after sixth grade. At first she charged $1.50 to do a bundle of laundry, later $10.00. But she always managed to save. By age 86, she'd accumulated $250,000. In 1995 she decided to donate $150,000 to endow a scholarship. Miss McCarty said, "The secret to building a fortune is compounding interest. You've got to leave your investment alone long enough for it to increase."

2·4 EXERCISES

Multiply.

1. $\frac{3}{7} \times \frac{2}{3}$
2. $\frac{1}{2} \times \frac{8}{9}$
3. $\frac{2}{5} \times 3$
4. $4\frac{1}{5} \times \frac{5}{6}$
5. $3\frac{2}{3} \times 6$
6. $1\frac{5}{6} \times \frac{3}{4}$
7. $1\frac{3}{4} \times 2\frac{1}{7}$
8. $3\frac{2}{5} \times 2\frac{1}{2}$
9. $6 \times 2\frac{1}{2}$
10. $\frac{5}{6} \times 1\frac{1}{5}$
11. $2\frac{2}{3} \times 3\frac{1}{2}$
12. $4\frac{3}{8} \times 1\frac{3}{5}$

Find the reciprocal of each number.

13. $\frac{5}{8}$
14. 2
15. $3\frac{1}{5}$
16. $2\frac{2}{5}$
17. $\frac{7}{9}$

Divide.

18. $\frac{3}{4} \div \frac{3}{2}$
19. $\frac{1}{3} \div 2$
20. $\frac{1}{2} \div 1\frac{1}{2}$
21. $\frac{1}{4} \div \frac{1}{2}$
22. $2 \div 2\frac{1}{3}$
23. $3\frac{1}{2} \div 1\frac{3}{4}$
24. $0 \div \frac{1}{4}$
25. $3\frac{1}{5} \div \frac{1}{10}$
26. $1\frac{2}{3} \div 3\frac{1}{5}$
27. $4\frac{2}{3} \div 7$
28. $\frac{2}{9} \div 2\frac{2}{3}$

29. Girls make up $\frac{5}{8}$ of the eighth-grade enrollment at Marshall Middle School. If $\frac{1}{5}$ of the girls try out for the basketball team, what fractional part of the entire class is this?

30. Of the cafeteria dessert selections, $\frac{1}{3}$ are baked goods. Each day before lunch, the cafeteria workers divide the desserts so that an equal number of baked goods is available on both serving lines. On each line, what fraction of the total desserts are baked goods?

2·5 Naming and Ordering Decimals

Decimal Place Value: Tenths and Hundredths

You can use what you know about **place value** for whole numbers to help you understand decimals.

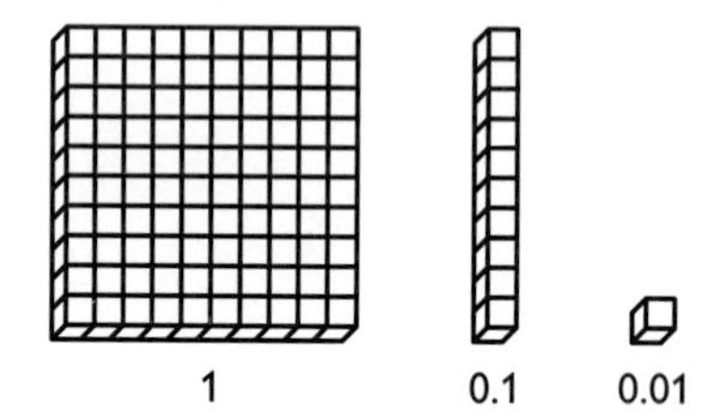

The base-ten blocks show that:
One whole (1) is 10 times greater than 1 tenth (0.1).
One tenth (0.1) is 10 times greater than 1 hundredth (0.01).

You can use a place-value chart to help you read and write decimal numbers.

ten thousands	thousands	hundreds	tens	ones	tenths	hundredths	
			3	5.	2	6	thirty-five and twenty-six hundredths
	1,	0	2	0.	0	2	one thousand twenty and two hundredths
			7	0.	7		seventy and seven tenths
			7	0.	7	0	seventy and seventy hundredths
				0.	3		three tenths

You can read the decimal by reading the whole number to the left of the decimal point as usual. You say "and" for the decimal point. Then find the place of the last decimal digit and use it to name the decimal.

You can write a decimal by writing the whole number, putting a decimal point, then placing the last digit of the decimal number in the place that names it.

1,000 + 20 + .02 is 1,020.02 written in expanded notation. The place-value chart can help you write decimals in expanded notation. You write each nonzero place as a number and add them together.

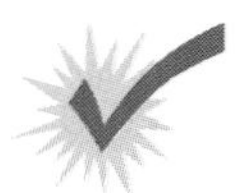

Check It Out

Write the decimal.

1. one and fifty hundredths
2. thirty-two hundredths
3. sixteen and sixty-three hundredths
4. three hundredths

Decimal Place Value: Thousandths

Thousandths is used as a more accurate measurement in sports statistics and scientific studies. The number 1 is 1,000 times 1 thousandth. The number 1 hundredth is equal to 10 thousandths ($\frac{1}{100} = \frac{10}{1,000}$).

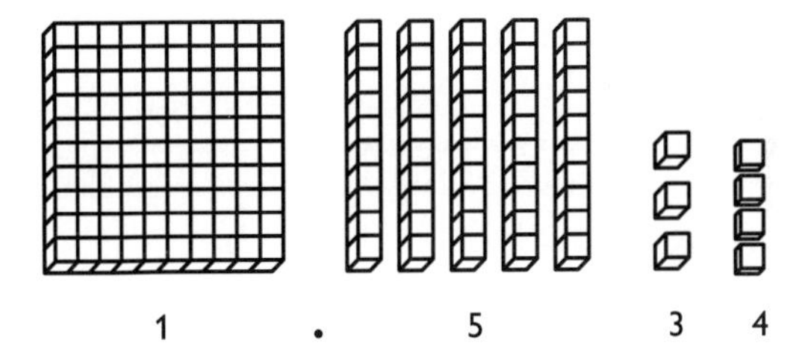

The base-ten blocks show the decimal number 1.534.

- Read the number:
  "one *and* five hundred thirty-four thousandths"
- The decimal number in expanded form is :
  1 + 0.5 + 0.03 + 0.004

Check It Out

Write the decimal in words.

5. 0.365
6. 1.102
7. 0.054

Naming Decimals Greater Than and Less Than One

Decimal numbers are based on units of ten.

ten thousands	thousands	,	hundreds	tens	ones	.	tenths	hundredths	thousandths	ten thousandths	hundred thousandths
10,000	1,000	COMMA (for thousands period)	100	10	1	DECIMAL POINT	0.1	0.01	0.001	0.0001	0.00001
							$\frac{1}{10}$	$\frac{1}{100}$	$\frac{1}{1,000}$	$\frac{1}{10,000}$	$\frac{1}{100,000}$

The chart shows place value for some of the digits of a decimal. You can use a place-value chart to help you name decimals greater than and less than one.

NAMING DECIMAL NUMBERS

Find the value of the digits in 36.7542.

- Values to the left of the decimal point are greater than one.

    36 means 3 tens and 6 ones.

- Read the decimal. The word name of a decimal is determined by the place value of the digit in the last place.

    The last digit (2) is in the ten-thousandth place.

36.7542 is read as thirty-six *and* seven thousand five hundred forty-two ten thousandths.

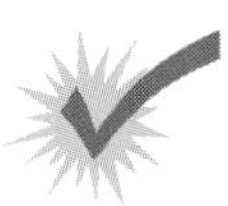

Check It Out

Use the place-value chart to tell what each boldfaced digit means. Then write the numbers in words.

8. **5**.306
9. 0.05**8**
10. 6.00**1**5
11. 0.0020**6**

Comparing Decimals

Zeros can be added to the right of the decimal in the following manner without changing the value of the number.

1.039 = 1.0390 = 1.03900 = 1.039000...

You can compare decimals by comparing their place value.

COMPARING DECIMALS

Compare 21.6032 and 21.6029.

- Start at the left. Find the first place where the numbers are different.

  21.6032 and 21.6029

  The thousandths place is different.

- Compare the digits that are different.

  $3 > 2$

- The numbers compare the same way the digits compare.

21.6032 > 21.6029

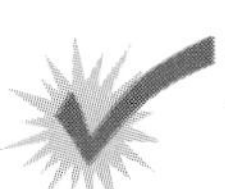

Check It Out

Write >, <, or =.

12. 0.2678 □ 0.2695
13. 24.95 □ 23.95
14. 0.007 □ 0.070

Ordering Decimals

To write decimals from least to greatest and vice versa, you need to first compare the numbers two at a time.

For example, to order 2.143; 0.214; and 2.14:

- Compare the numbers two at a time.
  2.143 > 2.140   2.140 > 0.214
- List the decimals least to greatest.
  0.214; 2.14; 2.143

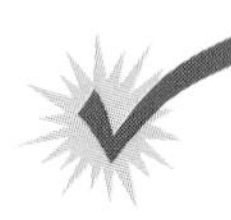

Check It Out

Write in order from least to greatest.

15. 0.7539; 0.754; 0.753; 0.759
16. 12.427; 12.0427; 12.4273; 12.00427

Rounding Decimals

Rounding decimals is similar to rounding whole numbers.

ROUNDING DECIMALS

Round 15.067 to the nearest hundredth.

- Find the rounding place. 15.067
  ↑
  hundredths
- Look at the digit to the right of the rounding place. 15.067
- If it is less than 5, leave the digit in the rounding place unchanged. If it is more than or equal to 5, increase the digit in the rounding place by 1. 7 > 5
- Write the rounded number.

15.067 rounded to the nearest hundredth is 15.07.

Check It Out

Round each decimal to the nearest hundredth.

17. 2.115   18. 38.412

2·5 EXERCISES

Give the value of the boldfaced digit in each decimal.

1. 7.0**8**9
2. 4.699**9**5
3. 1.**2**34
4. 34.49**8**

Write the decimal.

5. three and fifty-six thousandths
6. nine tenths
7. eight hundred ninety-four ten thousandths
8. twenty-three and six hundred forty-three hundred thousandths

Write the decimal in words.

9. 0.342
10. 43.9
11. 0.9999

Compare. Use <, >, or =.

12. 1.407 ☐ 1.470
13. 276.4 ☐ 276.40
14. 0.82991 ☐ 0.82909
15. 3.966 ☐ 3.960

List in order from least to greatest.

16. 12.444; 12.140; 12.404; 12,400
17. 0.96; 10.96; 0.96666; 109.6
18. 0.5; 0.55; 0.505; 0.055
19. 5.01; 50.1; 0.51; 0.15

Round each number to the place named.

20. 7.931, tenths
21. 1.9316, thousandths
22. 67.006, hundredths
23. 4.98745, ten thousandths
24. The Wong family traveled 433.44 mi on Saturday, 403.41 mi on Sunday, and 433.43 mi on Monday. Which day did they travel the farthest?
25. Sonja deposited the following amounts into her savings account: $484.59; $386.90; $566.89; and $345.45. Order the deposits from least to greatest.

2·6 Decimal Operations

Adding and Subtracting Decimals

Adding and subtracting decimals is similar to adding and subtracting whole numbers.

ADDING AND SUBTRACTING DECIMALS

Add 6.75 + 29.49 + 16.9.

- Line up the decimal points.

$$\begin{array}{r} 6.75 \\ +29.49 \\ \underline{16.9} \end{array}$$

- Add or subtract the place farthest right. Rename, if necessary.

$$\begin{array}{r} 1 \\ 6.75 \\ 29.49 \\ \underline{+16.9} \\ 4 \end{array}$$

- Add or subtract the next place left. Rename, if necessary.

$$\begin{array}{r} 2 \\ 6.75 \\ 29.49 \\ \underline{+16.9} \\ 14 \end{array}$$

- Continue through the whole numbers. Place the decimal point in the result.

$$\begin{array}{r} 6.75 \\ 29.49 \\ \underline{+16.9} \\ 53.14 \end{array}$$

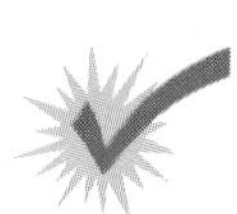

Check It Out

Solve.

1. 1.387 + 2.3444 + 3.45
2. 0.7 + 87.8 + 8.174
3. 56.13 − 17.59
4. 826.7 − 24.6444

Estimating Decimal Sums and Differences

One way that you can **estimate** decimal sums and differences is to use compatible numbers. Compatible numbers are close to the numbers in the problem and are easy to work with mentally.

ESTIMATING DECIMAL SUMS AND DIFFERENCES

Estimate the sum of $4.344 + 7.811$.

- Replace the numbers with compatible numbers.

  $4.344 \rightarrow 4$

  $7.811 \rightarrow 8$

- Add the numbers.

  $4 + 8 = 12$

Estimate the difference of $19.8 - 11.2$.

$19.8 \rightarrow 20$

$11.2 \rightarrow 10$

$20 - 10 = 10$

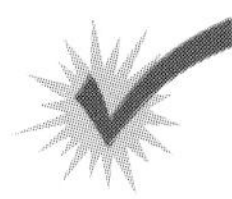

Check It Out

Estimate each sum or difference.

5. $4.63 + 7.71$
6. $12.4 - 10.66$
7. $19.055 - 4.41$
8. $124.95 + 59.50 + 100.40$

Multiplying Decimals

Multiplying decimals is much the same as multiplying whole numbers. You can model the multiplication of decimals with a 10×10 grid. Each tiny square is equal to one hundredth.

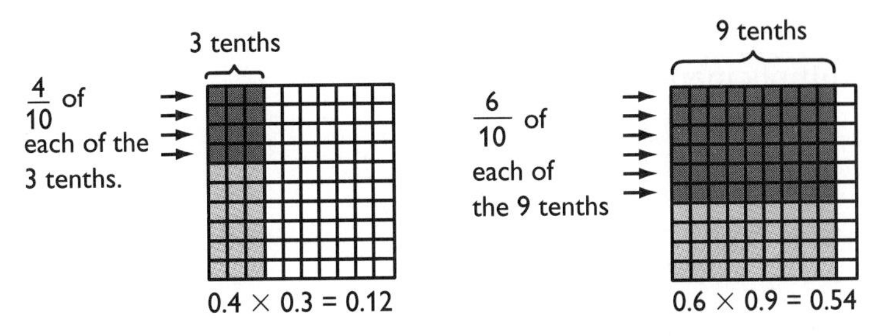

MULTIPLYING DECIMALS

Multiply 42.8×0.06.

- Multiply as with **whole numbers.**

$$\begin{array}{r} 42.8 \\ \underline{\times\ 0.06} \\ 2568 \end{array} \qquad \begin{array}{r} 428 \\ \underline{\times\ 6} \\ 2568 \end{array}$$

- Add the number of decimal places for the factors.

$$\begin{array}{rl} 42.8 & \leftarrow 1 \text{ decimal place} \\ \underline{\times\ 0.06} & \leftarrow 2 \text{ decimal places} \\ & \quad 1 + 2 = 3 \text{ decimal places} \end{array}$$

- Place the decimal point in the product.

$$\begin{array}{rl} 42.8 & \leftarrow 1 \text{ decimal place} \\ \underline{\times\ 0.06} & \leftarrow 2 \text{ decimal places} \\ 2.568 & \leftarrow 3 \text{ decimal places} \end{array}$$

$42.8 \times 0.06 = 2.568$

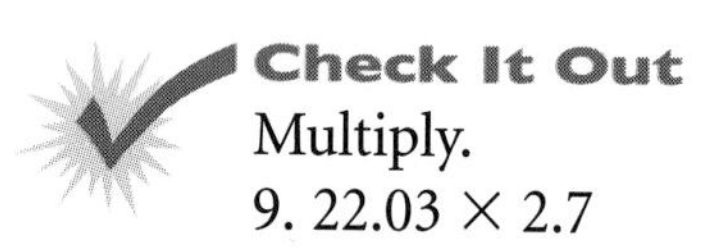

Check It Out

Multiply.

9. 22.03×2.7

10. 9.655×8.33

Estimating Decimal Products

To estimate decimal products, you can replace given numbers with compatible numbers. Compatible numbers are estimates you choose because they are easier to work with mentally.

Estimate 26.2×52.3.

- Replace the factors with compatible numbers.

  $26.2 \rightarrow 30 \qquad 52.3 \rightarrow 50$

- Multiply mentally.

  $30 \times 50 = 1{,}500$

Check It Out

Estimate each product by using simpler estimates.

11. 12.75 × 91.3
12. 3.76 × 0.61

Multiplying Decimals with Zeros in the Product

Sometimes when you are multiplying decimals, you need to add zeros in the product.

ZEROS IN THE PRODUCT

Multiply 0.2 × 0.375.

- Multiply as with whole numbers.

$$\begin{array}{r} 0.375 \\ \underline{\times\ 0.2} \end{array} \qquad \begin{array}{r} 375 \\ \underline{\times\ 2} \\ 750 \end{array}$$

- Count the number of decimal places in the factors.

$$\begin{array}{rl} 0.375 & \leftarrow 3 \text{ decimal places} \\ \underline{\times\ 0.2} & \leftarrow 1 \text{ decimal place} \\ 750 & \end{array}$$

- The product should have the same number of decimal places as the factors. Add zeros in the product as necessary.

  Since 4 places are needed in the product, write a zero to the left of the 7.

0.2 × 0.375 = 0.0750 = 0.075

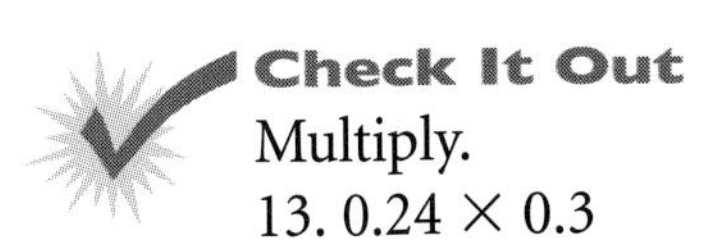

Check It Out

Multiply.

13. 0.24 × 0.3
14. 0.0007 × 4.033

Dividing Decimals

Dividing decimals is similar to dividing whole numbers. You can use a model to help you understand dividing decimals. For example, $0.8 \div 0.2$ means, how many groups of 0.2 in 0.8? There are 4 groups of 0.2 in 0.8, so $0.8 \div 0.2 = 4$.

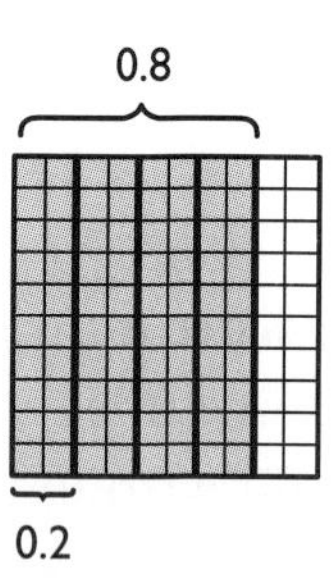

DIVIDING DECIMALS

Divide $38.35 \div 6.5$.

- Multiply the divisor by a power of ten to make it a whole number. — $6.5\overline{)38.35}$; $6.5 \times 10 = 65$
- Multiply the dividend by the same power of ten. — $65.\overline{)383.5}$; $38.35 \times 10 = 383.5$
- Place the decimal point in the quotient. — $65.\overline{)383.5}$ with the decimal point placed above
- Divide.

$$\begin{array}{r} \mathbf{5.9} \\ 65.\overline{)383.5} \\ \underline{325} \\ 58\,5 \\ \underline{58\,5} \\ 0 \end{array}$$

$38.35 \div 6.5 = 5.9$

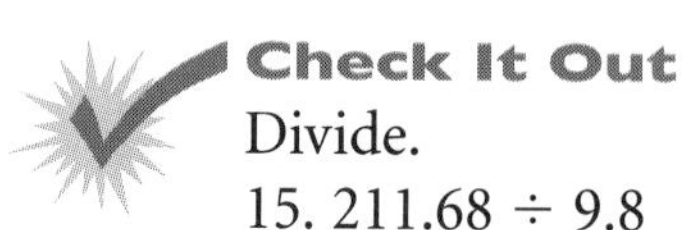
Check It Out

Divide.

15. $211.68 \div 9.8$
16. $42.363 \div 8.1$
17. $444.36 \div 4.83$
18. $1.548 \div 0.06$

Zeros in Division

You can use zeros to hold places in the dividend when you are dividing decimals.

ZEROS IN DIVISION

Divide 375.1 ÷ 6.2.

- Multiply the divisor and the dividend by a power of ten. Place the decimal point.

  6.2)375.1
  $6.2 \times 10 = 62$
  $375.1 \times 10 = 3751$

- Divide.

```
      60.
62.)3751.
    372
      31
       0
      31
```

- Use zeros to hold places in the dividend. Continue to divide until the remainder is zero.

```
      60.5
62.)3751.0
    372
      31
       0
      310
      310
        0
```

375.1 ÷ 6.2 = 60.5

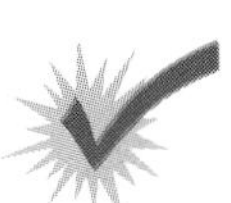

Check It Out

Divide until the remainder is zero.

19. 0.7042 ÷ 0.07

20. 37.2 ÷ 1.5

Rounding Decimal Quotients

You can use a calculator to divide decimals and round quotients.

Divide 6.3 by 2.6. Round to the nearest hundredth.

- Use your calculator to divide.

  6.3 [÷] 2.6 [=] [2.4230769]

- To round the quotient, look at one place to the right of the rounding place. Then round.

  2.4230769 rounds to 2.42.

Some calculators have a "fix" function. Press [FIX] and the number of decimal places you want. The calculator will then display all numbers rounded to that number of places.

Check It Out

Solve with a calculator. Round to the nearest hundredth.

21. 0.0258 ÷ 0.345

22. 0.817 ÷ 1.25

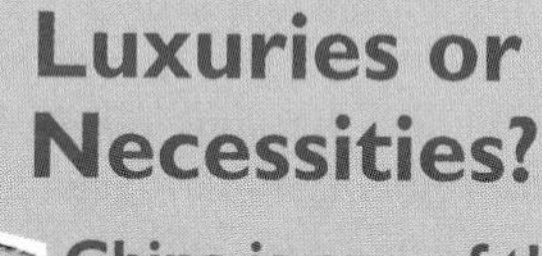

China is one of the fastest growing economies in the world. But after years of hardship, the country hasn't yet caught up in providing luxuries for its vast population of approximately 1,200,000,000.

	China	United States
Number of people per telephone	36.4	1.3
Number of people per TV	6.7	1.2

When China has the same number of telephones per person as the United States has, how many will it have? See Hot Solutions for answer.

2·6 EXERCISES

Estimate each sum or difference.

1. 7.61 − 0.82
2. 9.34 − 5.82
3. \$25.55 + \$195.38
4. 4.972 + 3.548
5. 6.42 − 0.81

Add.

6. 256.3 + 0.624
7. 78.239 + 38.6
8. 7.02396 + 4.88
9. \$250.50 + \$385.16
10. 2.9432 + 1.9 + 3 + 1.975

Subtract.

11. 43 − 28.638
12. 58.543 − 0.768
13. 435.2 − 78.376
14. 38.3 − 16.254
15. 11.01 − 2.0063

Multiply.

16. 0.66 × 17.3
17. 0.29 × 6.25
18. 7.526 × 0.33
19. 37.82 × 9.6
20. 22.4 × 9.4

Divide until the remainder is zero.

21. 29.38 ÷ 0.65
22. 62.55 ÷ 4.5
23. 84.6 ÷ 4.7
24. 0.657 ÷ 0.6

Divide. Round to the nearest hundredth.

25. 142.7 ÷ 7
26. 2.55 ÷ 1.6
27. 22.9 ÷ 6.2
28. 15.25 ÷ 2.3

29. The moon orbits the Earth in 27.3 days. How many orbits does the moon make in 365.25 days? Round your answer to the nearest hundredth.
30. *Apollo 11* astronauts Scott and Irwin drove the lunar rover about 26.4 km on the moon. Their average speed was 3.3 km/hr. How long did they drive the lunar rover?

2·7 Meaning of Percent

Naming Percents

A ratio of a number to 100 is called a **percent.** Percent means *per hundred* and is represented by the symbol %.

You can use graph paper to model percents. There are 100 squares in a 10 by 10 grid of graph paper. So the grid can be used to represent 100%. Since percent means how many out of 100, it is easy to tell what percent of the 100-square grid is shaded.

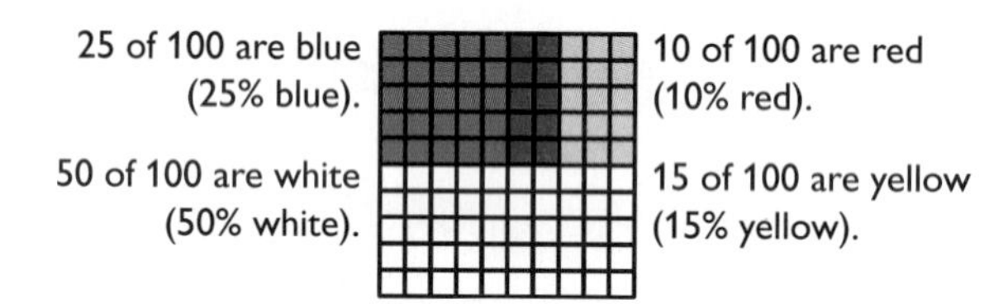

Check It Out

What percent of each square is shaded?

1.

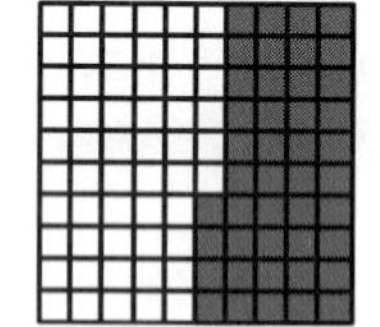

2.

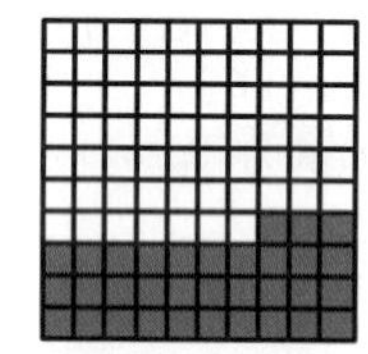

Understanding the Meaning of Percent

Any *ratio* with 100 as the second number can be expressed in three ways. You can write the ratio as a fraction, a decimal, and a percent.

A quarter is 25% of \$1.00. You can express a quarter as 25¢, \$0.25, $\frac{1}{4}$ of a dollar, $\frac{25}{100}$ of a dollar, and 25% of a dollar.

One way to think about percents is to become very comfortable with a few. You build what you know about percents based on these few **benchmarks.** You can use these benchmarks to help you estimate percents of other things.

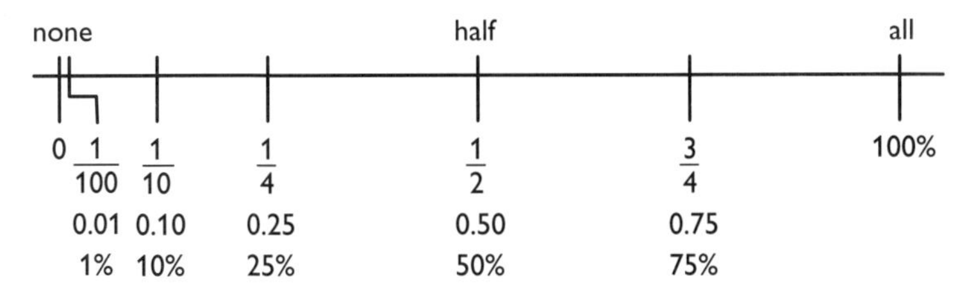

ESTIMATING PERCENTS

Estimate 26% of 200.

- Choose a benchmark, or combination of benchmarks, close to the target percent.

  26% is close to 25%.

- Find the fraction or decimal equivalent to the benchmark percent.

  $\frac{1}{4}$ is equal to 25%.

- Use the benchmark equivalent to estimate the percent.

  $\frac{1}{4}$ of 200 is 50.

26% of 200 is about 50.

Check It Out

Use fractional benchmarks to estimate the percents.

3. 47% of 300
4. 22% of 400
5. 72% of 200
6. 99% of 250

Using Mental Math to Estimate Percent

You can use fraction or decimal benchmarks to help you quickly estimate the percent of something, such as a tip in a restaurant.

Estimate a 15% tip for a bill of \$5.45.

- Round to a convenient number.

  \$5.45 rounds to \$5.50.

- Think of the percent as a combination of benchmarks.

  $15\% = 10\% + 5\%$ $10\% = 0.10 = \frac{1}{10}$ $5\% =$ half of 10%

- Multiply mentally.

  10% of \$5.50 $= 0.10 \times 5.50 = 0.55$

  5% of \$5.50 = half of 10% = half of 0.55 = about 0.25

  $0.55 + 0.25 =$ about 0.80

The tip is about \$0.80.

Check It Out

Estimate the amount of each tip.

7. 20% of \$4.75
8. 15% of \$40

Honesty Pays

David Hacker, a cabdriver, found a wallet in the back seat of the cab that contained \$25,000—about a year's salary for him.

The owner's name was in the wallet and Hacker remembered where he had dropped him off. He went straight to the hotel and found the man. The owner, a businessman, had already realized he had lost his wallet and figured he would never see it again. He didn't believe anyone would be that honest! On the spot, he handed the cabdriver fifty \$100 bills.

What percent of the money did Hacker receive as a reward? See Hot Solutions for answer.

2•7 EXERCISES

Write the percent of each square that is shaded.

1. 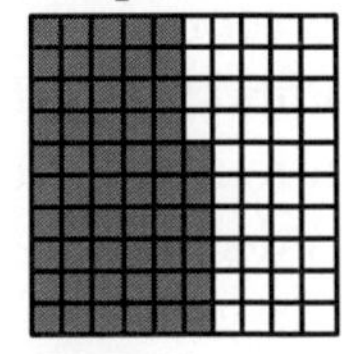
2. 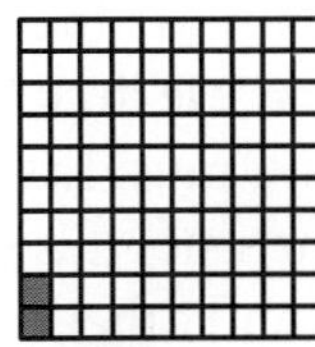
3. 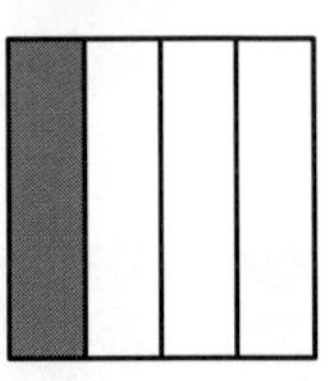
4. 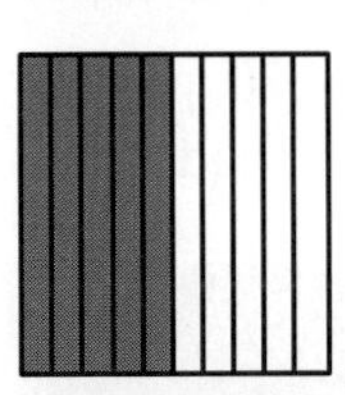
5. 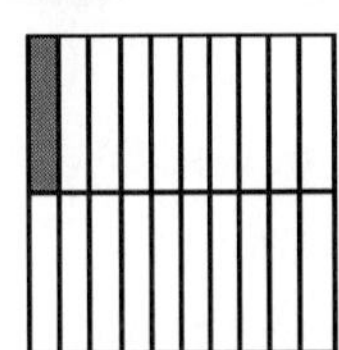
6. 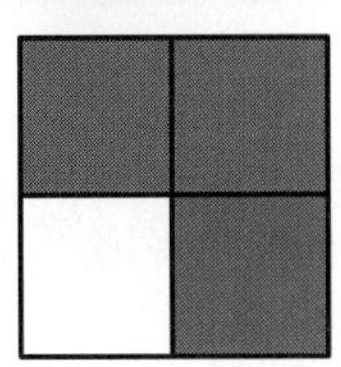
7. If you made half of your free throws, what percent would you be shooting?

Use fractional benchmarks to estimate the percent of each number.

8. 15% of 200
9. 49% of 800
10. 2% of 50
11. 76% of 200
12. Estimate a 15% tip for a bill of $65.
13. Estimate a 20% tip for a bill of $49.
14. Estimate a 10% tip for a bill of $83.
15. Estimate an 18% tip for a bill of $79.

2·8 Using and Finding Percents

Finding a Percent of a Number

There are several ways that you can find the percent of a number. You can use decimals or fractions. To find the percent of a number, you must first change the percent to a decimal or a fraction. Sometimes it is easier to change to a decimal representation and other times to a fractional one.

To find 30% of 80, you can use either the fraction method or the decimal method.

FINDING THE PERCENT OF A NUMBER: TWO METHODS

Find 30% of 80.

DECIMAL METHOD

- Change percent to decimal.

  $30\% = 0.3$

- Multiply.

  $80 \times .3 = 24$

FRACTION METHOD

- Change percent to fraction in lowest terms.

  $30\% = \frac{30}{100} = \frac{3}{10}$

- Multiply.

  $80 \times \frac{3}{10} = 24$

So 30% of 80 = 24.

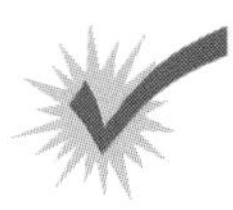

Check It Out

Give the percent of each number.

1. 80% of 75
2. 95% of 700
3. 21% of 54
4. 75% of 36

Finding the Percent of a Number: Proportion Method

You can use proportions to help you find the percent of a number.

FINDING THE PERCENT OF A NUMBER: PROPORTION METHOD

Pei works in a sporting goods store. He receives a commission of 12% on his sales. Last month he sold $9,500 worth of sporting goods. What was his commission?

- Use a proportion to find the percent of a number.

  P = Part (of the base or total) R = Rate (percentage)

  B = Base (total) $\frac{P}{R} = \frac{B}{100}$

- Identify the given items before trying to find the unknown.

P is	R is	B is
unknown, call x.	12%.	$9,500.

- Set up the proportion.

  $\frac{P}{R} = \frac{B}{100}$ $\frac{x}{12} = \frac{9{,}500}{100}$

- Cross multiply.

  $100x = 114{,}000$

- Divide both sides of the equation by the value of x.

  $\frac{114{,}000}{100} = \frac{100x}{100}$ $1{,}140 = x$

Pei received $1,140 in commission.

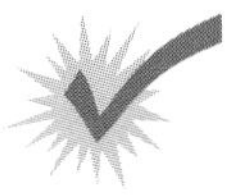

Check It Out

Use a proportion to find the percent of each number.

5. 95% of 700
6. 150% of 48
7. 65% of 200
8. 85% of 400

Finding Percent and Base

Finding what percent a number is of another number and finding what number is a certain percent of another number can be confusing. Setting up and solving a proportion can make the process less confusing.

Use the ratio $\frac{P}{B} = \frac{R}{100}$ where P = Part (of base), B = Base (total), and R = Rate (percentage).

FINDING THE PERCENT

What percent of 70 is 14 ?

- Set up a proportion, using this form.

  $\frac{\text{Part}}{\text{Base}} = \frac{\text{Rate}}{100}$

  $\frac{14}{70} = \frac{n}{100}$

  (The number after the word *of* is the base.)
- Show the cross products of the proportion.

  $14 \times 100 = 70 \times n$
- Find the products.

  $1{,}400 = 70n$
- Divide both sides of the equation by the coefficient of n.

  $\frac{1{,}400}{70} = \frac{70n}{70}$

  $n = 20$

14 is 20% of 70.

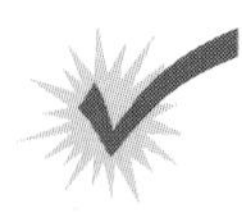

Check It Out

Solve.

9. What percent of 240 is 80?
10. What percent of 64 is 288?
11. What percent of 2 is 8?
12. What percent of 55 is 33?

FINDING THE BASE

12 is 48% of what number?

- Set up a percent proportion using this form.

  $\frac{\text{Part}}{\text{Base}} = \frac{\text{Percent}}{100}$

  (The phrase *what number* after the word *of* is the base.)

  $\frac{12}{n} = \frac{48}{100}$

- Show the cross products of the proportion.

  $12 \times 100 = 48 \times n$

- Find the products.

  $1200 = 48n$

- Divide both sides of the equation by the coefficient of n.

  $\frac{1200}{48} = \frac{48n}{48}$

  $n = 25$

12 is 48% of 25.

Check It Out

13. 52 is 50% of what number?
14. 15 is 75% of what number?
15. 40 is 160% of what number?
16. 84 is 7% of what number?

Percent of Increase or Decrease

Sometimes it is helpful to keep a record of your monthly expenses. Keeping a record allows you to see the actual percent of increase or decrease in your expenses. You can make a chart to record expenses.

EXPENSES	SEPTEMBER	OCTOBER	INCREASE OR DECREASE (AMOUNT OF)	(% OF)
Food	225	189	36	16%
Travel	75	93	18	
Rent	360	375	15	4%
Clothing	155	62	93	60%
Miscellaneous	135	108	27	20%
Entertainment	80	44		
Total	1,030	871	159	15%

You can use a calculator to find the percent of increase or decrease.

FINDING THE PERCENT OF INCREASE

During September $75 was spent on travel. In October the amount spent on travel was $93.

- Use a calculator to key in the following.

  new amount [−] original amount [=] amount of increase

  93 [−] 75 [=] [18.]

- Leave the amount of increase on display.

  [18.]

- Use your calculator to divide the amount of increase by the original amount.

  $\frac{\text{amount of increase}}{\text{original amount}}$ [=] percent of increase

  [18.] [÷] 75 [=] [0.24]

- Round to nearest hundredth and convert to percent.

  0.24 = 24%

The percent of increase from $75 to $93 is 24%.

Check It Out

Use a calculator to find the percent of increase.

17. 56 to 70
18. 20 to 39
19. 45 to 99
20. 105 to 126

FINDING THE PERCENT OF DECREASE

During September $80 was spent on entertainment. In October $44 was spent on entertainment. You can use a calculator to find the percent of decrease.

- Use a calculator to key in the following.

  original amount [−] new amount [=] amount of decrease

  80 [−] 44 [=] [36.]

- Leave the amount of decrease on display.

  [36.]

- Use your calculator to divide the amount of decrease by the original amount.

  $\frac{\text{amount of decrease}}{\text{original amount}}$ [=] percent of decrease

  [36.] [÷] 80 [=] [0.45]

- Round to nearest hundredth and convert from decimal to percent.

  0.45 = 45%

The percent of decrease from $80 to $44 is 45%.

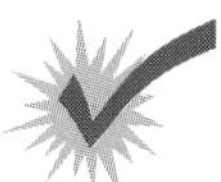

Check It Out

Use your calculator to find the percent of decrease.

21. 72 to 64
22. 46 to 23
23. 225 to 189
24. 120 to 84

Discounts and Sale Prices

A **discount** is the amount that an item is reduced from the regular price. The sale price is the regular price minus the discount. Discount stores have regular prices marked below the suggested retail price. You can use percent to find discounts and resulting sale prices.

A CD player has a regular price of $109.99. It is on sale for 25% off the regular price. How much money will you save by buying the item on sale?

You can use a calculator to help you find the discount and resulting sale price of an item.

FINDING DISCOUNTS AND SALE PRICES

The regular price of the item is $109.99. It is marked 25% off. Find the discount and the sale price.

- Use your calculator to multiply the regular price times the discount percent.

  regular price [×] discount percent [=] discount

  109.99 [×] 25 [%] [=] [27.4975]

- If necessary, round the discount to the nearest hundredth.

  27.4975 = 27.50

  The discount is $27.50.

- Use your calculator to subtract the discount from the regular price. This will give you the sale price.

  regular price [−] discount [=] sale price

  109.99 [−] 27.50 [=] [82.49]

The sale price is $82.49.

Check It Out

Use a calculator to find the discount and sale price.

25. Regular price: $813.25, Discount percent: 20%
26. Regular price: $18.90, Discount percent: 30%

Estimating a Percent of a Number

You can use what you know about compatible numbers and simple fractions to estimate a percent of a number. You can use the table to help you estimate the percent of a number.

Percent	1%	5%	10%	20%	25%	$33\frac{1}{3}\%$	50%	$66\frac{2}{3}\%$	75%	100%
Fraction	$\frac{1}{100}$	$\frac{1}{20}$	$\frac{1}{10}$	$\frac{1}{5}$	$\frac{1}{4}$	$\frac{1}{3}$	$\frac{1}{2}$	$\frac{2}{3}$	$\frac{3}{4}$	1

ESTIMATING A PERCENT OF A NUMBER

Estimate 17% of 46.

- Find the percent that is closest to the percent you are asked to find.

  17% is about 20%.
- Find the fractional equivalent for the percent.

  20% is equivalent to $\frac{1}{5}$.
- Find a compatible number for the number you are asked to find the percent of.

  46 is about 50.
- Use the fraction to find the percent.

  $\frac{1}{5}$ of 50 is 10.

17% of 46 is about 10.

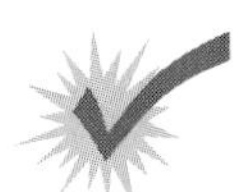

Check It Out

Use compatible numbers to estimate.

27. 67% of 150
28. 35% of 6
29. 27% of 54
30. 32% of 89

Finding Simple Interest

When you have a savings account, the bank pays you for the use of your money. With a loan, you pay the bank for the use of their money. In both situations, the payment is called *interest.* The amount of money you borrow or save is called the *principal.*

You want to borrow \$5,000 at 7% interest for 3 years. To find out how much interest you will pay, you can use the formula $I = P \times R \times T$. The chart can help you understand the formula.

P	Principal — the amount of money you borrow or save
R	Interest Rate — a percent of the principal you pay or earn
T	Time — the length of time you borrow or save
I	Total Interest — interest you pay or earn for the entire time
A	Amount — total amount (principal plus interest) you pay or earn

FINDING SIMPLE INTEREST

You can use a calculator to help find the interest that you will pay if you borrow \$5,000 at 7% interest for 3 years.

- Multiply the principal (*P*) by the interest rate (*R*) by the time (*T*) to find the interest (*I*) you will pay.

  $P \times R \times T = I$

  5000 [×] 7 [%] [×] 3 [=] 1050.

  \$1,050 is the interest.

- To find the total amount you will pay back, add the principal and the interest.

  $P + I = A$

  5000 [+] 1050 [=] 6050.

\$6,050 is the total amount of money to be paid back.

Check It Out

Find the interest (*I*) and the total amount (*A*).

31. Principal: \$4,800
    Rate: 12.5%
    Time: 3 years

32. Principal: \$2,500
    Rate: 3.5%
    Time: $1\frac{1}{2}$ years

2·8 EXERCISES

Find the percent of each number.

1. 2% of 50
2. 42% of 700
3. 125% of 34
4. 4% of 16.3

Solve.

5. What percent of 60 is 48?
6. 14 is what percent of 70?
7. 3 is what percent of 20?
8. What percent of 8 is 6?

Solve.

9. 82% of what number is 492?
10. 24% of what number is 18?
11. 3% of what number is 4.68?
12. 80% of what number is 24?

Find the percent of increase or decrease to the nearest percent.

13. 20 to 39
14. 175 to 91
15. 112 to 42

Estimate the percent of each number.

16. 48% of 70
17. 34% of 69

18. Mariko needed a helmet to snowboard on the half-pipe at Holiday Mountain. She bought a helmet for 45% off the regular price of $39.50. How much did she save? How much did she pay?

19. A snowboard is on sale for 20% off the regular price of $389.50. Find the discount and the sale price of the snowboard.

Find the discount and sale price.

20. Regular Price: $80
    Discount Percent: 20%
21. Regular Price: $17.89
    Discount Percent: 10%
22. Regular Price: $1,200
    Discount Percent: 12%
23. Regular Price: $250
    Discount Percent: 18%

Find the interest. Use a calculator.

24. $P = \$9{,}000$
    $R = 7.5\%$ per year
    $T = 2\frac{1}{2}$ years
25. $P = \$1{,}500$
    $R = 9\%$ per year
    $T = 2$ years

2·9 Fraction, Decimal, and Percent Relationships

Percents and Fractions

Percents and fractions can both describe a ratio out of 100. The chart will help you understand the relationship between percents and fractions.

Percent	Fraction
50 out of 100 = 50%	$\frac{50}{100} = \frac{1}{2}$
$33\frac{1}{3}$ out of 100 = $33\frac{1}{3}$%	$\frac{33.\overline{3}}{100} = \frac{1}{3}$
25 out of 100 = 25%	$\frac{25}{100} = \frac{1}{4}$
20 out of 100 = 20%	$\frac{20}{100} = \frac{1}{5}$
10 out of 100 = 10%	$\frac{10}{100} = \frac{1}{10}$
1 out of 100 = 1%	$\frac{1}{100} = \frac{1}{100}$
$66\frac{2}{3}$ out of 100 = $66\frac{2}{3}$%	$\frac{66.\overline{6}}{100} = \frac{2}{3}$
75 out of 100 = 75%	$\frac{75}{100} = \frac{3}{4}$

You can write fractions as percents and percents as fractions.

CONVERTING A FRACTION TO A PERCENT

Express $\frac{2}{5}$ as a percent.

- Set up a proportion. $\frac{2}{5} = \frac{n}{100}$
- Solve the proportion. $2 \times 100 = 5n$

  $\frac{2 \times 100}{5} = n$

  $n = 40$
- Express as a percent. 40%

  $\frac{2}{5} = 40\%$

Check It Out

Change each fraction to a percent.

1. $\frac{4}{5}$
2. $\frac{13}{20}$
3. $\frac{180}{400}$
4. $\frac{19}{50}$

Changing Percents to Fractions

To change from a percent to a fraction, write the percent as the numerator of a fraction with a denominator of 100, and express in lowest terms.

CHANGING PERCENTS TO FRACTIONS

Express 45% as a fraction.

- Change the percent directly to a fraction with a denominator of 100. The number of the percent becomes the numerator of the fraction.

  $45\% = \frac{45}{100}$
- Simplify if possible (p. 106).

  $\frac{45}{100} = \frac{9}{20}$

45% expressed as a fraction is $\frac{9}{20}$.

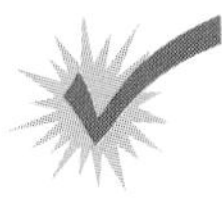

Check It Out

Change each percent to a fraction in lowest terms.

5. 55%
6. 29%
7. 85%
8. 92%

Changing Mixed Number Percents to Fractions

To change the *mixed number* (p. 109) $54\frac{1}{2}\%$ to a fraction, first change the mixed number to an *improper fraction* (p. 110).

- Change the mixed number to an improper fraction.
  $54\frac{1}{2}\% = \frac{109}{2}\%$
- Multiply the percent by $\frac{1}{100}$.
  $\frac{109}{2} \times \frac{1}{100} = \frac{109}{200}$
- Simplify, if possible.
  $54\frac{1}{2}\% = \frac{109}{200}$

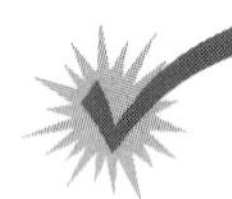

Check It Out

Change each mixed number percent to a fraction.

9. $44\frac{1}{2}\%$
10. $34\frac{2}{5}\%$

Percents and Decimals

Percents can be expressed as decimals and decimals can be expressed as percents. *Percent* means part of a hundred or hundredths.

CHANGING DECIMALS TO PERCENTS

Change 0.8 to a percent.

- Multiply the decimal by 100.
  $0.8 \times 100 = 80$
- Add the percent sign.
  $0.8 \rightarrow 80\%$

A Shortcut for Changing Decimals to Percents

Change 0.5 to a percent.

- Move the decimal point two places to the right. Add zeros, if necessary. $0.5 \rightarrow 50.$
- Add the percent sign. $0.5 \rightarrow 50\%$

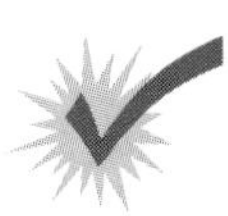

Check It Out

Write each decimal as a percent.

11. 0.08
12. 0.66
13. 0.398
14. 0.74

Since *percent* means part of a hundred, percents can be converted directly to decimals.

CHANGING PERCENTS TO DECIMALS

Change 3% to a decimal.

- Express the percent as a fraction with 100 as the denominator.

  $3\% = \frac{3}{100}$

- Divide the numerator by 100.

  $3 \div 100 = 0.03$

$3\% = 0.03$

A Shortcut for Changing Percents to Decimals

Change 8% to a decimal.

- Move the decimal point two places to the left.

  $8\% \rightarrow .\,8.$

- Add zeros, if necessary.

  $8\% = 0.08$

Check It Out

Express each percent as a decimal.

15. 14.5%
16. 0.01%
17. 23%
18. 35%

The Ups and Downs of Stocks

A corporation raises money by selling stocks—certificates that represent shares of ownership in the corporation. The stock page of a newspaper lists the high, low, and ending prices of the stock for the previous day. It also shows the overall fractional amount by which the price changed. A (+) sign indicates that the value of the stock increased; a (–) sign indicates the value decreased.

Suppose you see a listing on the stock page that shows the closing price of a stock was $21\frac{3}{4}$ with $+\frac{1}{4}$ next to it. What do those fractions mean? First it tells you that the price of the stock was $21\frac{3}{4}$ dollars or \$21.75. The $+\frac{1}{4}$ means the price went up $\frac{1}{4}$ of a dollar from the day before. Since $\frac{1}{4} \times \$1.00 = \$0.25$, the stock went up 25¢. To find the percent increase in the price of the stock, you have to first determine the original price of the stock. The stock went up $\frac{1}{4}$, so the original price is $21\frac{3}{4} - \frac{1}{4} = 21\frac{1}{2}$. What is the percent of increase to the nearest whole percent? See Hot Solutions for answer.

Fractions and Decimals

Fractions can be written as either **terminating** or **repeating decimals.**

Fractions	Decimals	Terminating or repeating
$\frac{1}{2}$	0.5	terminating
$\frac{1}{3}$	$0.333333\overline{3}$	repeating
$\frac{1}{6}$	$0.16666\overline{6}$	repeating
$\frac{2}{3}$	$0.6666\overline{6}$	repeating
$\frac{1}{11}$	$0.0909\overline{09}$	repeating
$\frac{3}{22}$	$0.13636\overline{36}$	repeating

CHANGING FRACTIONS TO DECIMALS

Write $\frac{3}{25}$ as a decimal.

- Divide the numerator of the fraction by the denominator.

  $3 \div 25 = 0.12$

$\frac{3}{25} = 0.12$. The remainder is zero. The decimal is a terminating decimal.

Write $\frac{1}{6}$ and $\frac{5}{22}$ as decimals.

- Divide the numerator of the fraction by the denominator.

  $1 \div 6 = 0.1666\ldots \quad 5 \div 22 = 0.22727\ldots$

- Place a bar over any digit or digits that repeat.

  $0.1\overline{6} \qquad 0.2\overline{27}$

$\frac{1}{6} = 0.1\overline{6}$ and $\frac{5}{22} = 0.2\overline{27}$. Both decimals are repeating decimals.

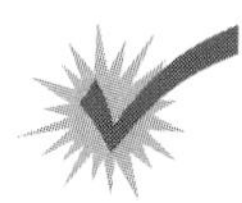

Check It Out

Use a calculator to find a decimal for each fraction.

19. $\frac{4}{5}$
20. $\frac{11}{20}$
21. $\frac{28}{32}$
22. $\frac{5}{12}$

CHANGING DECIMALS TO FRACTIONS

Write 0.55 as a fraction.

- Write the decimal as a fraction.

$$0.55 = \frac{55}{100}$$

- Write the fraction in lowest terms (p. 106).

$$\frac{55}{100} = \frac{55 \div 5}{100 \div 5} = \frac{11}{20}$$

$$0.55 = \frac{11}{20}$$

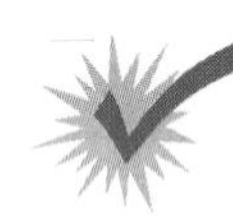

Check It Out

Write each decimal as a fraction.

23. 2.4
24. 0.056
25. 0.14
26. 1.2

2·9 EXERCISES

Change each fraction to a percent.

1. $\frac{17}{100}$
2. $\frac{19}{20}$
3. $\frac{13}{100}$
4. $\frac{19}{50}$
5. $\frac{24}{25}$

Change each percent to a fraction in lowest terms.

6. 42%
7. 60%
8. 44%
9. 12%
10. 80%

Write a decimal as a percent.

11. 0.4
12. 0.41
13. 0.105
14. 0.83
15. 3.6

Write each percent as a decimal.

16. 35%
17. 13.6%
18. 18%
19. 4%
20. 25.4%

Change each fraction to a decimal. Use bar notations to show repeating decimals.

21. $\frac{3}{18}$
22. $\frac{30}{111}$
23. $\frac{4}{18}$
24. $\frac{7}{15}$

Write each decimal or mixed number as a fraction.

25. 0.4
26. 2.004
27. 3.42
28. 0.27

29. One middle-school survey said 40% of eighth grade students preferred pizza for lunch. Another survey said $\frac{2}{5}$ of eighth grade students preferred pizza for lunch. Could both surveys be correct? Explain.

30. Blades on Second is advertising $109 skateboards at 33% off. Skates on Seventh is advertising the same skateboard at $\frac{1}{3}$ off. Which is the better buy?

What have you learned?

You can use the problems and the list of words that follow to see what you have learned in this chapter. You can find out more about a particular problem or word by referring to the boldfaced topic number (for example, **2•2**).

Problem Set

1. Of the 16 girls on the softball team, 12 play regularly. What percent play regularly? **2•8**
2. Itay missed 6 questions on a 25-question test. What percent did he get correct? **2•8**
3. Fenway Park in Boston has a seating capacity of 34,450 seats. 27% of the seats are held by season-ticket holders. How many seats are taken by season-ticket holders? **2•8**
4. Which fraction is equivalent to $\frac{14}{21}$? **2•1**

   A. $\frac{2}{7}$ B. $\frac{7}{7}$ C. $\frac{2}{3}$ D. $\frac{3}{2}$
5. Which fraction is greater, $\frac{1}{12}$ or $\frac{3}{35}$? **2•2**

Add or subtract. Write your answers in lowest terms. **2•3**

6. $\frac{5}{8} + \frac{3}{4}$
7. $2\frac{1}{5} - 1\frac{1}{2}$
8. $3 - 1\frac{1}{8}$
9. $7\frac{3}{4} + 2\frac{7}{8}$
10. Write the improper fraction $\frac{11}{4}$ as a mixed number. **2•1**

In items 11–14, multiply or divide as indicated. **2•4**

11. $\frac{4}{5} \times \frac{5}{6}$
12. $\frac{3}{10} \div 4\frac{1}{2}$
13. $2\frac{5}{8} \times \frac{4}{7}$
14. $5\frac{1}{3} \div 2\frac{1}{6}$
15. Give the place value of the 2 in 455.021. **2•5**
16. Write 6.105 in expanded form. **2•5**
17. Write as a decimal: Three hundred two and twenty-three thousandths. **2•5**
18. Write the following numbers in order from least to greatest: 0.990; 0.090; 0.099; 0.909. **2•5**

Solve. **2•6**

19. 10.55 + 3.884
20. 13.4 − 2.08
21. 8.05 × 6.4
22. 69.69 ÷ 11.5

Use a calculator. Round answers to the nearest tenth. **2•8**

23. What percent of 125 is 30?
24. Find 18% of 85.
25. 36 is 40% of what number?

hot words

WRITE DEFINITIONS FOR THE FOLLOWING WORDS.

benchmark **2•7**
common denominator **2•2**
cross product **2•1**
denominator **2•1**
discount **2•8**
equivalent fractions **2•1**
estimate **2•6**
factor **2•4**
fraction **2•1**
greatest common factor **2•1**
improper fraction **2•1**
mixed number **2•1**
numerator **2•1**
percent **2•7**
place value **2•5**
product **2•4**
reciprocal **2•4**
repeating decimal **2•9**
terminating decimal **2•9**
whole number **2•1**

hot topics 3

Powers and Roots

3•1	Powers and Exponents	168
3•2	Square and Cube Roots	176
3•3	Scientific Notation	182
3•4	Laws of Exponents	188

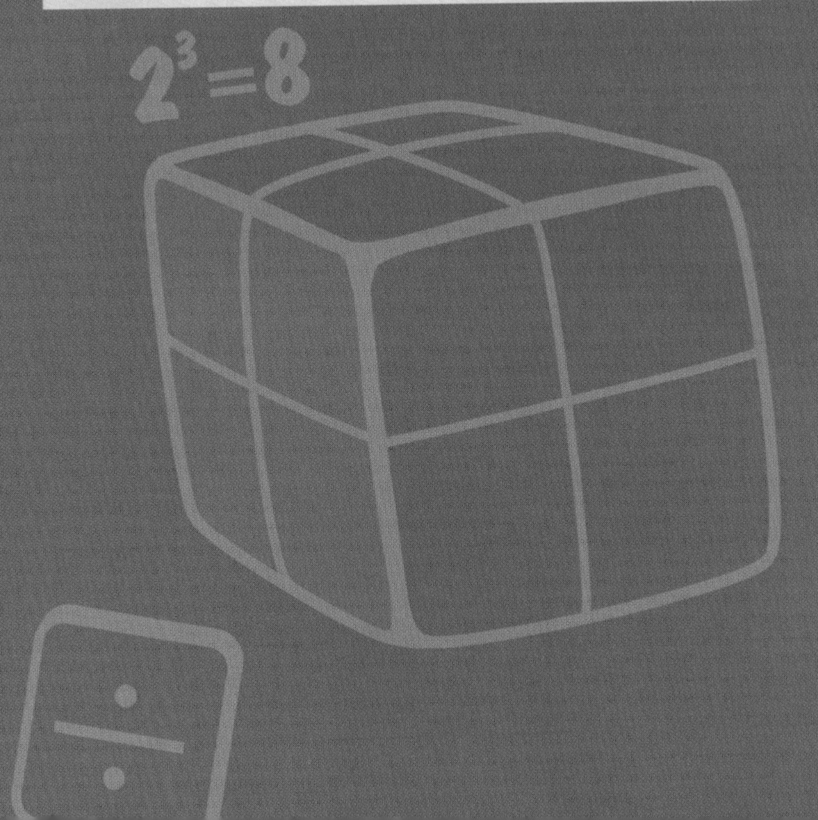

What do you already know?

You can use the problems and the list of words that follow to see what you already know about this chapter. The answers to the problems are in Hot Solutions at the back of the book, and the definitions of the words are in Hot Words at the front of the book. You can find out more about a particular problem or word by referring to the boldfaced topic number (for example, **3•2**).

Problem Set

Write each multiplication using an exponent. **3•1**

1. $5 \times 5 \times 5 \times 5 \times 5 \times 5 \times 5$
2. $a \times a \times a \times a \times a$

Evaluate each square. **3•1**

3. 2^2
4. 9^2
5. 6^2

Evaluate each cube. **3•1**

6. 2^3
7. 5^3
8. 7^3

Evaluate each power. **3•1**

9. 6^4
10. 3^7
11. 2^9

Evaluate each power of 10. **3•1**

12. 10^3
13. 10^7
14. 10^{11}

Evaluate each square root. **3•2**

15. $\sqrt{16}$
16. $\sqrt{49}$
17. $\sqrt{121}$

Estimate each square root between two consecutive whole numbers. **3•2**

18. $\sqrt{33}$
19. $\sqrt{12}$
20. $\sqrt{77}$

Estimate each square root to the nearest thousandth. **3•2**
21. $\sqrt{15}$
22. $\sqrt{38}$

Evaluate each cube root. **3•2**
23. $\sqrt[3]{8}$
24. $\sqrt[3]{64}$
25. $\sqrt[3]{343}$

Identify each number as very large or very small. **3•3**
26. 0.00014
27. 205,000,000

Write each number in scientific notation. **3•3**
28. 78,000,000
29. 200,000
30. 0.0028
31. 0.0000302

Write each number in standard form. **3•3**
32. 8.1×10^6
33. 2.007×10^8
34. 4×10^3
35. 8.5×10^{-4}
36. 9.06×10^{-6}
37. 7×10^{-7}

Evaluate each expression. **3•4**
38. $8 + (9 - 5)^2 - 3 \cdot 4$
39. $3^2 + 6^2 \div 9$
40. $(10 - 8)^3 + 4 \cdot 3 - 2$

CHAPTER 3

hot words

area **3•1**
base **3•1**
cube **3•1**
cube root **3•2**
exponent **3•1**
factor **3•1**
order of operations **3•4**
perfect squares **3•2**
power **3•1**
scientific notation **3•3**
square **3•1**
square root **3•2**
volume **3•1**

3·1 Powers and Exponents

Exponents

Multiplication, as you know, is the shortcut for showing a repeated addition: $5 \times 3 = 3 + 3 + 3 + 3 + 3$. A shortcut for showing the repeated multiplication $3 \times 3 \times 3 \times 3 \times 3$ is to write 3^5. The 3, the factor to be multiplied, is called the **base.** The 5 is the **exponent,** which tells how many times the base is to be multiplied. The expression can be read as "3 to the fifth **power.**" When you write an exponent, it is written slightly higher than the base and the size is usually a little smaller.

MULTIPLICATION USING EXPONENTS

Write the multiplication $2 \times 2 \times 2 \times 2 \times 2 \times 2 \times 2$ using an exponent.

- Check that the same **factor** is being used in the multiplication.

  All the factors are 2.

- Count the number of times 2 is being multiplied.

  There are 7 factors of 2.

- Write the multiplication using an exponent.

Since the factor 2 is being multiplied 7 times, write 2^7.

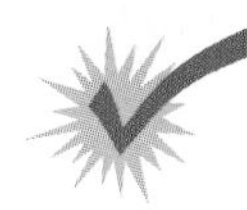

Check It Out

Write each multiplication using an exponent.

1. $4 \times 4 \times 4$
2. $6 \times 6 \times 6 \times 6 \times 6 \times 6 \times 6 \times 6 \times 6$
3. $x \times x \times x \times x$
4. $y \times y \times y \times y \times y \times y$

Evaluating the Square of a Number

The **square** of a number means to apply the exponent 2 to a base. The square of 4, then, is 4^2. To evaluate 4^2, identify 4 as the base and 2 as the exponent. Remember, the exponent tells you how many times to use the base as a factor. So 4^2 means to use 4 as a factor 2 times:

$$4^2 = 4 \times 4 = 16$$

The expression 4^2 can be read as "4 to the second power." It can also be read as "4 squared."

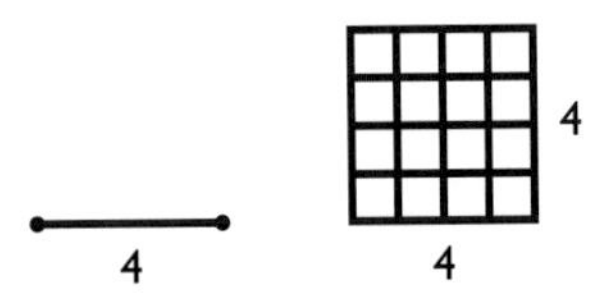

When a square is made from a segment whose length is 4, the **area** of the square is $4 \times 4 = 4^2 = 16$.

EVALUATING THE SQUARE OF A NUMBER

Evaluate 9^2.

- Identify the base and the exponent.
  The base is 9 and the exponent is 2.
- Write the expression as a multiplication.
  $9^2 = 9 \times 9$
- Evaluate.
  $9 \times 9 = 81$

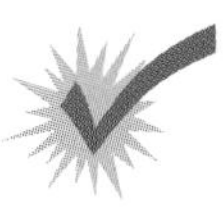

Check It Out

Evaluate each square.

5. 5^2
6. 10^2
7. 3 squared
8. 7 squared

Evaluating the Cube of a Number

The **cube** of a number means to apply the exponent 3 to a base. The cube of 2, then, is 2^3. Evaluating cubes is very similar to evaluating squares. For example, if you wanted to evaluate 2^3, notice that 2 is the base and 3 is the exponent. Remember, the exponent tells you how many times to use the base as a factor. So 2^3 means to use 2 as a factor 3 times:

$$2^3 = 2 \times 2 \times 2 = 8$$

The expression 2^3 can be read as "2 to the third power." It can also be read as "2 cubed."

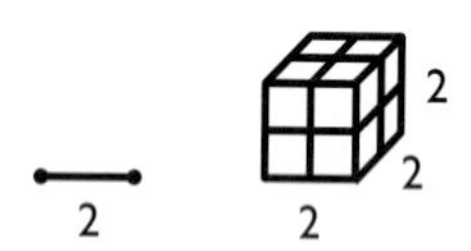

When a cube has edges of length 2, the **volume** of the cube is $2 \times 2 \times 2 = 2^3 = 8$.

EVALUATING THE CUBE OF A NUMBER

Evaluate 5^3.

- Identify the base and the exponent.

  The base is 5 and the exponent is 3.
- Write the expression as a multiplication.

  $5^3 = 5 \times 5 \times 5$
- Evaluate.

  $5 \times 5 \times 5 = 125$

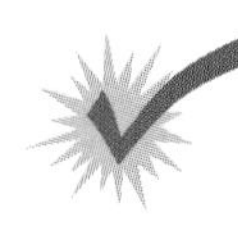

Check It Out

Evaluate each cube.

9. 4^3
10. 10^3
11. 3 cubed
12. 8 cubed

The Future of the Universe

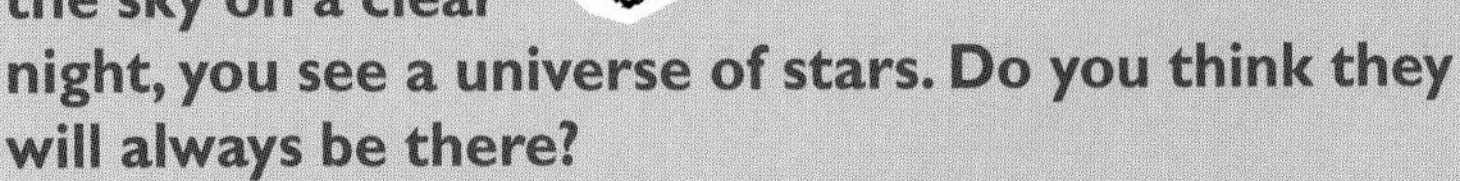

When you look at the sky on a clear night, you see a universe of stars. Do you think they will always be there?

One current hypothesis has it that eventually all the stars will die and the universe will be dark. Our planet's star, the sun, is expected to die in about 5 billion years. All the stars in the universe will have died in about 100 trillion years.

For a while after that, there will be a kind of twilight—ghosts of dying stars. Finally even these will be gone and the universe will become totally dark. Astrophysicists (astronomers who study events that take place on stars) predict that the dark era will begin about 10,000 trillion trillion trillion trillion trillion trillion trillion trillion years from now.

If a trillion is 10^{12}, express the number of years until the dark era begins using powers of ten. Why do you think scientists use exponents to talk about the timing of events in the universe? See Hot Solutions for answer.

Evaluating Higher Powers

You have evaluated the second power of numbers (squares) and the third power of numbers (cubes). You can evaluate higher powers of numbers as well.

To evaluate 5^4, identify 5 as the base and 4 as the exponent. The exponent tells you how many times to use the base as a factor. So 5^4 means to use 5 as a factor 4 times:

$$5^4 = 5 \times 5 \times 5 \times 5 = 625$$

The expression 5^4 can be read as "5 to the fourth power." There isn't any special name for the fourth power, or any higher power, because you cannot draw a figure with four or more dimensions.

EVALUATING HIGHER POWERS

Evaluate 4^6.

- Identify the base and the exponent.
  The base is 4 and the exponent is 6.
- Write the expression as a multiplication.
  $4^6 = 4 \times 4 \times 4 \times 4 \times 4 \times 4$
- Evaluate.
  $4 \times 4 \times 4 \times 4 \times 4 \times 4 = 4{,}096$

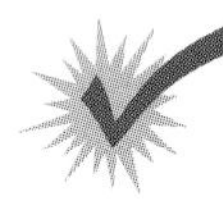

Check It Out

Evaluate each power.

13. 2^7
14. 9^5
15. 3 to the fourth power
16. 5 to the eighth power

Powers of Ten

Our decimal system is based on 10. For each factor of 10, the decimal point moves one place to the right.

3.15 → 31.5 (×10)   14.25 → 1,425 (×100)   3. → 30 (×10)

When the decimal point is at the end of a number and the number is multiplied by 10, a zero is added at the end of the number.

Try to discover a pattern for the powers of 10.

Powers	As a Multiplication	Result	Number of Zeros
10^2	10×10	100	2
10^4	$10 \times 10 \times 10 \times 10$	10,000	4
10^5	$10 \times 10 \times 10 \times 10 \times 10$	100,000	5
10^8	$10 \times 10 \times 10 \times 10 \times 10 \times 10 \times 10 \times 10$	100,000,000	8

Notice that the number of zeros after the 1 is the same as the power of 10. This means that if you want to evaluate 10^7, you simply write a 1 followed by 7 zeros: 10,000,000.

Check It Out

Evaluate each power of 10.

17. 10^3
18. 10^6
19. 10^9
20. 10^{14}

Using a Calculator to Evaluate Powers

You can use a calculator to evaluate powers. On any calculator, you can always multiply a number any number of times just by using the [×] key.

Many calculators have the key [x^2]. This key is used to calculate the square of a number. Enter the number that is the base, then press [x^2]. The display will show the square of the number. (For some calculators, you will have to press [=] or [ENTER] to see the answer.)

Some calculators have the key [y^x] or [x^y]. This key is used to calculate any power of a number. Enter the number that is the base. Press [y^x] or [x^y]. Then enter the number that is the exponent and press [=].

Other calculators use the key [∧] for calculating powers. Enter the number that is the base. Press [∧]. Then enter the number that is the exponent and press [ENTER].

For more information on calculators, see Topics 9•1 and 9•2.

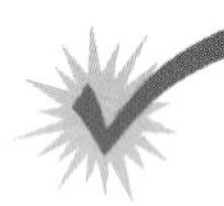

Check It Out

Use a calculator to evaluate each power.

21. 18^2
22. 5^{10}
23. 2^{25}
24. 29^5

3•1 EXERCISES

Write each multiplication using an exponent.

1. $7 \times 7 \times 7$
2. $9 \times 9 \times 9 \times 9 \times 9 \times 9 \times 9$
3. $a \times a \times a \times a \times a \times a$
4. $w \times w \times w \times w \times w \times w \times w \times w \times w \times w$
5. 16×16

Evaluate each square.

6. 8^2
7. 15^2
8. 7^2
9. 1 squared
10. 20 squared

Evaluate each cube.

11. 7^3
12. 11^3
13. 6^3
14. 3 cubed
15. 9 cubed

Evaluate each higher power.

16. 6^4
17. 2^{12}
18. 5^5
19. 4 to the seventh power
20. 1 to the fifteenth power

Evaluate each power of 10.

21. 10^2
22. 10^8
23. 10^{13}

Use a calculator to evaluate each power.

24. 8^6
25. 6^{10}

3·2 Square and Cube Roots

Square Roots

In mathematics, certain operations are opposites of each other. That is, one operation "undoes" the other. Addition undoes subtraction: $9 - 5 = 4$, so $4 + 5 = 9$. Multiplication undoes division: $12 \div 4 = 3$, so $3 \times 4 = 12$.

The opposite, or undoing, of squaring a number is finding the **square root.** You know that 5 squared $= 5^2 = 25$. The square root of 25 is the number that can be multiplied by itself to get 25, which is 5. The symbol for square root is $\sqrt{\ }$. Therefore $\sqrt{25} = 5$.

FINDING THE SQUARE ROOT

Find $\sqrt{81}$.

- Think, what number times itself makes 81?

  $9 \times 9 = 81$

- Find the square root.

  Since $9 \times 9 = 81$, the square root of 81 is 9.

So $\sqrt{81} = 9$.

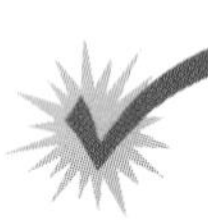

Check It Out

Find each square root.

1. $\sqrt{16}$
2. $\sqrt{49}$
3. $\sqrt{100}$
4. $\sqrt{144}$

Estimating Square Roots

The table shows the first ten **perfect squares** and their square roots.

Perfect square	1	4	9	16	25	36	49	64	81	100
Square root	1	2	3	4	5	6	7	8	9	10

So then, how much is $\sqrt{40}$? In this problem, 40 is the square. In the table, 40 lies between 36 and 49, which tells you $\sqrt{40}$ must be between $\sqrt{36}$ and $\sqrt{49}$—a number between 6 and 7. You can estimate the value of a square root by finding the two consecutive numbers that the square root must be between.

ESTIMATING A SQUARE ROOT

Estimate $\sqrt{70}$.

- Identify the perfect squares that 70 is between.
  70 is between 64 and 81.
- Find the square roots of the perfect squares.
  $\sqrt{64} = 8$ and $\sqrt{81} = 9$.
- Estimate the square root.
  $\sqrt{70}$ is between 8 and 9.

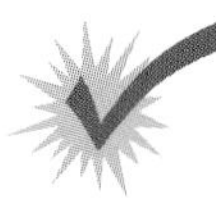

Check It Out

Estimate each square root.

5. $\sqrt{55}$
6. $\sqrt{18}$
7. $\sqrt{7}$
8. $\sqrt{95}$

Better Estimates of Square Roots

If you want to know a better estimate for the value of a square root, you will want to use a calculator. Most calculators have a key [√] for finding square roots.

On some calculators, the $\sqrt{\ }$ function is shown not on a key but above the [x^2] key on the calculator's surface. If this is true for your calculator, you should then see a key that has either [INV] or [2nd] on it. To use the $\sqrt{\ }$ function, you would press [INV] or [2nd], then the key with $\sqrt{\ }$ above it.

When finding the square root of a number that is not a perfect square, the answer will be a decimal, and the entire calculator display will be used. Generally you should round square roots to the nearest thousandth. Remember that the thousandths place is the third place after the decimal point.

See topics 9•1 and 9•2 for more about calculators.

ESTIMATING THE SQUARE ROOT OF A NUMBER

Estimate $\sqrt{42}$.

- Use a calculator.

  Press 42 [√], or 42 [INV] [x^2],
  or press [2nd] [x^2] 42 [ENTER].
- Read the display.

  6.4807407 if your calculator shows 8 digits, or
  6.480740698 if your calculator shows 10 digits
- Round to the nearest thousandth.

  Locate the digit in the third place after the decimal, which is 0. Then look at the digit to its right, which is 7. Since this digit is 5 or more, round up.
- Estimate the square root.

  $\sqrt{42} = 6.481$

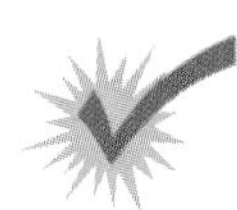

Check It Out

Estimate each square root to the nearest thousandth.

9. $\sqrt{2}$
10. $\sqrt{50}$
11. $\sqrt{75}$
12. $\sqrt{99}$

Cube Roots

In the same way that finding a square root "undoes" the squaring of a number, finding a **cube root** undoes the cubing of a number. Finding a cube root answers the question "What number times itself three times makes the cube?" Since 2 cubed $= 2 \times 2 \times 2 = 2^3 = 8$, the cube root of 8 is 2. The symbol for cube root is $\sqrt[3]{\ }$. Therefore $\sqrt[3]{8} = 2$.

FINDING THE CUBE ROOT OF A NUMBER

Find $\sqrt[3]{216}$.

- *Think:* What number times itself three times will make 216?

  $6 \times 6 \times 6 = 216$

- Find the cube root.

  $\sqrt[3]{216} = 6$

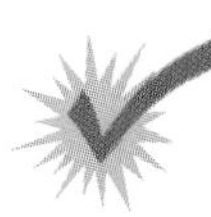

Check It Out

Find the cube root of each number.

13. $\sqrt[3]{64}$
14. $\sqrt[3]{343}$
15. $\sqrt[3]{1000}$
16. $\sqrt[3]{125}$

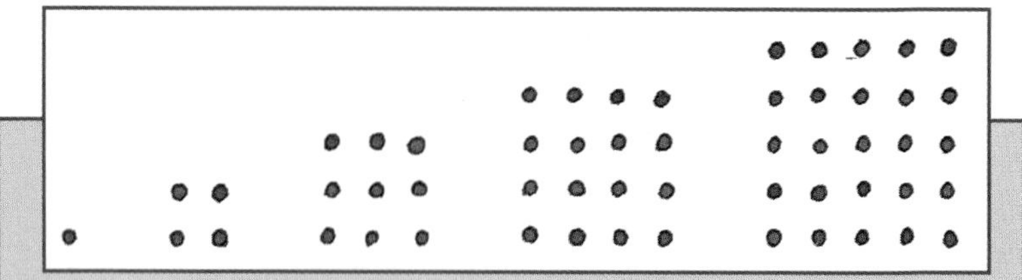

Squaring Triangles

As you can see, some numbers can be pictured using arrays of dots that form geometric figures. You might have already noticed that this sequence shows the first five square numbers: 1^2, 2^2, 3^2, 4^2, and 5^2.

Can you think of places where you have seen numbers that form a triangular array? Think

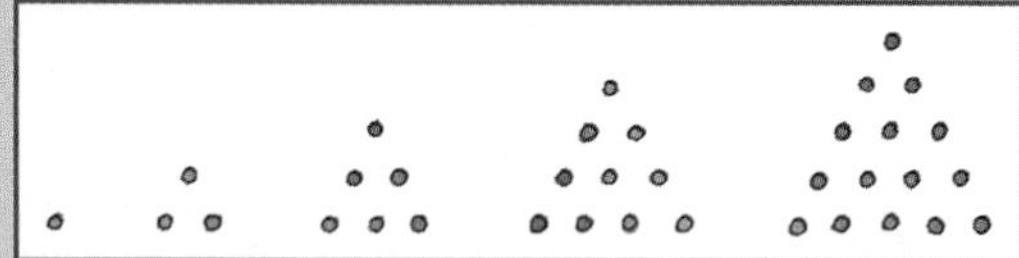

of cans stacked in a pyramid supermarket display, bowling pins, and 15 pool balls before the break. What are the next two triangular numbers?

Add each pair of consecutive triangular numbers to form a new sequence as shown here. What do you notice about this sequence?

```
1   3   6  10  15   ...
 \ / \ / \ / \ / \ /  ...
  4   9
```

Think about how you could use the dot arrays for the square numbers to show the same result. *Hint*: What line could you draw in each array? See Hot Solutions for answers.

3•2 EXERCISES

Find each square root.

1. $\sqrt{9}$
2. $\sqrt{64}$
3. $\sqrt{121}$
4. $\sqrt{25}$
5. $\sqrt{196}$

6. $\sqrt{30}$ is between which two numbers?
   A. 3 and 4 B. 5 and 6
   C. 29 and 31 D. None of these
7. $\sqrt{84}$ is between which two numbers?
   A. 4 and 5 B. 8 and 9
   C. 9 and 10 D. 83 and 85
8. $\sqrt{21}$ is between what two consecutive numbers?
9. $\sqrt{65}$ is between what two consecutive numbers?
10. $\sqrt{106}$ is between what two consecutive numbers?

Estimate each square root to the nearest thousandth.

11. $\sqrt{3}$
12. $\sqrt{10}$
13. $\sqrt{47}$
14. $\sqrt{86}$
15. $\sqrt{102}$

Find the cube root of each number.

16. $\sqrt[3]{27}$
17. $\sqrt[3]{512}$
18. $\sqrt[3]{1331}$
19. $\sqrt[3]{1}$
20. $\sqrt[3]{8000}$

3·3 Scientific Notation

Using Scientific Notation

Often, in science and in mathematics, numbers are used that are very large or very small. Large numbers often have many zeros at the end. Small numbers often have many zeros in the beginning.

Large number: 450,000,000
many zeros at the end

Small number: 0.000000032
many zeros at the beginning

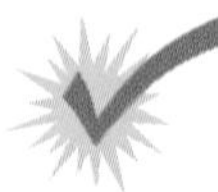

Check It Out

Identify each number as very large or very small.

1. 0.000015
2. 6,000,000

Bugs

Insects are the most successful form of life on Earth. About one million have been classified and named. It is estimated that there are up to four million more. That's not total insects we are talking about; that's different *kinds* of insects!

Estimates are that there are 200,000,000 insects for each person on the planet. Given a world population of approximately 6,000,000,000, just how many insects do we share with the earth? Use a calculator to arrive at an estimate. Express the number in scientific notation. See Hot Solutions for answer.

Writing Large Numbers Using Scientific Notation

Instead of writing large numbers with all the zeros, and possibly forgetting one of them, you can use **scientific notation.**

Scientific notation uses *powers of 10* (p. 173). To write a number in scientific notation, move the decimal point in the number so that only one digit is to the left of the decimal. Count the number of decimal places that the decimal has to move to the right to get the original number. Recall that each factor of 10 moves the decimal point one place to the right. Then multiply the number by the correct power of 10.

WRITING A LARGE NUMBER IN SCIENTIFIC NOTATION

Write 4,250,000,000 in scientific notation.

- Move the decimal point so that only one digit is to the left of the decimal.

4.250000000.

- Count the number of decimal places that the decimal has to be moved to the right.

4.250000000.
9 places

- Write the number without the ending zeros, and multiply by the correct power of 10.

4.25×10^9

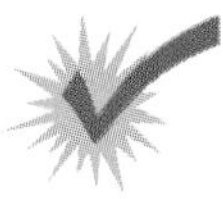

Check It Out

Write each number in scientific notation.

3. 68,000
4. 7,000,000
5. 30,500,000,000
6. 73,280,000

Writing Small Numbers Using Scientific Notation

To write a small number in scientific notation, move the decimal again so that there is one nonzero digit to the left of the decimal. Count the number of places that the decimal has to move to the left to get the original number. To move the decimal one place to the left, you still have to use factors of 10. When you determine the correct number of places, use a negative number for the exponent.

WRITING A SMALL NUMBER IN SCIENTIFIC NOTATION

Write 0.0000000425 in scientific notation.

- Move the decimal point so that only one nonzero digit is to the left of the decimal.

  0.00000004.25

- Count the number of decimal places that the decimal has to be moved to the left.

  0.00000004.25
  8 places

- Write the number without the beginning zeros, and multiply by the correct power of 10. Use a negative exponent to move the decimal to the left.

  $$4.25 \times 10^{-8}$$

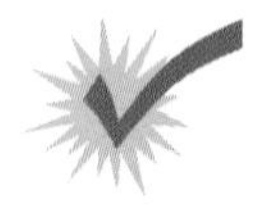

Check It Out

Write each number in scientific notation.

7. 0.0038
8. 0.0000004
9. 0.00000000000603
10. 0.0007124

Converting from Scientific Notation to Standard Form

Converting to Standard Form When the Exponent Is Positive

When the power of 10 is positive, each factor of 10 moves the decimal point one place to the right. When the last digit of the number is reached, there may still be some factors of 10 remaining. Add a zero at the end of the number for each remaining factor of 10.

CONVERTING TO A NUMBER IN STANDARD FORM

Write 7.035×10^6 in standard form.

- Notice the exponent.

  The exponent is positive. The decimal point moves to the right 6 places.

- Move the decimal point the correct number of places to the right. Add necessary zeros at the end of the number to fill to the decimal point.

  7.035000.

  Move the decimal point to the right 6 places.

- Write the number in standard form.

  $7.035 \times 10^6 = 7{,}035{,}000$

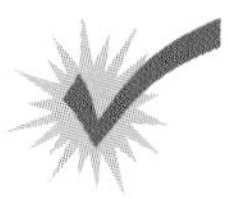

Check It Out

Write each number in standard form.

11. 5.3×10^4
12. 9.24×10^8
13. 1.205×10^5
14. 8.84073×10^{12}

Converting to Standard Form When the Exponent Is Negative

When the power of 10 is negative, each factor of 10 moves the decimal point one place to the left. Since there is only one digit to the left of the decimal, you will have to add zeros at the beginning of the number.

CONVERTING TO A NUMBER IN STANDARD FORM

Write 4.16×10^{-5} in standard form.

- Notice the exponent.

  The exponent is negative. The decimal point moves to the left 5 places.

- Move the decimal point the correct number of places to the left. Add zeros at the beginning of the number to fill to the decimal point.

  0.00004.16

  Move the decimal point to the left 5 places.

- Write the number in standard form.

  $4.16 \times 10^{-5} = 0.0000416$

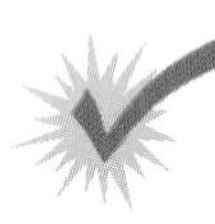

Check It Out

Write each number in standard form.

15. 7.1×10^{-4}
16. 5.704×10^{-6}
17. 8.65×10^{-2}
18. 3.0904×10^{-11}

3·3 EXERCISES

Identify each number as very large or very small.

1. 0.000034
2. 83,900,000
3. 0.000245
4. 302,000,000,000

Write each number in scientific notation.

5. 420,000
6. 804,000,000
7. 30,000,000
8. 13,060,000,000,000
9. 0.00037
10. 0.0000506
11. 0.002
12. 0.000000005507

Write each number in standard form.

13. 2.4×10^7
14. 7.15×10^4
15. 4.006×10^{10}
16. 8×10^8
17. 4.9×10^{-7}
18. 2.003×10^{-3}
19. 5×10^{-5}
20. 7.0601×10^{-10}

21. Which of the following expresses the number 5,030,000 in scientific notation?
    A. 5×10^6 B. 5.03×10^6
    C. 5.03×10^{-6} D. 50.3×10^5
22. Which of the following expresses the number 0.0004 in scientific notation?
    A. 4×10^4 B. 0.4×10^{-3}
    C. 4×10^{-4} D. 4×10^{-3}
23. Which of the following expresses the number 3.09×10^7 in standard form?
    A. 30,000,000 B. 30,900,000
    C. 0.000000309 D. 3,090,000,000
24. Which of the following expresses the number 5.2×10^{-5} in standard form?
    A. 0.000052 B. 0.0000052
    C. 520,000 D. 5,200,000
25. When written in scientific notation, which of the following numbers will have the greatest power of 10?
    A. 93,000 B. 408,000
    C. 5,556,000 D. 100,000,000

3·4 Laws of Exponents

Exponents Within the Order of Operations

You know that when you evaluate expressions using the **order of operations,** you do the operations within parentheses first. Then you do the multiplications and divisions. Lastly you do the additions and subtractions.

Remember that exponents represent a repeated multiplication. When there are exponents involved in the expression, this repeated multiplication has to take place before other numbers are multiplied. So evaluating powers comes after the operations within parentheses but before the multiplications and divisions.

EVALUATING EXPRESSIONS WITH EXPONENTS

Evaluate the expression $3(6 - 2) + 4^3 \div 8 - 3^2$.

$= 3(4) + 4^3 \div 8 - 3^2$	• Do the operations within parentheses first.
$= 3(4) + 64 \div 8 - 9$	• Evaluate the powers.
$= 12 + 8 - 9$	• Do the multiplications and divisions in order from left to right.
$= 11$	• Do the additions and subtractions in order from left to right.

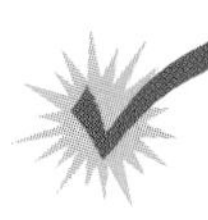

Check It Out

Evaluate each expression.

1. $5^2 - 8 \div 4$
2. $(7 - 3)^2 + 16 \div 2^4$
3. $5 + (3^2 - 2 \cdot 4) + 12$
4. $16 - (4 \cdot 3 - 7) + 2^3$

3·4 EXERCISES

Evaluate each expression.

1. $4^2 \div 2^3$
2. $(5 - 3)^5 - 4 \cdot 5$
3. $7^2 - 3(5 + 3^2)$
4. $8^2 \div 4 \cdot 2$
5. $15 \div 3 + (10 - 7)^2 \cdot 2$
6. $7 \cdot 3 - (8 - 2 \cdot 3)^3 - 1$
7. $5^2 - 2 \cdot 3^2$
8. $2 \cdot 5 + 3^4 \div (4 + 5)$
9. $(7 - 3)^2 - (9 - 6)^3 \div 9$
10. $3 \cdot 4^2 \div 6 + 2(3^2 - 5)$

Parentheses
Exponents
Multiplication &
Division
Addition &
Subtraction

What have you learned?

You can use the problems and the list of words that follow, to see what you have learned in this chapter. You can find out more about a particular problem or word by referring to the boldfaced topic number (for example, **3•2**).

Problem Set

Write each multiplication using an exponent. **3•1**

1. $7 \times 7 \times 7 \times 7 \times 7 \times 7 \times 7 \times 7 \times 7$
2. $n \times n \times n \times n$

Evaluate each square. **3•1**

3. 3^2
4. 7^2
5. 12^2

Evaluate each cube. **3•1**

6. 4^3
7. 9^3
8. 5^3

Evaluate each power. **3•1**

9. 3^8
10. 7^4
11. 2^{11}

Evaluate each power of 10. **3•1**

12. 10^2
13. 10^5
14. 10^9

Evaluate each square root. **3•2**

15. $\sqrt{9}$
16. $\sqrt{64}$
17. $\sqrt{169}$

Estimate each square root between two consecutive numbers. **3•2**

18. $\sqrt{51}$
19. $\sqrt{18}$
20. $\sqrt{92}$

Estimate each square root to the nearest thousandth. **3•2**

21. $\sqrt{23}$
22. $\sqrt{45}$

Evaluate each cube root. **3•2**

23. $\sqrt[3]{27}$
24. $\sqrt[3]{125}$
25. $\sqrt[3]{729}$

Identify each number as very large or very small. **3•3**

26. 0.000063
27. 8,600,000

Write each number in scientific notation. **3•3**

28. 9,300,000
29. 800,000,000
30. 0.000054
31. 0.0605

Write each number in standard form. **3•3**

32. 3.4×10^4
33. 7.001×10^{10}
34. 9×10^6
35. 5.3×10^{-3}
36. 6.02×10^{-9}
37. 4×10^{-4}

Evaluate each expression. **3•4**

38. $3 \cdot 5^2 - 4^2 \cdot 2$
39. $6^2 - (8^2 \div 2^5 + 3 \cdot 5)$
40. $(1 + 2 \cdot 3)^2 - (2^3 - 4 \div 2^2)$

hot words WRITE DEFINITIONS FOR THE FOLLOWING WORDS.

area **3•1**
base **3•1**
cube **3•1**
cube root **3•2**
exponent **3•1**
factor **3•1**
order of operations **3•4**
perfect squares **3•2**
power **3•1**
scientific notation **3•3**
square **3•1**
square root **3•2**
volume **3•1**

hot topics 4

Data, Statistics, and Probability

4•1	Collecting Data	196
4•2	Displaying Data	202
4•3	Analyzing Data	214
4•4	Statistics	222
4•5	Combinations and Permutations	232
4•6	Probability	240

What do you already know?

You can use the problems and the list of words that follow to see what you already know about this chapter. The answers to the problems are in Hot Solutions at the back of the book, and the definitions of the words are in Hot Words at the front of the book. You can find out more about a particular problem or word by referring to the boldfaced topic number (for example, **4•2**).

Problem Set

Use the following for items 1–3. A student asked others riding on the school bus about their favorite PE time. The answers are shown below. **4•1**

FAVORITE PE TIME

	6th Graders	7th Graders	8th Graders
Early morning	///	~~////~~	
Late morning	////	~~////~~ ~~////~~	////
Early afternoon	~~////~~	~~////~~	/
Late afternoon	//	/	~~////~~ /

1. When is the favorite PE time among all students who gave answers?
2. Which grade had the most responses?
3. Is this a random sample?

Use the following graphs for items 4 and 5. These two circle graphs show whether cars made a right turn, a left turn, or drove straight ahead at an intersection near school. **4•2**

4. Between 8 A.M. and 9 A.M., what percent of the cars turn?
5. Do the graphs show that more cars go straight between 9 A.M. and 10 A.M. than between 8 A.M. and 9 A.M.?

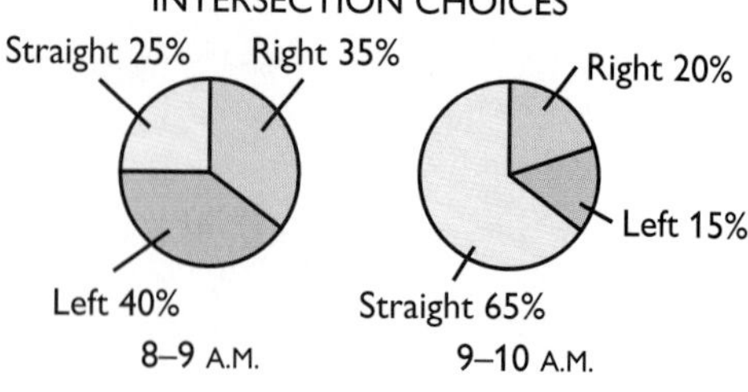

6. In a class election, tally marks were used to count votes. What is the bar graph called that is made from these marks? **4•2**

7. In a scatter plot of phone calls, the line of best fit rises from left to right. What kind of correlation is illustrated? **4•3**
8. In Mr. Dahl's class of 27 students, the lowest test grade was 58%, the highest was 92%, and the most common was 84%. What was the range of grades? **4•4**
9. In item 8, can you find the mean, median, or mode? **4•4**
10. $P(4, 3) = ?$ **4•5**
11. $C(7, 2) = ?$ **4•5**
12. Write all the combinations of the digits 3, 5, and 7 using only two numbers at a time. **4•5**

Use the following information to answer items 13–15. A bag contains 10 colored chips—3 red, 4 blue, 1 green, and 2 black. **4•6**

13. One chip is drawn. What is the probability it is blue or black?
14. Two chips are drawn. What is the probability that they are both green?
15. A chip is drawn. Then a second one is drawn without replacement. What is the probability that both are blue?

CHAPTER 4

hot words

average **4•4**
bimodal distribution **4•3**
box plot **4•2**
circle graph **4•2**
combination **4•5**
correlation **4•3**
dependent events **4•6**
double-bar graph **4•2**
event **4•6**
experimental probability **4•6**
factorial **4•5**
flat distribution **4•3**
histogram **4•2**
independent events **4•6**
leaf **4•2**
line graph **4•2**
line of best fit **4•3**
mean **4•4**
median **4•4**
mode **4•4**
normal distribution **4•3**
outcome **4•6**
outcome grid **4•6**
permutation **4•5**
population **4•1**
probability **4•6**
probability line **4•6**
random sample **4•1**
range **4•4**
sample **4•1**
sampling with replacement **4•6**
scatter plot **4•3**
skewed distribution **4•3**
spinner **4•5**
stem **4•2**
stem-and-leaf plot **4•2**
strip graph **4•6**
survey **4•1**
table **4•1**
tally marks **4•1**
theoretical probability **4•6**
tree diagram **4•5**
weighted average **4•4**

4·1 Collecting Data

Surveys

Have you ever been asked to name your favorite movie? Have you been asked what kind of pizza you like? These kinds of questions are often asked in **surveys.** A statistician studies a group of people or objects, called a **population.** They usually get information from a small part of the population, called a **sample.**

In a survey, eighth-grade students were chosen at random from three countries and asked if they spent three or more hours on a normal school day watching TV, hanging out with friends, playing sports, reading a book for fun, or studying. The following bar graph shows the percent of students who said yes in each category.

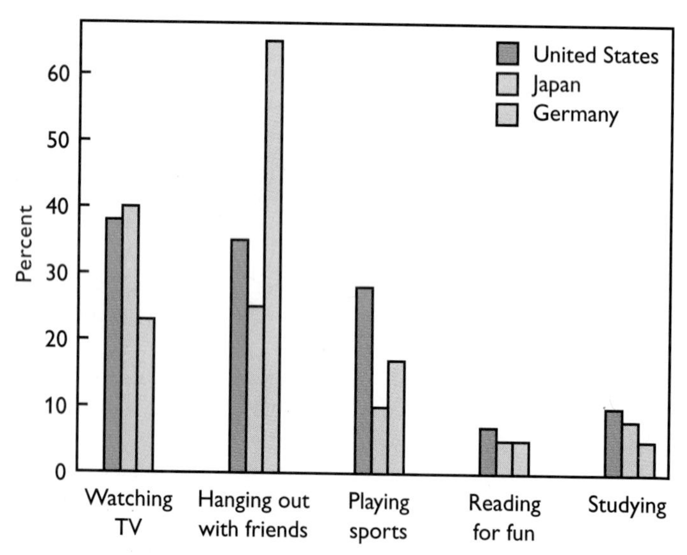

In this case, the population is all eighth-grade students in the United States, Japan, and Germany. The sample is the students who were actually asked the questions.

In any survey,

- The population consists of the people or objects about which information is desired.
- The sample consists of the people or objects in the population that are actually studied.

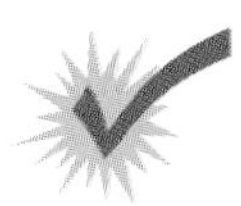

Check It Out

Identify the population and the size of the sample.

1. In a survey, 150,000 adults over the age of 45 were asked if they listened to radio station KROK.
2. Two hundred elk in Roosevelt National Forest.

Random Samples

When you choose a sample to survey for data, you want to be sure the sample is representative of the population. You also want to be sure it is a **random sample,** where each person in the population has an equal chance of being included.

Mr. Singh wanted to find out whether his students wanted pizza, chicken fingers, ice cream, or bagels at a class party. He picked a sample to question by writing the names of his students on cards and drawing ten cards from a bag.

DETERMINING WHETHER A SAMPLE IS RANDOM

Determine whether Mr. Singh's sample (above) is random.

- Determine the population.

  The population is Mr. Singh's class.
- Determine the sample.

  The sample consists of ten students.
- Determine if the sample is random.

  Since every student in Mr. Singh's class had the same chance of being chosen, the sample is random.

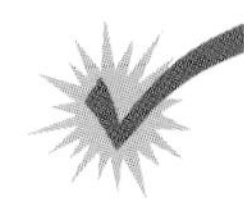

Check It Out

3. A student asked 20 of her parents' friends who they planned to vote for. Is the sample random?
4. A student assigns numbers to his 24 classmates and then uses a spinner divided into 24 equal parts to pick ten numbers. He asks those ten students what their favorite movie is. Is the sample random?

Questionnaires

When you write questions for a survey, it is important to be sure the questions are not biased. That is, the questions should not make assumptions or influence the answers. The following two questionnaires are designed to find out what kind of food your classmates like and what they do after school. The first questionnaire uses biased questions. The second questionnaire uses questions that are not biased.

Survey 1

A. What kind of pizza do you like?
B. What is your favorite afternoon TV program?

Survey 2

A. What is your favorite food?
B. What do you like to do after school?

When you are developing a questionnaire,

- Decide what topic you want to ask about.
- Define a population and decide how to select a sample from that population.
- Develop questions that are not biased.

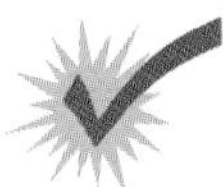

Check It Out

5. Why is question **A** in Survey 1 biased?
6. Why is question **B** in Survey 2 better than question **B** in Survey 1?
7. Write a question that asks the same thing as the following question but is not biased. Are you a caring citizen who recycles newspapers?

Compiling Data

After Mr. Singh collected the data from his students, he had to decide how to show the results. As he asked students their food preference, he used **tally marks** to tally the answers in a table. The following table shows their answers.

FOOD PREFERRED IN MR. SINGH'S CLASS

Preferred Food	Number of Students
Pizza	////
Chicken Fingers	//
Ice Cream	///
Bagels	/

Follow this procedure when you are making a table to compile data.

- List the categories or questions in the first column or row.
- Tally the responses in the second column or row.

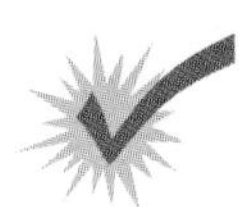

Check It Out

8. How many students chose chicken fingers?
9. What was the food least preferred by the students surveyed?
10. If Mr. Singh uses the survey to pick food to serve at the class party, what should he serve? Explain.

The WorldPOPClock

The U.S. Bureau of the Census estimates how many people are in the world each second on their WorldPOPClock. The estimate is based on projected births and deaths around the world.

At 2:00 A.M. EST on March 2, 1997, the WorldPOPClock estimate was 5,825,618,337. Using the table below, calculate the world population as of the date you read this page.

Time Unit	Projected Increase
Year	79,178,194
Month	6,598,183
Day	216,927
Hour	9,039
Minute	151
Second	2.5

You can check your answer on the WorldPOPClock on the Internet. Go to http://www.census.gov/ipc/www/popwnote.html Then click the link to WorldPOPClock.

4·1 EXERCISES

1. Three hundred eighth-grade Roddaville students were asked to name their favorite mall. Identify the population and the sample. How big is the sample?

2. Norma chose businesses to survey by obtaining a list of businesses in the city and writing each name on a slip of paper. She placed the slips of paper in a bag and drew 50 names. Is the sample random?
3. Jonah knocked on 25 doors in his neighborhood. He asked the residents who answered if they were in favor of the idea of the city building a swimming pool. Is the sample random?

Are the following questions biased? Explain.

4. Are you happy about the ugly building being built in your neighborhood?
5. How many hours do you watch TV each week?

Write unbiased questions to replace the following questions.

6. Do you prefer cute, cuddly kittens as pets, or do you like dogs better?
7. Are you thoughtful about not playing your stereo after 10 P.M.?

Ms. Chow asked her students which type of book they prefer to read and tallied the following data.

BOOK PREFERENCES OF MS. CHOW'S CLASS

Type of Book	Number of Seventh Graders	Number of Eighth Graders
Biography	卌 卌	卌 卌 II
Mystery	卌 I	III
Fiction	卌 卌 II	卌 卌
Science Fiction	卌 II	卌 I
Nonfiction	III	卌 I

8. Which type of book was most popular? How many students preferred that type?

9. Which type of book was preferred by 13 students?
10. How many students were surveyed?

4•2 Displaying Data

Interpret and Create a Table

You know that statisticians collect data about people or objects. One way to show the data is to use a **table.** Here are the number of letters in the words of the first two sentences in *Black Beauty.*

3 5 5 4 1 3 4 8 3 1 5 8 6 4 1 4 2 5 5 2 2 4 5 5 6 4 2 3 6 3 11 4 2 3 4 3

MAKING A TABLE

Make a table to organize the data about letters in the words.

- Name the first row or column *what* you are counting. Label the first row *No. of Letters.*
- Tally the amounts for each category in the second row or column.

No. of Letters	1	2	3	4	5	6	7	8	more than 8
No. of Words	III	~~IIII~~	~~IIII~~ II	~~IIII~~ III	~~IIII~~ II	III		II	I

- Count the tallies and record the number in the second row or column.

No. of Letters	1	2	3	4	5	6	7	8	more than 8
No. of Words	3	5	7	8	7	3	0	2	1

The most common number of letters in a word is 4. Three words have 1 letter.

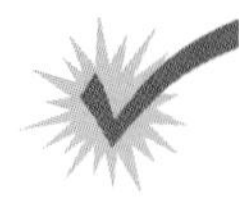

Check It Out

1. What information is lost by using the category "more than 8"?
2. Use the data below to make a table to show the number of gold medals won by countries in the 1994 Winter Olympics.
   10 9 11 7 6 3 3 2 4 0 1 0 0 2 1 0 0 1 0 0 1 0

Interpret a Box Plot

A **box plot** shows data using the middle value of the data and the quartiles, or 25% divisions of the data. The box plot shows exam scores on a math test for a class of 8th graders.

On a box plot, 50% of the scores are above the middle score and 50% are below it. The first quartile score is the middle score of the bottom half of the scores. The third quartile score is the middle score in the top half of the scores.

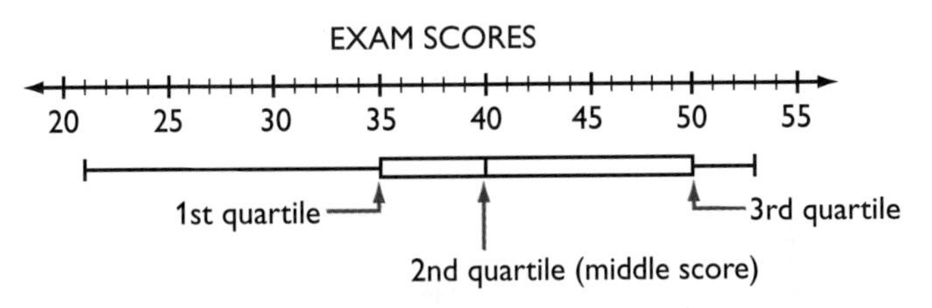

Here's what we can tell about the exam scores.

- The high score is 53. The low score is 21.
- The middle score is 40. The first quartile score is 35 and the third quartile score is 50.
- 50% of the scores are between 35 and 50.

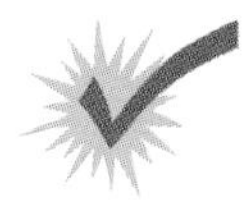

Check It Out

Use the following box plot to answer these questions.

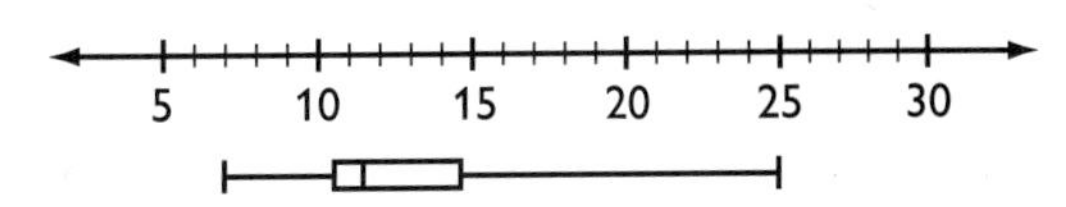

3. What is the greatest amount of fat in a fast-food milkshake?
4. What is the middle amount of fat in a fast-food milkshake?
5. What percent of the milkshakes contain between 7 and 11.5 g of fat?

Interpret and Create a Circle Graph

Another way to show data is to use a **circle graph.** A circle graph can be used to show parts of a whole. Arturo conducted a survey to find out what kind of solid waste was thrown away. He came up with the following data: 39% of the trash was paper; 6%, glass; 8%, metals; 9%, plastic; 7%, wood; 7%, food; 15%, yard waste; the remaining 9%, miscellaneous waste. Arturo wants to make a circle graph to show his data.

To make a circle graph,

- Find what percent of the whole each part of the data is.
  In this case, the percents are given.
- Multiply each percent by 360°, the number of degrees in a circle.

$360° \times 39\% = 140.4°$ $\quad$ $360° \times 6\% = 21.6°$

$360° \times 8\% = 28.8°$ $\quad$ $360° \times 9\% = 32.4°$

$360° \times 7\% = 25.2°$ $\quad$ $360° \times 15\% = 54°$

- Draw a circle, measure each central angle, and complete the graph.

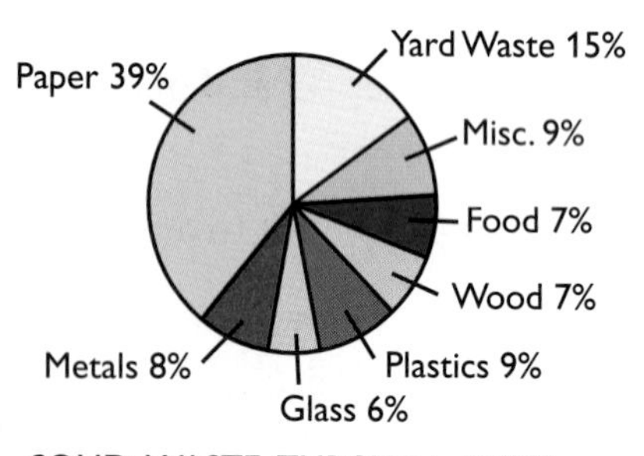

SOLID WASTE THROWN AWAY

From the graph, you can see that more than half of the solid waste is made up of paper and yard waste. Equal amounts of food and wood are discarded.

Check It Out

Use the circle graph to answer items 6 and 7.

6. About what fraction of people buy used cars from a dealership?

Dealership
Other
Family
Private Owner
Used Car Lot

WHERE WE BUY USED CARS

7. About what fraction of people buy used cars from private owners?

8. Students earned money. Make a circle graph to show the results.
   Car wash: $335 Bake sale: $128
   Recycling: $155 Book sale: $342

And the Winner Is...

At the end of February 1996, the performer Seal won three Grammys.

Album Sales
75,000
50,000
25,000
0
4 11 18 25 3 10 17 24 31
February March

Based on the graph, how did winning the Grammy awards affect Seal's album sales? How often were the sales tabulated? What kind of graph is this? See Hot Solutions for answers.

Interpret and Create a Frequency Graph

You have used tally marks to show data. A frequency graph is a vertical graph of the tally marks you make in collecting data. Suppose you collect the following information about the times your friends get up on a school day.

5:30, 6, 5:30, 8, 7:30, 8, 7:30, 9, 8, 8, 6, 6:30, 6, 8

You can make a frequency graph by placing X's above a number line.

To make a frequency graph:

- Draw a number line showing the numbers in your data set.
- Place an X to represent each result above the number line for each piece of data you have.
- Title the graph.

Your frequency graph should look like this:

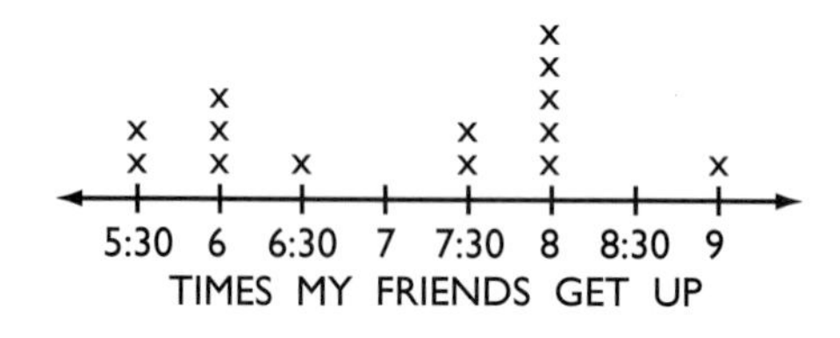

You can tell from the frequency graph that your friends get up anywhere from 5:30 to 9:00 on school days.

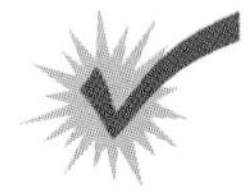

Check It Out

9. What is the most common time for your friends to get up?
10. How many friends get up before 7:00 A.M.?
11. Make a frequency graph to show the number of letters in the words of the first two sentences in *Black Beauty* (p. 202).

Interpret a Line Graph

You know that a *line graph* can be used to show changes in data over time. The following line graph compares the monthly average vault scores of two gymnasts.

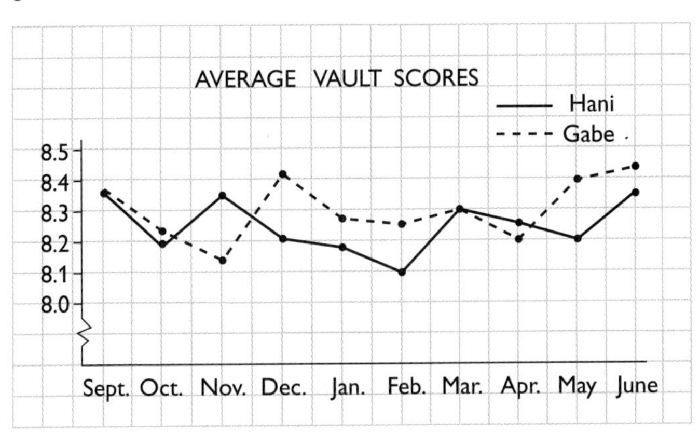

From the graph, you can see that Hani and Gabe had the same average scores in two months, September and March.

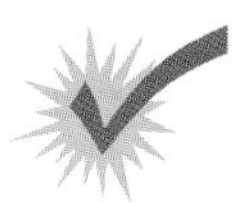

Check It Out

12. In December which gymnast had better scores?
13. Which gymnast typically scores higher on the vault?

Interpret a Stem-and-Leaf Plot

The following numbers show the ages of the students in a T'ai Chi class.

8 12 78 34 38 15 18 9 45 24 39 28 20 66 68 75 45 52 18 56

It is hard to tell much about the ages when they are displayed like this. You know that you could make a table, a box chart, or a line graph to show this information. Another way to show the information is to make a **stem-and-leaf plot.** The following stem-and-leaf plot shows the ages of the students.

Notice that the tens digits appear in the left-hand column. These are called **stems.** Each digit on the right is called a **leaf.** From looking at the plot, you can tell that more students are in their teens than in their twenties or thirties and that two students are younger than ten.

Stem	Leaves
0	8 9
1	2 5 8 8
2	0 4 8
3	4 8 9
4	5 5
5	2 6
6	6 8
7	5 8

1|2 means 12 years old

Check It Out

The stem-and-leaf plot shows the average points per game of high-scoring players over several years.

14. How many players scored an average number of points between 30 and 31?
15. What was the highest average number of points scored? the lowest?

Stem	Leaves
27	2
28	4
29	3 6 8
30	1 3 4 6 6 7 8
31	1 1 5
32	3 5 6 9
33	1 6
34	0 5
35	0
36	
37	1

30|1 means 30.1 points

Interpret and Create a Bar Graph

Another type of graph you can use to show data is called a *bar graph.* In this graph, either horizontal or vertical bars are used to show data. Consider the data showing Kirti's earnings from mowing lawns.

May	$78
June	$92
July	$104
August	$102
September	$66

You can make a bar graph to show Kirti's earnings.

To make a bar graph:

- Choose a vertical scale and decide what to place along the horizontal scale.
- For each item on horizontal scale, draw a bar of the appropriate height.
- Write a title for the graph.

A bar graph of Kirti's earnings is shown below.

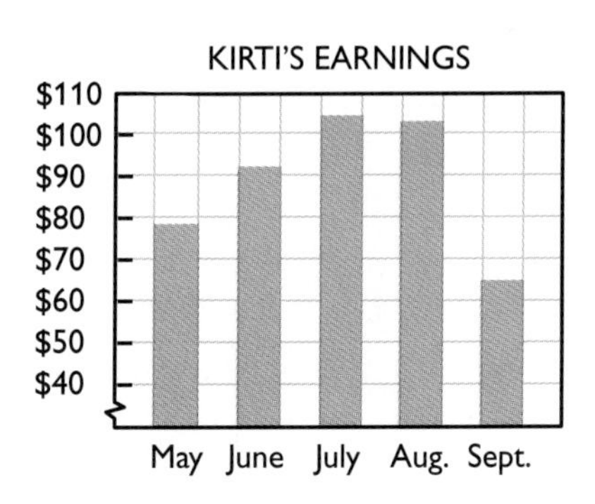

From the graph, you can see that his earnings were highest in July.

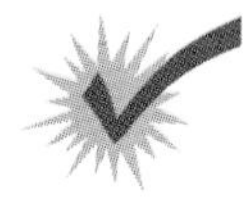

Check It Out

16. During which month were Kirti's earnings the lowest?

17. Write a sentence describing Kirti's earnings.

18. Use the data to make a bar graph to show the number of middle-school students on the honor role.

Sixth Grade	144
Seventh Grade	182
Eighth Grade	176

Interpret a Double-Bar Graph

If you want to show information about two or more things, you can use a **double-bar graph.** The following graph shows the sources of revenue for public schools for the past few years.

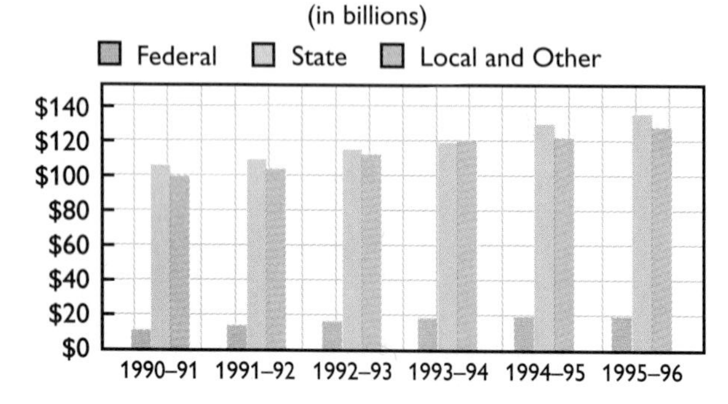

You can see from the graph that the states usually contribute more toward public schools than do local and other sources. Note that the amounts are given in billions. That means $20 on the graph represents $20,000,000,000.

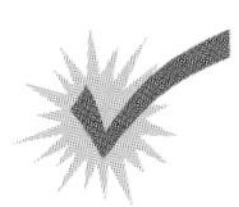

Check It Out

19. About how much did states contribute to public schools in 1993–94?
20. Write a sentence that describes the federal contribution during the years shown.

Interpret and Create a Histogram

A **histogram** is a special kind of bar graph that shows frequency of data. Suppose you asked several classmates how many hours, to the nearest hour, they talked on the telephone last week and collected the following numbers.

4 3 2 3 1 2 0 2 1 3 4 2 1 0 1 6

To create a histogram,

- Make a table showing frequencies.

Hours	Tally	Frequency
0	//	2
1	////	4
2	////	4
3	///	3
4	//	2
5		0
6	/	1

- Make a bar graph showing the frequencies.
- Title the graph.

In this case, you might call it "Hours Spent on the Telephone."
Your frequency diagram might look like this.

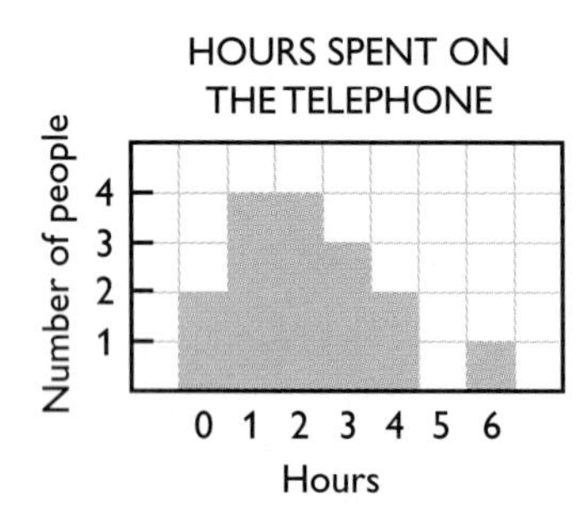

You can see from the diagram that as many students spent 1 hour on the phone as spent 2 hours.

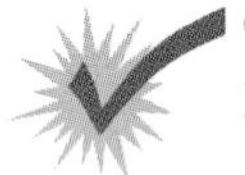

Check It Out

21. How many classmates were surveyed?
22. Make a histogram from the data about *Black Beauty* (p. 202).

4·2 EXERCISES

1. Make a table and a histogram to show the following data.
   Hours Spent Each Week Reading for Pleasure
   3 2 5 4 3 1 5 0 2 3 1 4 3 5 1 7 0 3 0 2
2. Which was the most common amount of time spent each week reading for pleasure?
3. Make a frequency graph to show the data in item 1.
4. Use your frequency graph to describe the hours spent reading for pleasure.
5. Of the first ten presidents, two were born in Massachusetts, one in New York, one in South Carolina, and six in Virginia. Make a circle graph to show this information and write a sentence about your graph.
6. The stem-and-leaf plot shows the heights of 19 girls.

Stem	Leaves
5	3 4 4 4 6 8
6	0 0 3 4 4 4 5 6 8 8
7	0 1 2

5 | 3 means 53 inches

   What can you say about the height of most of the girls?

7. The eighth grade classes collected 56 lb of aluminum in September, 73 lb in October, 55 lb in November, and 82 lb in December. Make a bar graph to show the data.
8. The box plot shows the daily high temperatures in Seaside in July. What is the middle temperature? 50% of the temperatures are between 65° and what temperature?

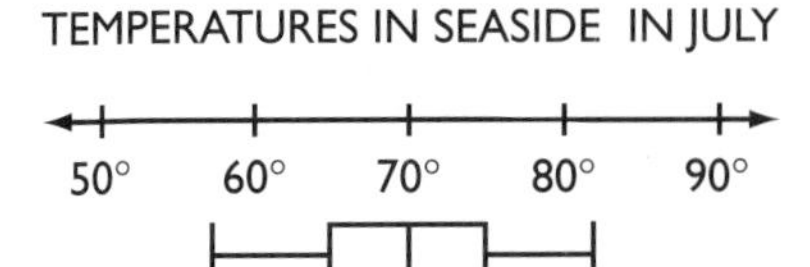

4·3 Analyzing Data

Scatter Plots

Once you have collected data, you will want to analyze and interpret it. You can plot points on a *coordinate graph* (p. 316) to make **scatter plots** of data and determine if the data are related.

Chilled to the Bone

Wind carries heat away from the body, increasing the cooling rate. So whenever the wind blows, you feel cooler. If you live in an area where the temperature drops greatly in winter, you know you may feel much, much colder on a blustery winter day than the temperature indicates.

Wind Speed (mi/hr)	Air Temperature (°F) 35	30	25	20	15	10	5	0
Calm	35	30	25	20	15	10	5	0
5	32	27	22	16	11	6	0	−5
10	22	16	10	3	−3	−9	−15	−22
15	16	9	2	−5	−11	−18	−25	−31
20	12	4	−3	−10	−17	−24	−31	−39
25	8	1	−7	−15	−22	−29	−36	−44
30	6	−2	−10	−18	−25	−33	−41	−49

This wind-chill table shows the effects of the cooling power of the wind in relation to temperature under calm conditions (no wind).

Listen to or read your local weather report each day for a week or two in the winter. Record the daily average temperature and wind speed. Use the table to determine how chilly it felt each day.

Samuel collected information showing the number of candy boxes sold by each person in his soccer club and the number of years each person had been in the club.

Years in Club	4	3	6	2	3	4	1	2	1	3	4	5	2	2
Boxes Sold	23	18	30	26	22	20	20	20	15	19	23	26	22	18

Let's make a scatter plot to see if there was any relationship between the two. Making a scatter plot is like drawing a graph on a coordinate plane. First you write the data as ordered pairs and then you graph the ordered pairs.

To make a scatter plot,

- Collect two sets of data that you can graph as ordered pairs.
- Label the vertical and horizontal axes and graph the ordered pairs.

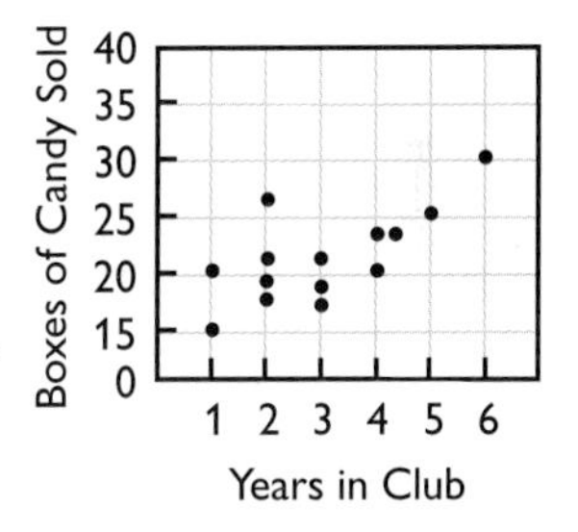

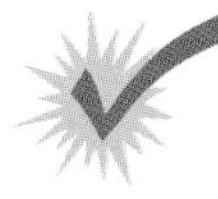

Check It Out

Make a scatter plot showing the following data.

1. Winning Times for the Men's 100-meter Run, Summer Olympics

Year	1900	1912	1924	1936	1948	1960	1972	1984	1996
Time (in sec)	11.0	10.8	10.6	10.3	10.3	10.2	10.1	9.99	9.84

2. Relationship between Shoe Size and Number of siblings

Shoe Size	5	$5\frac{1}{2}$	7	9	9	8	$7\frac{1}{2}$	5	8	$9\frac{1}{2}$	10	7	6	$9\frac{1}{2}$	6
Number of Siblings	5	3	0	0	1	1	4	2	3	6	5	2	1	2	6

Correlation

The following scatter plots have slightly different appearances.

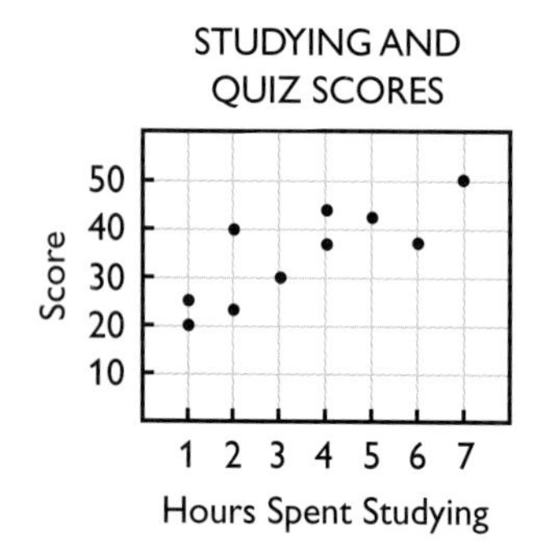

The studying and quiz scores scatter plot shows the relationship between the hours spent studying and quiz scores. There is an upward trend in the scores. You call this a positive **correlation.**

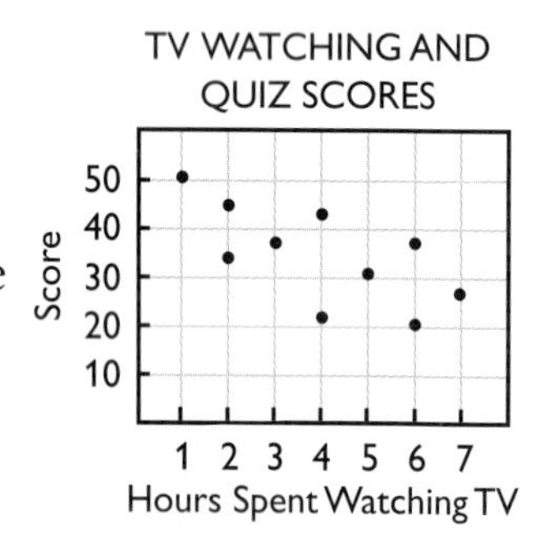

This scatter plot shows the relationship between hours spent watching TV and quiz scores. There is a downward trend in the scores. You call this a negative correlation.

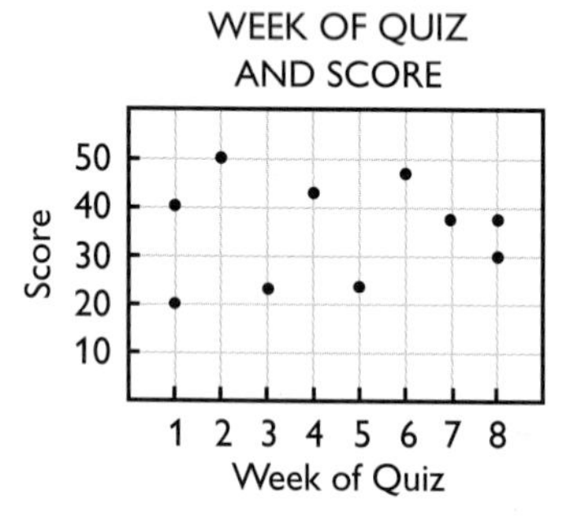

The third scatter plot shows the relationship between the week a quiz was taken and the score. There does not appear to be any relationship. You call this no correlation.

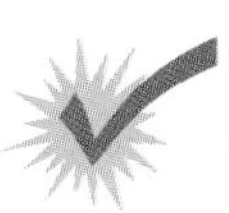

Check It Out

3. Which of the following scatter plots shows no relationship?

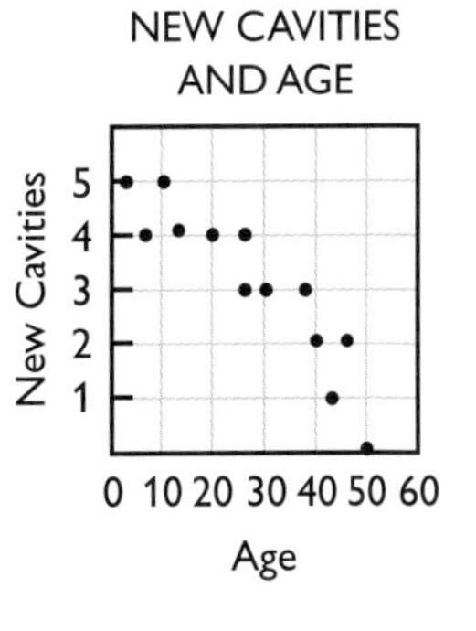

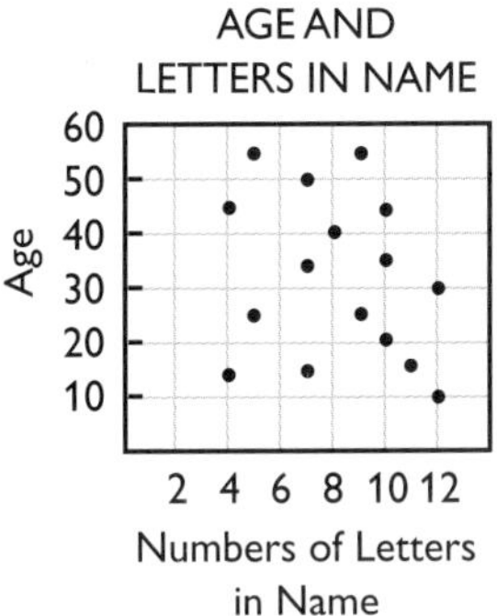

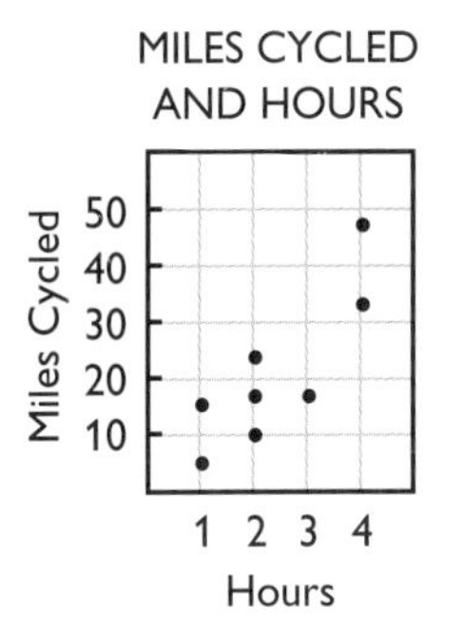

4. Describe the correlation in the scatter plot showing the relationship between age and number of cavities.
5. Which scatter plot shows a positive correlation?

Birthday Surprise

How likely do you think it is that two people in your class have the same birthday? With 365 days in a year, you might think the chances are very slim. After all, the probability that a person is born on any given day is $\frac{1}{365}$, or about 0.3%.

It might surprise you to learn that in a group of 23 people, the chances that two share the same birthday is just a slight bit more than 50%. With 30 people, the likelihood increases to 71%. And with 50 people, you can be 97% sure that two of them were born on the same day.

Line of Best Fit

When the points you plot on a scatter plot are either positively or negatively related, you can sometimes draw a **line of best fit.** Consider again the graph showing the relationship between age and number of cavities.

To draw a line of best fit,

- Decide if the points on the scatter plot show a trend.
  The points on this graph show a negative correlation.
- Draw the line that seems to run through the center of the group of points.

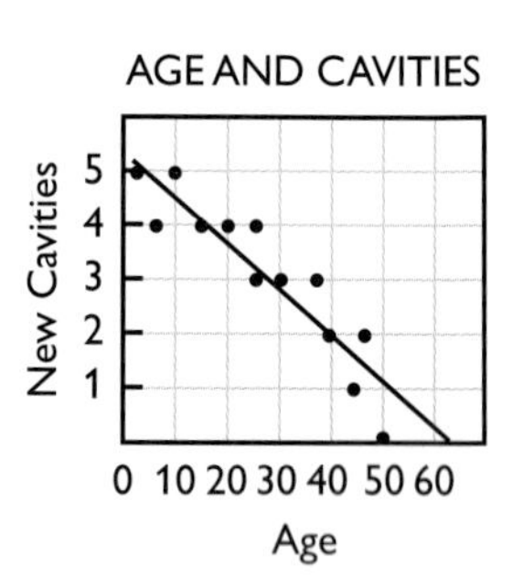

You can use the line to predict information. From the line it appears that people of age 60 would be expected to have fewer than one new cavity, and people of age 70 would be predicted to have no new cavities.

You draw the line to help you predict. But the line can show data that couldn't be real. Always think about whether your prediction is reasonable. For example, people of age 60 would not get $\frac{1}{4}$ cavities. You would probably predict one cavity.

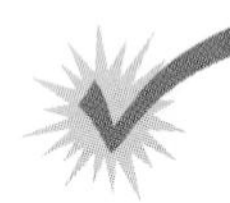

Check It Out

6. Use the following data to make a scatter plot and draw a line of best fit.

Latitude (°N)	35	34	39	42	35	42	33	42	21
Mean April Temperature (°F)	55	62	54	49	61	49	66	47	76

7. Predict the mean April temperature of Juneau, which has a latitude of 58°N.

Distribution of Data

A veterinarian measured the weights of 25 cats to the nearest pound and recorded the data on the following histogram. Notice how symmetrical the histogram is. If you draw a curve over the histogram, the curve illustrates a **normal distribution.**

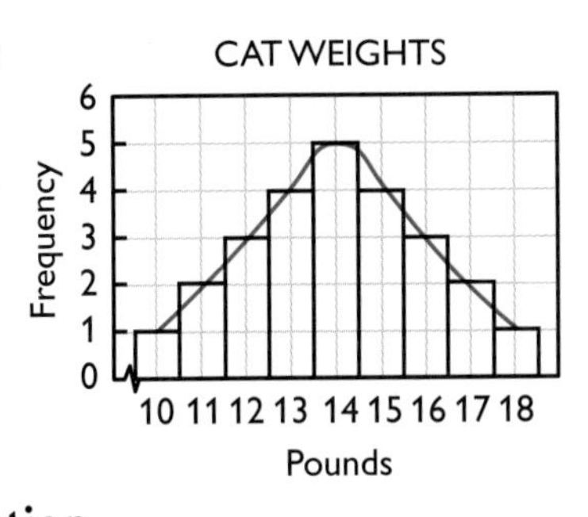

Often a histogram has a **skewed distribution.** These two histograms show the heights of students on the gymnastic team and basketball team. Again you can draw a curve to show the shape of the histogram. The one showing the gymnastic team heights is skewed to the left. The one showing the basketball team heights is skewed to the right.

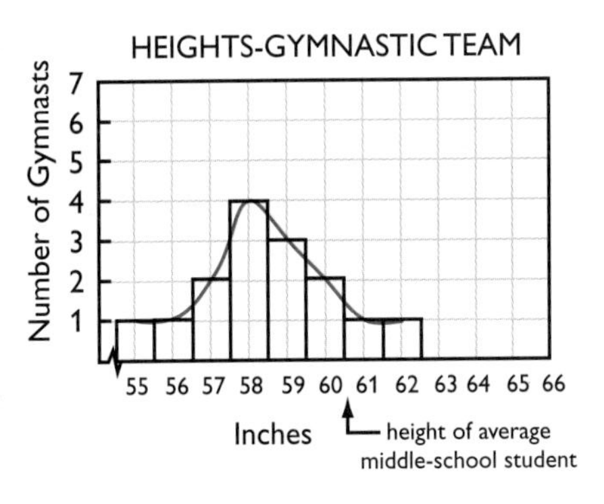

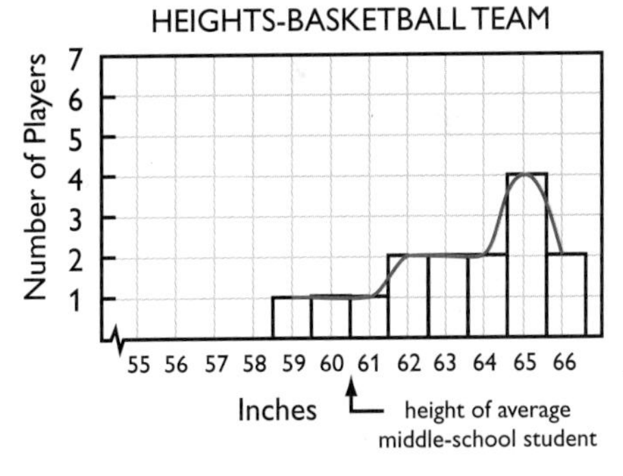

The histogram on the left below illustrates heights of adults. This histogram has two peaks, one for female heights and one for male heights. This kind of distribution is called a **bimodal distribution.** The one on the right shows the number of dogs boarded each week at the pet kennel. It is called a **flat distribution.**

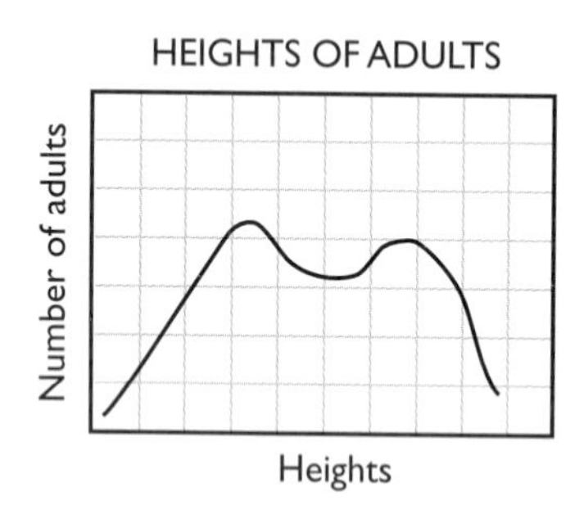

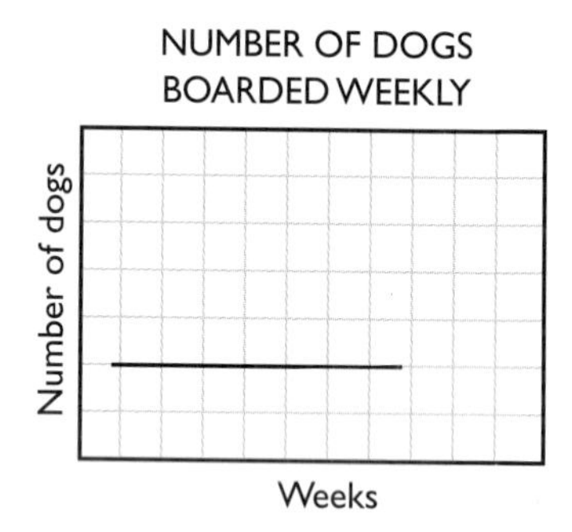

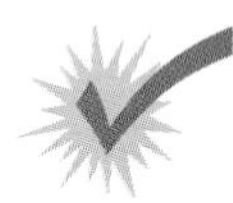

Check It Out

Identify each type of distribution as normal, skewed to the right, skewed to the left, bimodal, or flat.

8.

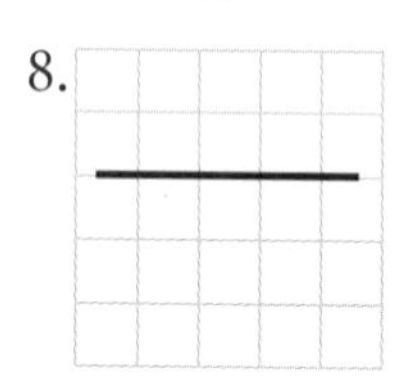

9.

10.

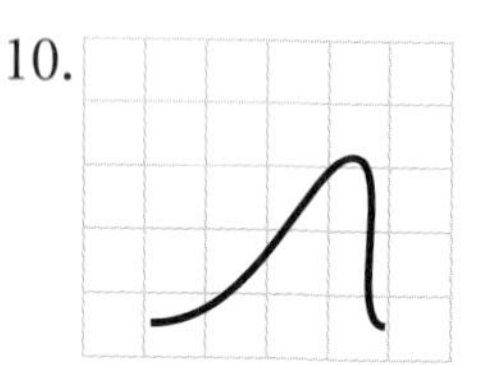

11.

12.

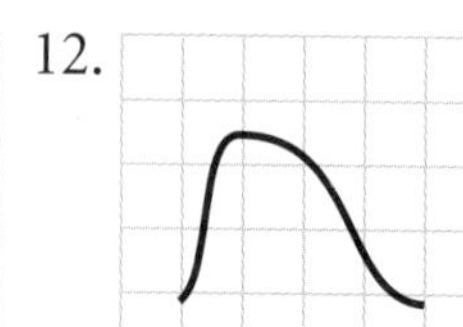

4•3 EXERCISES

1. Make a scatter plot of the following data.

Times at Bat	5	2	4	1	5	6	1	3	2	6
Hits	4	0	2	0	2	4	1	1	2	3

2. Describe the correlation in the scatter plot in item 1.
3. Draw a line of best fit for the scatter plot in item 1. Use it to predict the number of hits for 8 times at bat.

Describe the correlation in each of the following scatter plots.

4. 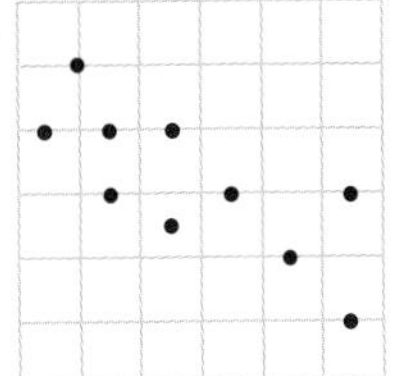

5. 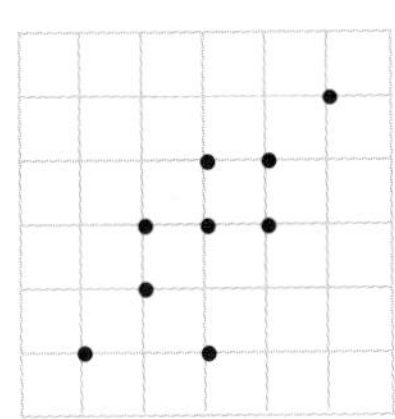

6. 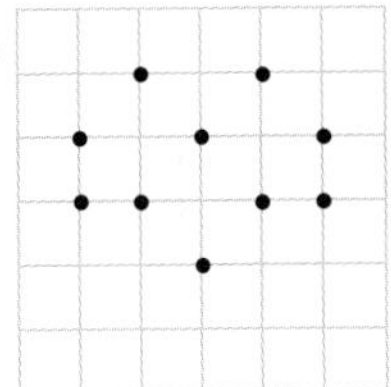

Tell whether each of the following distributions is normal, skewed to the right, skewed to the left, bimodal, or flat.

7. 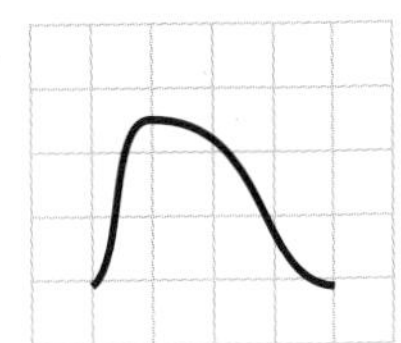

8. 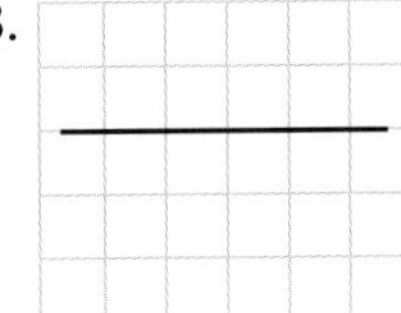

9. 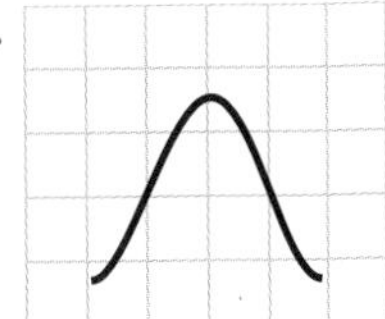

10. 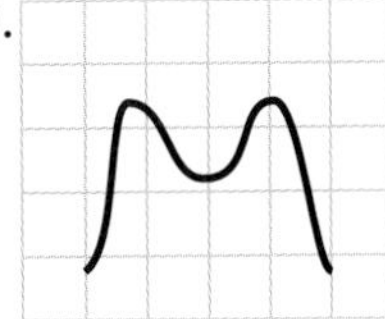

4•4 Statistics

Laila collected the following data about the amount her classmates spend on CDs each month.

$15, $15, $15, $15, $15, $15, $15
$25, $25, $25, $25, $25
$30, $30, $30
$45
$145

Laila said her classmates typically spend $15 per month, but Jacy disagreed. He said the typical amount was $25. A third classmate, Maria, said they were both wrong—the typical expenditure was $30. Each was correct because each was using a different common measure.

Mean

One measure of data is the **mean.** To find the mean, or **average,** add the amounts the students spend and divide by the number of amounts.

FINDING THE MEAN

Find the mean of money spent monthly on CDs by students in Laila's class.

- Add the amounts.

  $15 + $15 + $15 + $15 + $15 + $15 + $15 + $25 + $25 + $25 + $25 + $25 + $30 + $30 + $30 + $45 + $145 = $510

- Divide the total by the number of amounts.

  In this case, there are 17 amounts.

  $510 ÷ 17 = $30

The mean amount each student spends on CDs is $30. Maria used the mean to describe the amounts when she said each typically spent $30.

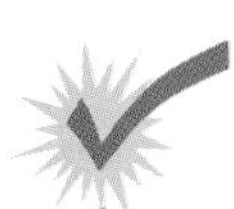

Check It Out

Find the mean:

1. 15, 12, 6, 4.5, 12, 2, 11.5, 1, 8
2. 100, 79, 88, 100, 45, 92
3. 125, 136, 287, 188, 201, 245
4. The low temperatures in Pinetop the first week in February were 38°, 25°, 34°, 28°, 25°, 15°, and 24°. Find the mean temperature.
5. Ling averaged 86 points on five tests. What would she have to score on the sixth test to bring her average up one point on the six tests?

Graphic Impressions

Humans can live longer than 100 years. But not all animals can live that long. A mouse, for example, has a maximum life span of 3 years, while a toad may live for 36 years.

Both these graphs compare the maximum life span of guppies, giant spiders, and crocodiles.

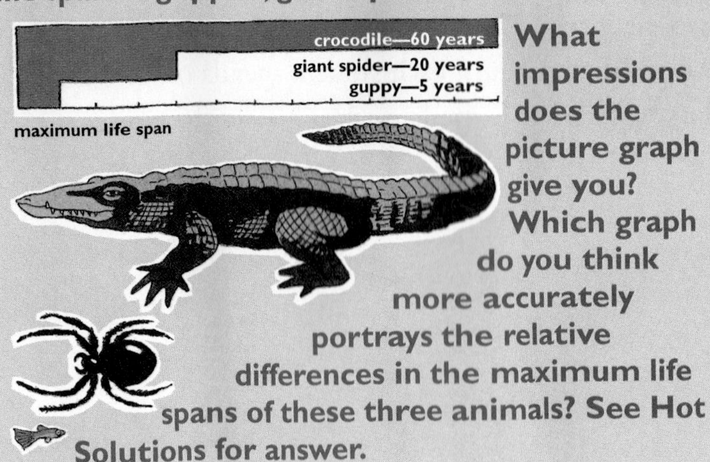

What impressions does the picture graph give you? Which graph do you think more accurately portrays the relative differences in the maximum life spans of these three animals? See Hot Solutions for answer.

Median

You see that you can find the mean by adding all the amounts and dividing by the number of amounts. Another way to look at numbers is to find the median. The **median** is the middle number in the data when the numbers are arranged in order. Let's look again at the amounts spent on CDs.

$15, $15, $15, $15, $15, $15, $15
$25, $25, $25, $25, $25
$30, $30, $30
$45
$145

FINDING THE MEDIAN

Find the median of amounts spent on CDs.

- Arrange the data in numerical order from least to greatest or greatest to least.

  Looking at the amounts spent on CDs, we can see they are already arranged in order.

- Find the middle number.

  There are 17 numbers. The middle number is $25 because there are eight numbers above $25 and eight below it.

The median amount each student spends on CDs is $25.

Jacy was using the median when he said the typical amount spent on CDs was $25.

When the number of amounts is even, you can find the median by finding the mean of the two middle numbers. So to find the median of the numbers 1, 6, 4, 2, 5, and 8, you must find the two numbers in the middle.

- Find the median of an even number of data.

  1 2 4 5 6 8
- Arrange the numbers in order from least to greatest or greatest to least.

  1 2 4 5 6 8 or 8 6 5 4 2 1
- Find the mean of the two middle numbers.

  The two middle numbers are 4 and 5.

  $(4 + 5) \div 2 = 4.5$

The median is 4.5. Half the numbers are greater than 4.5 and half the numbers are less than 4.5.

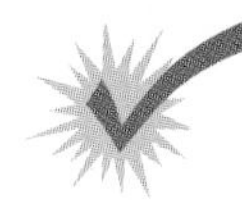

Check It Out

Find the median:

6. 11, 15, 10, 7, 16, 18, 9
7. 1.4, 2.8, 5.7, 0.6
8. 11, 27, 16, 48, 25, 10, 18
9. The top ten scoring totals in the NBA are: 24,489; 31,419; 23,149; 25,192; 20,880; 20,708; 23,343; 25,389; 26,710; and 14,260 points. Find the median scoring total.

Mode

You can describe a set of numbers by using the mean or by using the median, which is the middle number. Another way to describe a set of numbers is to use the mode. The **mode** is the number in the set that occurs most often. Let's look again at the amounts spent on CDs.

$15, $15, $15, $15, $15, $15, $15
$25, $25, $25, $25, $25
$30, $30, $30
$45
$145

To find the mode, group like numbers together and look for the one that appears most frequently.

FINDING THE MODE

- Arrange the numbers in order or make a frequency table of the numbers.

Amount	Frequency
$ 15	7
$ 25	5
$ 30	3
$ 45	1
$145	1

- Select the number that appears most frequently.
  The most frequent amount spent is $15.

The mode of the amount each student spends on CDs is $15.

So Laila was using the mode when she said $15 was the typical amount students spent on CDs.

A group of numbers may have no mode or more than one mode. Data that has two modes is called *bimodal*.

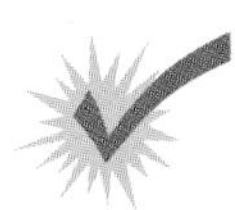

Check It Out

Find the mode:

10. 1, 3, 3, 9, 7, 2, 7, 7, 4, 4
11. 1.6, 2.7, 5.3, 1.8, 1.6, 1.8, 2.7, 1.6
12. 2, 10, 8, 10, 4, 2, 8, 10, 6
13. In 1961, Roger Maris hit 61 home runs. The top 25 home run hitters of 1961 hit the following numbers of home runs in one season: 61, 49, 54, 49, 49, 60, 52, 50, 49, 52, 59, 54, 51, 49, 58, 54, 56, 54, 51, 52, 51, 49, 51, 58, 49. Find the mode.

Olympic Decimals

In Olympic gymnastics, the competitors perform a set of specific events. Scoring is based on a 10-point scale, where 10 is a perfect score. Marks may be given in decimal numbers. After the high and low scores have been eliminated, the remaining marks are averaged.

For some of the events, gymnasts are judged on their technical merit and for composition and style.

	Technical Merit	Composition and Style
USA	9.4	9.8
China	9.6	9.7
France	9.3	9.9
Germany	9.5	9.6
Australia	9.6	9.7
Canada	9.5	9.6
Japan	9.7	9.8
Russia	9.6	9.5
Sweden	9.4	9.7
England	9.6	9.7

Marks for technical merit are based on the difficulty and variety of the routines and the skills of the gymnasts. Marks for composition and style are based on the originality and artistry of the routines.

Use these marks to determine the mean scores for technical merit and for composition and style. See Hot Solutions for answer.

Range

Another measure used with numbers is the range. The **range** tells how far apart the greatest and least numbers in a set are. Consider the following miles of coastline on the Pacific Coast in the United States.

State	Miles of Coastline
California	840
Oregon	296
Washington	157
Hawaii	750
Alaska	5,580

To find the range, you must subtract the least number of miles from the greatest.

FINDING THE RANGE

Find the range of miles of Pacific coastline.

- Find the greatest and least values.

  The greatest value is 5,580 mi and the least value is 157 mi.

- Subtract.

  5,580 mi − 157 mi = 5,423 mi

The range is 5,423 miles.

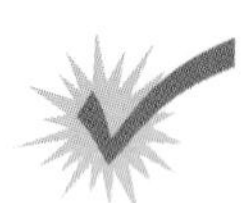

Check It Out

Find the range.

14. 100, 700, 800, 500, 50, 300
15. 1.4, 2.8, 5.7, 0.6
16. 56°, 43°, 18°, 29°, 25°, 70°
17. The winning scores for the Candlelights basketball team are 78, 83, 83, 72, 83, 61, 75, 91, 95, and 72. Find the range in the scores.

Weighted Averages

The amounts spent on CDs by Laila's classmates were \$15, \$25, \$30, \$45, and \$145. To find the mean amount spent, you cannot just find the mean of those numbers because, for example, more people spent \$15 than spent \$45. You must find the **weighted average.**

FINDING THE WEIGHTED AVERAGE

Find the weighted average of amounts spent on CDs.

- Determine each amount and the number of times it occurs in the set.

  \$15—7 times
  \$25—5 times
  \$30—3 times
  \$45—1 time
  \$145—1 time

- Multiply each amount by the number of times it occurs.

  $\$15 \times 7 = \$105$
  $\$25 \times 5 = \$125$
  $\$30 \times 3 = \$90$
  $\$45 \times 1 = \$45$
  $\$145 \times 1 = \$145$

- Add the products and divide by the total of the weights.

  $(\$105 + \$125 + \$90 + \$45 + \$145) \div (7 + 5 + 3 + 1 + 1) = \$510 \div 17 = \$30$

The weighted average spent on CDs was \$30.

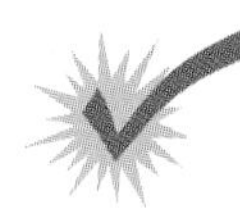

Check It Out

Find the weighted average.

18. 45 occurs 5 times, 36 occurs 10 times, and 35 occurs 15 times.
19. The average number of checkout lanes in a Well-made department store is 8, and the average number in a Cost-easy store is 5. If there are 12 Well-made stores and 8 Cost-easy stores, find the average number of checkout lanes.

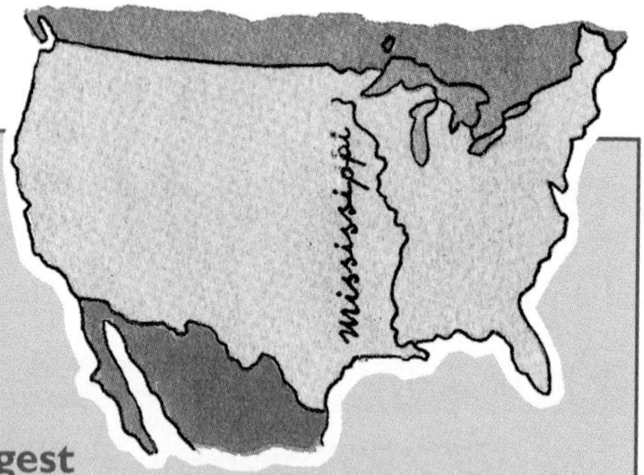

How Mighty Is the Mississippi?

The legendary Mississippi is the longest river in the United States, but not in the world. Here's how it compares to the world's 12 longest rivers.

River	Location	Length (miles)
Nile	Africa	4,145
Amazon	South America	4,000
Yangtze	Asia	3,915
Yellow	Asia	2,903
Congo	Africa	2,900
Irtysh	Asia	2,640
Mekong	Asia	2,600
Niger	Africa	2,600
Yenisey	Asia	2,543
Parana	South America	2,485
Mississippi	North America	2,348
Missouri	North America	2,315

In this set of data, what's the mean length, the median length, and the range? See Hot Solutions for answer.

4·4 EXERCISES

Find the mean, median, mode, and range.

1. 2, 2, 4, 4, 6, 6, 8, 8, 8, 8, 10, 10, 12, 14, 18
2. 5, 5, 5, 5, 5, 5, 5, 5, 5
3. 50, 80, 90, 50, 40, 30, 50, 80, 70, 10
4. 271, 221, 234, 240, 271, 234, 213, 253, 196
5. Are any of the sets of data above bimodal? Explain.
6. Find the weighted average: 15 occurs 3 times, 18 occurs 1 time, 20 occurs 5 times, and 80 occurs 1 time.
7. Kelly had 85, 83, 92, 88, and 69 on her first five math tests. She needs an average of 85 to get a B. What score must she get on her last test to get a B?
8. Which measure—mean, median, or mode—must be a member of the set of data?
9. The following times represent the lengths of phone calls, in minutes, made by an eighth grader one weekend.
   10 2 16 8 55 2 18 11 9 5 4 7
   Find the mean, median, and mode of the calls. Which measure best represents the data? Explain.
10. The price of a house is higher than half of the other houses in the area. Would you use the mean, median, mode, or range to describe it?

4·5 Combinations and Permutations

Tree Diagrams

You often need to be able to count outcomes. For example, suppose you have two **spinners.** One has the numbers 1 through 3 and the other spinner has the numbers 1 and 2. Suppose you want to find out how many different two-digit numbers you can make by spinning the first spinner and then the second one. You can make a **tree diagram.**

To make a tree diagram, you list what can happen with the first spinner.

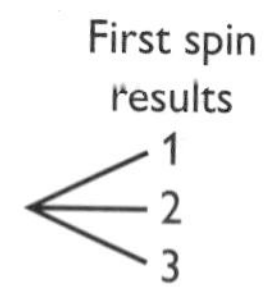

Then, by each one, you list what can happen with the second spinner.

First spin results	Second spin results	Possible different numbers
1	1	11
	2	12
2	1	21
	2	22
3	1	31
	2	32

After listing the possibilities, you can count to see how many there are. In this case, there are six possible numbers.

MAKING A TREE DIAGRAM

Make a tree diagram to find out how many possible ways three coins can land if you toss them into the air.

- List what happens with the first trial.

  The first coin can come up heads or tails.

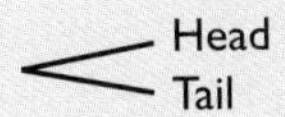

- List what happens with the second and third (and so on) trials.

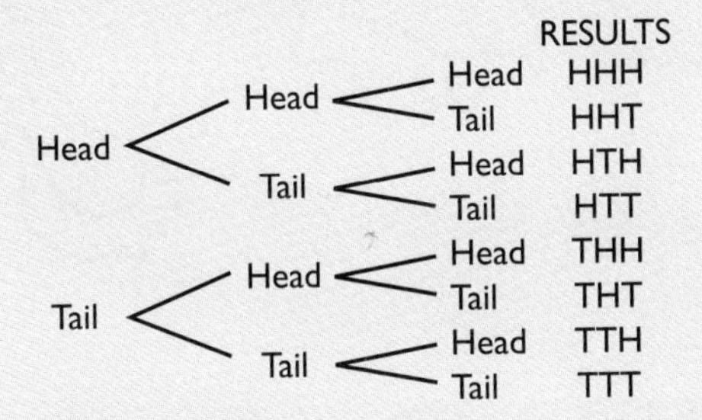

- Draw lines and list the options.

  The results are listed above. There are eight ways the coins can land.

You can find the number of possibilities by multiplying the number of choices at each step. For the three coins problem, $2 \times 2 \times 2 = 8$. This represents two possibilities for coin one, two possibilities for coin two, and two possibilities for coin three.

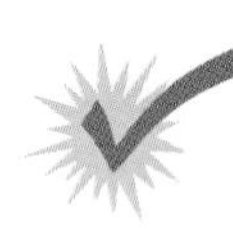

Check It Out

Use a tree diagram to find the answers. Check by multiplying.

1. If you toss three number cubes, each showing the numbers 1–6, how many possible three-digit numbers can you form?
2. How many possible routes are there from Creekside to Mountainville?

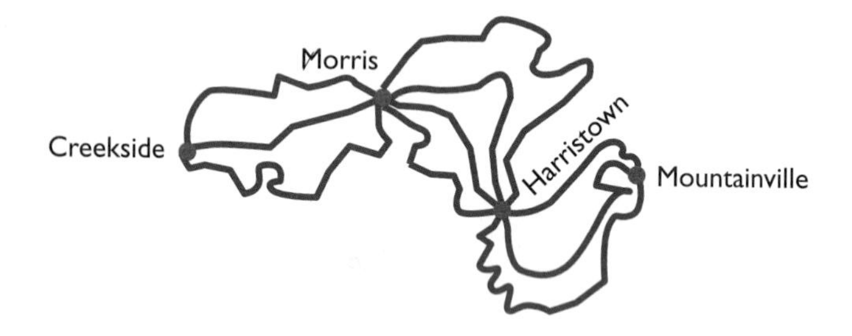

Monograms

What are your initials? Do you have anything with your monogram on it? A *monogram* is a design that is made up of one or more letters, usually the initials of a name. Monograms often appear on stationary, towels, shirts, or jewelry.

How many different three-letter monograms can you make with the letters of the alphabet? Use a calculator to compute the total number. Don't forget to allow for repeat letters in the combination. See Hot Solutions for answer.

Permutations

You know that you can use tree diagrams to count the ways something can happen. The tree diagram also shows ways things can be arranged or listed. A listing in which the order is important is called a **permutation.** Suppose you want to line up Rita, Jacob, and Zhao for a photograph. You can use a tree diagram to show all the different ways they could line up.

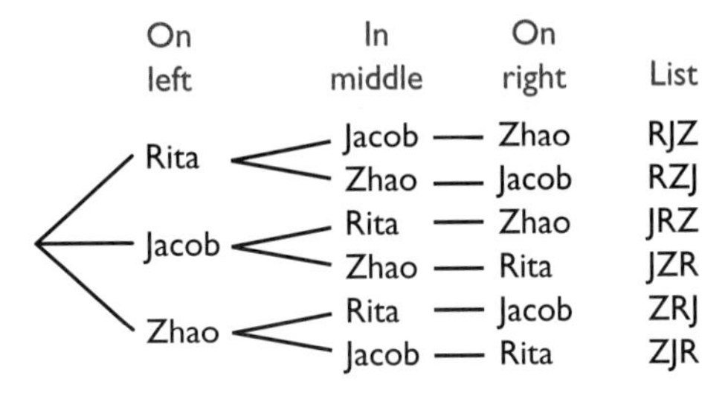

There are 3 ways to choose the first person, 2 ways to choose the second, and 1 way to chose the third, so the total number of permutations is $3 \times 2 \times 1 = 6$. Remember that Rita, Jacob, Zhao is a different permutation from Zhao, Jacob, Rita.

$P(3, 3)$ represents the number of permutations of 3 things taken 3 at a time. Thus $P(3, 3) = 6$.

FINDING PERMUTATIONS

Find $P(6, 5)$.

- Determine how many choices there are for each place.

  There are 6 choices for the first place, 5 for the second, 4 for the third, 3 for the fourth, and 2 for the last.

- Find the product.

  $6 \times 5 \times 4 \times 3 \times 2 = 720$

So $P(6, 5) = 720$.

Factorial Notation

You saw that to find the number of permutations of 8 things, you found the product $8 \times 7 \times 6 \times 5 \times 4 \times 3 \times 2 \times 1$. The product $8 \times 7 \times 6 \times 5 \times 4 \times 3 \times 2 \times 1$ is called 8 **factorial.** The shorthand notation for a factorial is an exclamation point. So $8! = 8 \times 7 \times 6 \times 5 \times 4 \times 3 \times 2 \times 1$.

Check It Out

Find each value.

3. $P(15, 2)$
4. $P(6, 6)$
5. The Grandview Middle School has a speech contest. There are 8 finalists. In how many different orders can the speeches be given?
6. One person from a class of 35 students is to be chosen as a delegate to Government Day, and another person is to be chosen as an alternate. In how many ways can a delegate and an alternate be chosen?

Find the value. Use a calculator if available.

7. 9!

Combinations

When you find the number of ways to select a delegate and an alternate from a class of 35, the order in which you select the students is important. Suppose, instead, that you simply pick two delegates. Then the order is not important. That is, choosing Elena and Rahshan is the same as choosing Rahshan and Elena when picking two delegates.

You can use the number of permutations to find the number of **combinations.** Say you want to select 2 students to be delegates from a group of 6 students (Elena, Rahshan, Felicia, Hani, Toshi, and Kelly).

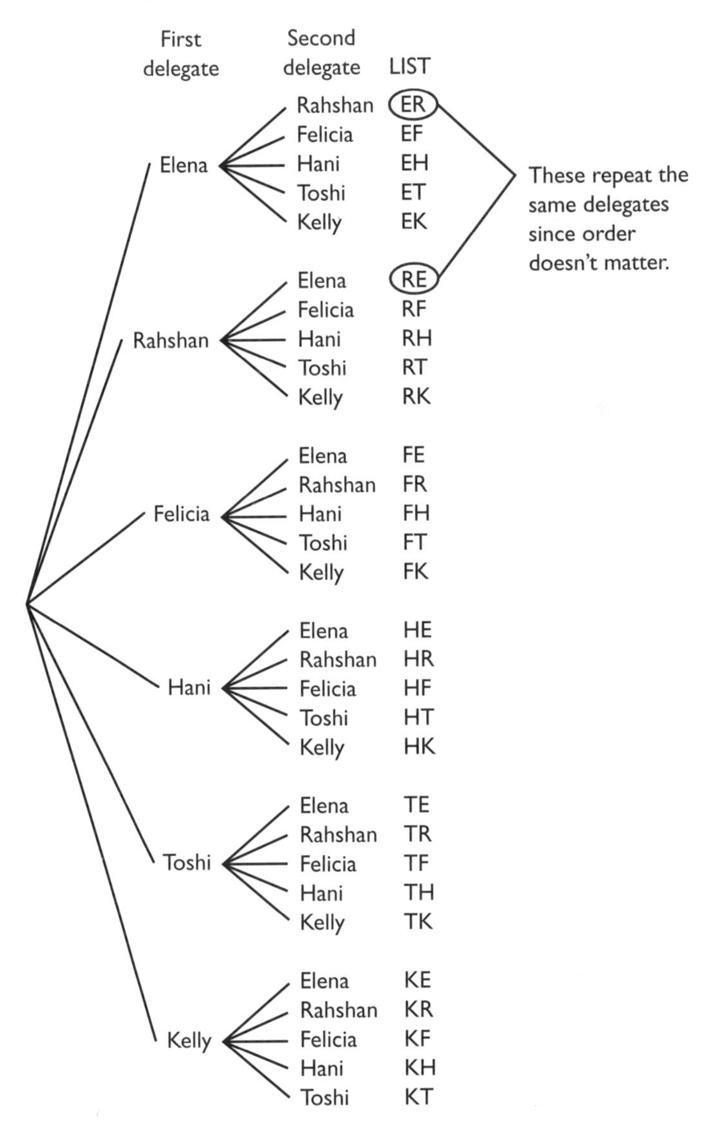

To find the number of combinations of six students taken two at a time, you start by finding the permutations. You have six ways to choose the first delegate and five ways to choose the second, so this is $6 \times 5 = 30$. But the order doesn't matter, so some combinations were counted too often! You need to divide by the number of different ways the two delegates can be arranged (2!).

$$C(6, 2) = \frac{P(6, 2)}{2!} = \frac{6 \times 5}{2 \times 1} = 15$$

FINDING COMBINATIONS

Find $C(6, 3)$.

- Find the number of permutations.

  $P(6, 3) = 6 \times 5 \times 4 = 120$

- Divide by the number of ways the objects can be arranged.

  $120 \div 3! = 120 \div 6 = 20$

So $C(6, 3) = 20$.

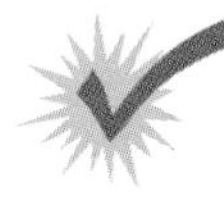

Check It Out

Find each value.

8. $C(9, 6)$
9. $C(14, 2)$
10. How many different combinations of three plants can you choose from a dozen plants?
11. Are there more combinations or permutations of two books from a total of four? Explain.

4•5 EXERCISES

1. Make a tree diagram to show the results when you toss a coin and roll a number cube containing the numbers 1 through 6.

Find each value.

2. $P(7, 5)$
3. $C(8, 8)$
4. $P(9, 4)$
5. $C(7, 3)$
6. $5! \times 4!$
7. $P(8, 8)$

Solve.

8. Eight friends want to play enough games of tennis (singles) to be sure that everyone plays everyone else. How many games will they have to play?
9. At a chess tournament, trophies are given for first, second, third, and fourth places. Twenty students enter the tournament. How many different arrangements of four winning students are possible?
10. Determine if the following is a permutation or a combination.
    a. Choosing a team of 5 players from 20 people
    b. Arranging 12 people in a line for a photograph
    c. Choosing first, second, and third places from 20 show dogs

4·6 Probability

If you and a friend want to decide who goes first in a game, you might flip a coin. You and your friend have an equal chance of winning the toss. The **probability** of an event is a number from 0 to 1 that measures the chance that an event will occur.

Experimental Probability

The probability of an event is a number from 0 to 1. One way to find the probability of an event is to conduct an experiment. Suppose you want to know the probability of winning a game of checkers played with your friend. You play 12 games and win 8 of them. You can compare the number of games you win to the number of games you play to find the probability of winning. In this case, the **experimental probability** that you will win is $\frac{8}{12}$, or $\frac{2}{3}$.

DETERMINING EXPERIMENTAL PROBABILITY

Find the experimental probability of drawing a red marble from a bag of 10 colored marbles.

- Conduct an experiment. Record the number of trials and the result of each trial.

  Choose a marble from the bag, record its color, and replace it. Repeat 10 times. Suppose you draw red, green, blue, green, red, blue, blue, red, green, blue.

- Compare the number of occurrences of one result to the number of trials. That is the probability for that result.

  Compare the number of red marbles to the total number of draws.

The experimental probability of drawing a red marble in this test is $\frac{3}{10}$.

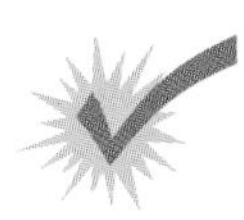

Check It Out

Three pennies are tossed 100 times. The results are shown on the circle graph.

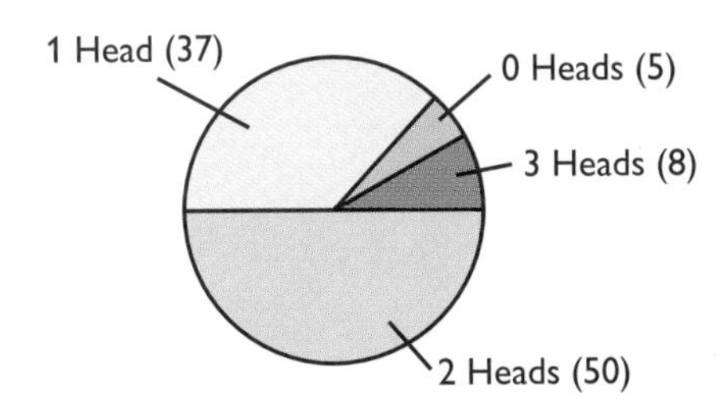

1. Find the experimental probability of getting two heads.
2. Find the experimental probability of getting no heads.
3. Drop a thumbtack 50 times and record the number of times it lands point up. Find the experimental probability of the tack landing point up. Compare your answers with other student answers.

Theoretical Probability

You know that you can find the experimental probability of tossing a head when you toss a coin by doing the experiment and recording the results. You can also find the **theoretical probability** by considering the outcomes of the experiment. The **outcome** of an experiment is a result. The outcomes when tossing a coin are head and tail. An **event** is a specific outcome, such as heads. So the probability of getting a head is

$$\frac{\text{number of ways an event occurs}}{\text{number of outcomes}} = \frac{1}{2}$$

DETERMINING THEORETICAL PROBABILITY

Find the probability of drawing a red marble from a bag containing 5 red, 8 blue, and 7 white marbles.

- Determine the number of ways the event occurs.

  In this case, the event is getting a red marble. There are 5 red marbles.

- Determine the total number of outcomes. Use a list, multiply, or make a *tree diagram* (p. 232).

  There are 20 marbles in the bag.

- Use the formula

  $$P(\text{event}) = \frac{\text{number of ways an event occurs}}{\text{number of outcomes}}$$

- Find the probability of the target event.

  In this case, drawing a red marble is represented by $P(\text{red})$.

  $P(\text{red}) = \frac{5}{20} = \frac{1}{4}$

The probability of drawing a red marble is $\frac{1}{4}$.

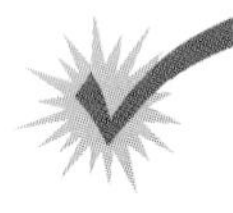

Check It Out

Find each probability. Use the spinner for items 4–5.

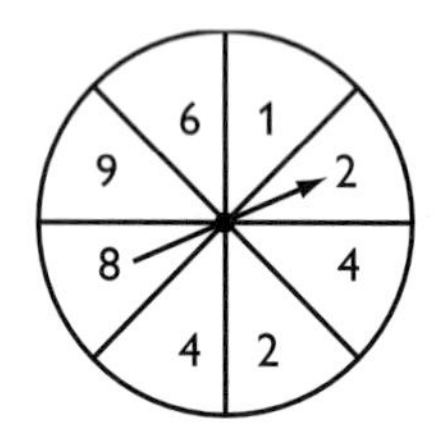

4. $P(\text{even number})$
5. $P(\text{number greater than 10})$
6. $P(2)$ when tossing a number cube
7. The letters of the word *Mississippi* are written on identical slips of paper and placed in a box. If you draw a slip at random, what is the probability that it will be a vowel?

Expressing Probabilities

You can express a probability as a fraction, as shown before. But, just as you can write a fraction as a decimal, ratio, or percent, you can also write a probability in any of those forms (p. 154).

The probability of getting a head when you toss a coin is $\frac{1}{2}$. You can also express the probability as follows:

Fraction	Decimal	Ratio	Percent
$\frac{1}{2}$	0.5	1:2	50%

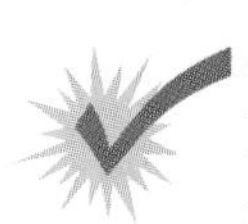

Check It Out

Express each of the following probabilities as a fraction, decimal, ratio, and percent.

8. the probability of drawing a red marble from a bag containing 4 red marbles and 12 green ones
9. the probability of getting an 8 when spinning a spinner divided into eight equal divisions numbered 1 through 8
10. the probability of getting a green gumball out of a machine containing 25 green, 50 red, 35 white, 20 black, 5 purple, 50 blue, and 15 orange gumballs
11. the probability of being chosen to do your oral report first if your teacher puts all 25 students' names in a bag and draws

Strip Graphs

When you conduct an experiment such as tossing a coin, you need to find a way to show the outcome of each toss. One way to show the outcomes of an experiment is to use a **strip graph.**

If you tossed a coin repeatedly, you could make the following strip graph.

H	H	H	T	T	H	T	T	

The strip graph shows that the following results occurred on the first eight tosses: head, head, head, tail, tail, head, tail, and tail.

To make a strip graph,
- Draw a series of boxes in a long strip.
- Enter each outcome in a box.

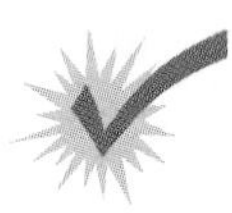

Check It Out

Consider the following strip graph.

2	5	4	2	3	1	6	3	

12. Describe the first eight outcomes.
13. What do you think the experiment might be?
14. Make a strip graph to show the outcomes if you toss a coin ten times. Compare your grids to others' grids.

Lottery Fever

You read the headline. You say to yourself, "Somebody's *bound* to win this time." But the truth is, you would be wrong! The chances of winning a Pick-6 lottery are always the same, and very, very, very small.

Start with the numbers from 1 to 7. There are always 7 different ways to choose 6 out of 7 things. (Try it for yourself.) So, your chances of winning a 6-out-of-7 lottery would be $\frac{1}{7}$ or about 14.3%. Suppose you try using 6 out of 10 numbers. There are 210 different ways you can do that, making the likelihood of winning a 6-out-of-10 lottery $\frac{1}{210}$ or 0.4%. For a 6-out-of-20 lottery, there are 38,760 possible ways to pick 6 numbers, and only 1 of these would be the winner. That's about a 0.003% chance of winning. Get the picture?

The chances of winning a 6-out-of-50 lottery are 1 in 15,890,700 or 1 in about 16 million. For comparison, think about the chances that you'll get struck by lightning—a rare occurrence. It is estimated that in the U. S. roughly 260 people are struck by lightning each year. Suppose the population of the U. S. is about 260 million. Would you be more likely to win the lottery or be struck by lightning? See Hot Solutions for answer.

Outcome Grids

You have seen how to use a tree diagram to show possible outcomes. Another way to show the outcomes in an experiment is to use an **outcome grid.** The following outcome grid shows the outcomes when rolling two number cubes and observing the sum of the two numbers.

2nd Number Cube

1st Number Cube	1	2	3	4	5	6
1	2	3	4	5	6	7
2	3	4	5	6	7	8
3	4	5	6	7	8	9
4	5	6	7	8	9	10
5	6	7	8	9	10	11
6	7	8	9	10	11	12

You can use the grid to find the sum that occurs most often, which is 7.

MAKING OUTCOME GRIDS

Make an outcome grid to show the results of tossing a coin and rolling a number cube.

- List the outcomes of the first type down the side. List the outcomes of the second type across the top.

Number Cube

Coin	1	2	3	4	5	6
Head						
Tail						

- Fill in the outcomes.

Number Cube

Coin	1	2	3	4	5	6
Head	H1	H2	H3	H4	H5	H6
Tail	T1	T2	T3	T4	T5	T6

Once you have completed the outcome grid, it is easy to count target outcomes and determine probabilities.

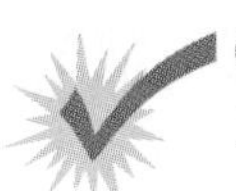

Check It Out

15. Make an outcome grid to show the two-letter outcomes when spinning the spinner twice.

Second Spin

First Spin	R	B	G	Y
R				
B				
G				
Y				

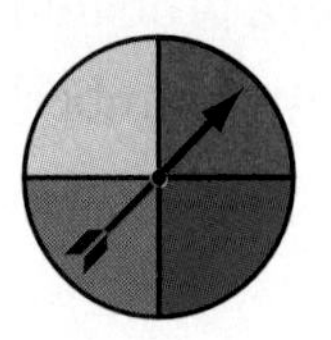

16. What is the probability of getting green as one color when you spin the spinner in item 15 twice?

Probability Line

You know that the probability of an event is a number from 0 to 1. One way to show probabilities and how they relate to each other is to use a **probability line.** The following probability line shows the possible ranges of probability values.

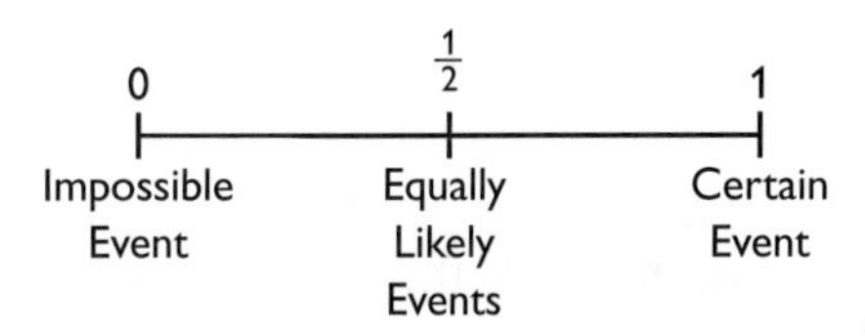

The line shows that events which are certain have a probability of 1. Such an event is the probability of getting a number between 0 and 7 when rolling a standard number cube. An event that cannot happen has a probability of zero. The probability of getting an 8 when spinning a spinner that shows 0, 2, and 4 is 0. Events that are equally likely, such as getting a head or a tail when you toss a coin, have a probability of $\frac{1}{2}$.

SHOWING PROBABILITY ON A PROBABILITY LINE

Suppose you roll two number cubes. Show the probabilities of getting a sum of 4 and of getting a sum of 7 on a probability line.

- Draw a number line and label it from 0 to 1.

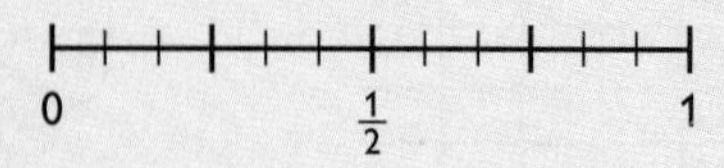

- Calculate the probabilities of the given events and show them on the probability line.

  From the outcome grid on page 246, you can see there are 3 sums of 4 and 6 sums of 7 out of 36 sums. So $P(\text{sum of } 4) = \frac{3}{36} = \frac{1}{12}$ and $P(\text{sum of } 7) = \frac{6}{36} = \frac{1}{6}$. The probabilities are shown on the following probability line.

  P(sum of 4) P(sum of 7)

  0 $\frac{1}{12}$ $\frac{1}{6}$ $\frac{1}{2}$ 1

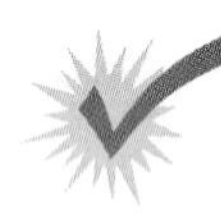

Check It Out

Draw a probability line. Then plot the following.

17. the probability of tossing a tail on one flip of a coin
18. the probability of rolling a 1 or a 2 on one roll of a die
19. the probability of being chosen if there are four people and an equal chance of any of them being chosen
20. the probability of getting a green gumball out of the machine if there are 25 each of green, yellow, red, and blue gumballs

Dependent and Independent Events

If you toss a coin and roll a number cube, the result of one does not affect the other. We call these events **independent events.** To find the probability that we get a head and then a 5, you can find the probability of each event and then multiply. The probability of getting a head is $\frac{1}{2}$ and the probability of getting a 5 on a roll of the number cube is $\frac{1}{6}$. So the probability of getting a head and a 5 is $\frac{1}{2} \times \frac{1}{6} = \frac{1}{12}$.

Suppose you have 4 oatmeal and 6 raisin cookies in a bag. The probability that you get an oatmeal cookie if you choose a cookie at random is $\frac{4}{10} = \frac{2}{5}$. Once you have taken an oatmeal cookie out, however, there are only 9 cookies left, 3 of which are oatmeal. So the probability that a friend gets an oatmeal cookie once you have drawn 1 out is $\frac{3}{9} = \frac{1}{3}$. These events are called **dependent events** because the probability of one depends on the other.

In the case of dependent events, you still multiply to get the probability of both events happening. So the probability that your friend gets an oatmeal cookie and you also get one is $\frac{2}{5} \times \frac{1}{3} = \frac{2}{15}$.

To find the probability of dependent and independent events,
- Find the probability of the first event.
- Find the probability of the second event.
- Find the product of the two probabilities.

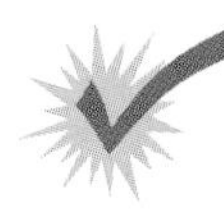

Check It Out

21. Find the probability of getting an even number and an odd number if you roll two number cubes. Are the events dependent or independent?
22. You draw two marbles from a bag containing six red marbles and fourteen white ones. What is the probability that you get two white marbles? Are the events dependent or independent?

Sampling With and Without Replacement

If you draw a card from a deck of cards, the probability that it is an ace is $\frac{4}{52}$, or $\frac{1}{13}$. If you put the card back in the deck and draw another card, the probability that it is an ace is still $\frac{1}{13}$, and the events are independent. This is called **sampling with replacement.**

If you do not put the card back in, the probability of drawing an ace the second time will depend on what you drew the first time. If you drew an ace, there will be only three aces left of 51 cards, so the probability of drawing a second ace will be $\frac{3}{51}$, or $\frac{1}{17}$. In sampling without replacement, the events are dependent.

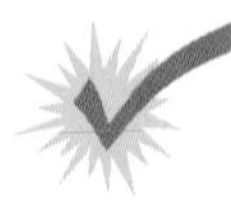

Check It Out

23. You draw a card from a deck of cards and then put it back. Then you draw another card. What is the probability you get a spade and then a heart?
24. Answer the question again if you do not replace the card.

4·6 EXERCISES

You spin the spinner shown for items 1 and 2. Find each probability as a fraction, decimal, ratio, and percent.

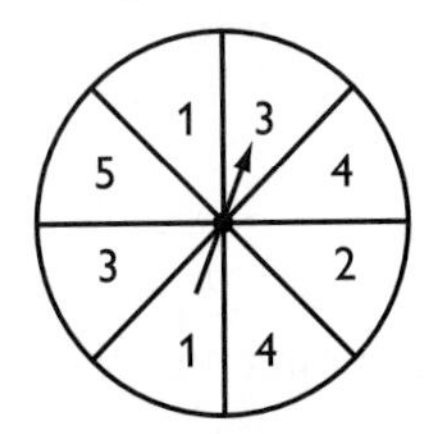

1. $P(4)$
2. $P(\text{odd number})$
3. If you toss a coin 48 times and get 26 heads, what is the probability of getting a head? Is this experimental or theoretical probability?
4. If you roll a number cube, what is the probability of getting a 6? Is this experimental or theoretical probability?
5. Draw a probability line to show the probability of getting a number greater than 6 when rolling a number cube numbered 1 through 6.
6. Make a strip graph to show the following outcomes when flipping a coin: heads, heads, tails, heads, tails, heads, tails, heads, heads, tails.
7. Make an outcome grid to show the outcomes of spinning two spinners divided into four equal sections labeled 1 through 4.
8. Find the probability of drawing two red kings from a deck of cards if you replace the card between drawings.
9. Find the probability of drawing two red kings from a deck of cards if you do not replace the card between drawings.
10. Look again at items 8 and 9. In which item are the events dependent?

What have you learned?

You can use the problems and the list of words that follow to see what you have learned in this chapter. You can find out more about a particular problem or word by referring to the boldfaced topic number (for example, **4•2**).

Problem Set

1. One Saturday in a shopping mall, Livna took a survey by asking one person every ten minutes, "How far did you travel to the mall today?" Was this a random sample? **4•1**
2. Taking a survey at the mall, Salvador asked, "What do you think of the beautiful new landscaping at the mall?" Was the question biased or unbiased? **4•1**
3. What kind of graph can be used to compare two sets of data on the same graph? **4•2**
4. On a circle graph, how many degrees must be in a sector to show 50%? **4•2**

For items 5–8, use the following stem-and-leaf plot, which shows the length of time in minutes that people stayed at the library on Monday morning.

Stem	Leaves
0	2 2 5 7 7
1	3 3 5 7 8 8 9
2	1 4 4 6 6 6 8 9
3	3 6 7 8
4	5

1|3 = 13 min

5. How many times are recorded in this plot? **4•2**
6. How many people stayed at the library for 15 minutes or less? **4•2**
7. What was the median time spent at the library? **4•4**
8. The person who stayed the longest arrived at 10:30 A.M. What time did that person leave the library? **4•2**

Use this information for items 9–10. A bookstore manager compared the prices of 100 new books to the number of pages in each book to see if there was a relationship between them. For each book, the manager made an ordered pair of the form (number of pages, price). **4•3**

9. What kind of graph will these data make?
10. On the graph, many of these 100 points seem to lie on a straight line. What is this line called?

11. True or false: The graph of a normal distribution rises from left to right in a smooth curve. **4•3**
12. Find the mean, median, mode, and range of the numbers 42, 43, 19, 16, 16, 36, and 17. **4•4**
13. $C(6, 3) = ?$ **4•5**

Use the following information to answer items 14–15. A bag contains 4 red, 3 blue, 2 green, and 1 black marble. **4•6**

14. One marble is drawn. What is the probability that it is red?
15. Three marbles are drawn. What is the probability that 2 are black and 1 is green?

hot words

WRITE DEFINITIONS FOR THE FOLLOWING WORDS.

average **4•4**
bimodal distribution **4•3**
box plot **4•2**
circle graph **4•2**
combination **4•5**
correlation **4•3**
dependent events **4•6**
double-bar graph **4•2**
event **4•6**
experimental probability **4•6**
factorial **4•5**
flat distribution **4•3**
histogram **4•2**
independent events **4•6**
leaf **4•2**
line graph **4•2**
line of best fit **4•3**
mean **4•4**
median **4•4**
mode **4•4**
normal distribution **4•3**
outcome **4•6**
outcome grid **4•6**
permutation **4•5**
population **4•1**
probability **4•6**
probability line **4•6**
random sample **4•1**
range **4•4**
sample **4•1**
sampling with replacement **4•6**
scatter plot **4•3**
skewed distribution **4•3**
spinner **4•5**
stem **4•2**
stem-and-leaf plot **4•2**
strip graph **4•6**
survey **4•1**
table **4•1**
tally marks **4•1**
theoretical probability **4•6**
tree diagram **4•5**
weighted average **4•4**

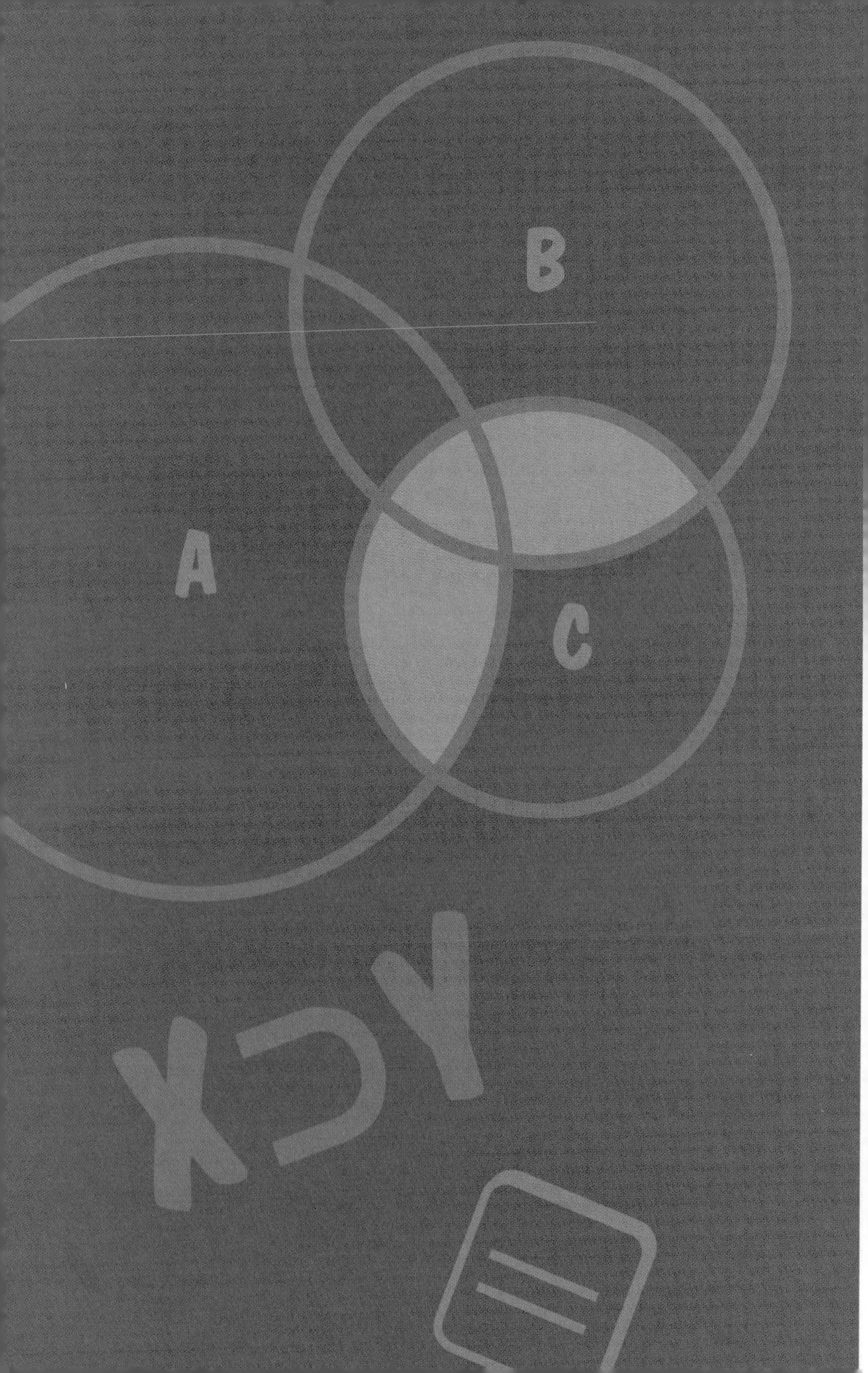
A
B
C

Logic

5·1	If/Then Statements	258
5·2	Counterexamples	264
5·3	Sets	266

What do you already know?

You can use the problems and the list of words that follow to see what you already know about this chapter. The answers to the problems are in Hot Solutions at the back of the book, and the definitions of the words are in Hot Words at the front of the book. You can find out more about a particular problem or word by referring to the boldfaced topic number (for example, **5•2**).

Problem Set

Tell whether each statement is true or false.

1. In a conditional statement, the *if*-clause is the hypothesis and the *then*-clause is the conclusion. **5•1**
2. You form the inverse of a conditional statement by switching the hypothesis and the conclusion. **5•1**
3. If a conditional statement is true, then its related converse is always false. **5•1**
4. If a conditional statement is true, then its related contrapositive is always true. **5•1**
5. Every set is a subset of itself. **5•3**
6. A counterexample shows that a statement is false. **5•2**
7. You form the union of two sets by combining all the elements in both sets. **5•3**
8. The negation of "It's Monday" is "It's Tuesday." **5•1**
9. The intersection of two sets can be the empty set. **5•3**

Write each conditional in if/then form. **5•1**

10. The jet flies to Belgium on Tuesday.
11. The bank is closed on Sunday.

Write the converse of each conditional statement. **5•1**

12. If $x = 7$, then $x^2 = 49$.
13. If an angle has a measure less than 90°, then the angle is acute.

Write the negation of each statement. **5•1**

14. The playground will close at sundown.
15. These two lines form an angle.

Write the inverse of each conditional statement. **5•1**

16. If you pass all your courses, then you will graduate.
17. If two lines intersect, then they form four angles.

Write the contrapositive of each conditional statement. **5•1**

18. If you are over 12 years old, then you buy an adult ticket.
19. If a pentagon has five equal sides, then it is equilateral.

Find a counterexample that shows that each of these statements is false. **5•2**

20. Tuesday is the only day of the week that begins with the letter T.
21. The legs of a trapezoid are equal.

Use the Venn diagram for items 22–25. **5•3**

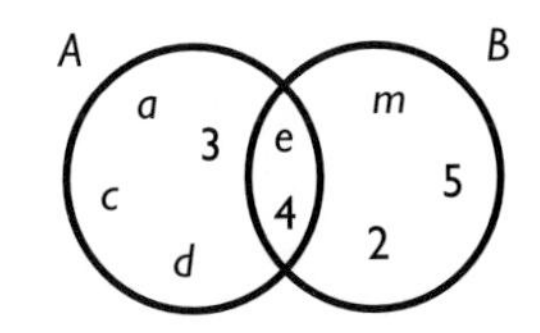

22. List the elements in set *A*.
23. List the elements in set *B*.
24. Find $A \cup B$.
25. Find $A \cap B$.

CHAPTER 5

hot words

contrapositive **5•1**
converse **5•1**
counterexample **5•2**
intersection **5•3**
inverse **5•1**
union **5•3**
Venn diagram **5•3**

5•1 If/Then Statements

Conditional Statements

A *conditional* is a statement that you can express in *if/then* form. The *if* part of a conditional is the *hypothesis,* and the *then* part is the *conclusion.* Often you can rewrite a statement that contains two related ideas as a conditional in if/then form by making one of the ideas the hypothesis and the other the conclusion.

Statement: All the members of the varsity swim team are seniors.

The conditional in if/then form:

If a person is a swim team member, then the person is a senior.

hypothesis: If a person is a swim team member

conclusion: then the person is a senior.

FORMING CONDITIONAL STATEMENTS

Write this conditional in if/then form:

Julie only goes swimming in water that is above 80°F.

- Find the two ideas.

  (1) Julie goes swimming (2) Water is above 80°F

- Decide which idea will be the hypothesis and which will be the conclusion.

  Hypothesis: Julie goes swimming

  Conclusion: Water is above 80°

- Place the hypothesis in the *if* clause and the conclusion in the *then* clause. If necessary, add words so that your sentence makes sense.

  If Julie is swimming, then the water is above 80°F.

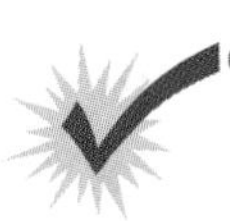

Check It Out

Write each statement in if/then form.

1. Perpendicular lines meet to form right angles.
2. An integer that ends in 0 or 5 is a multiple of 5.

Converse of a Conditional

When you switch the hypothesis and conclusion in a conditional statement, you form a new statement called the **converse.**

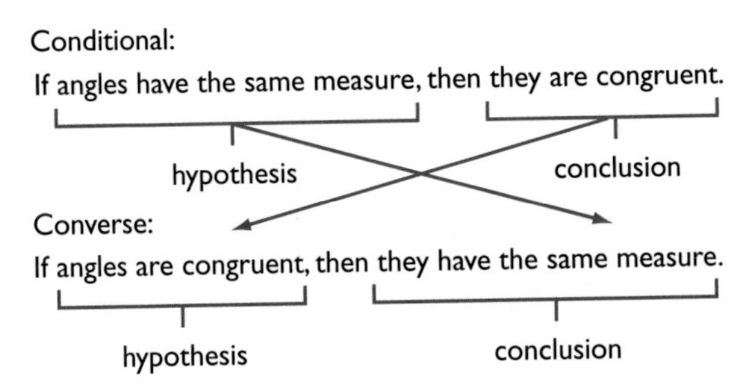

The converse of a conditional may or may not have the same truth value as the conditional on which it is based.

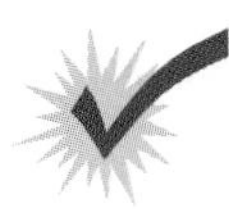

Check It Out

Write the converse of each conditional.

3. If an integer ends with 1, 3, 5, 7, or 9, then the integer is odd.
4. If Jacy is 15 years old, then he is too young to vote.

Negations and the Inverse of a Conditional

A *negation* of a given statement has the opposite value of the given statement. That means that if the given statement is true, the negation is false; if the given statement is false, the negation is true.

Statement: A square is a quadrilateral. (True)

Negation: A square is not a quadrilateral. (False)

Statement: A pentagon has four sides. (False)

Negation: A pentagon does not have four sides. (True)

When you negate the hypothesis and conclusion of a conditional, you form a new statement called the **inverse.**

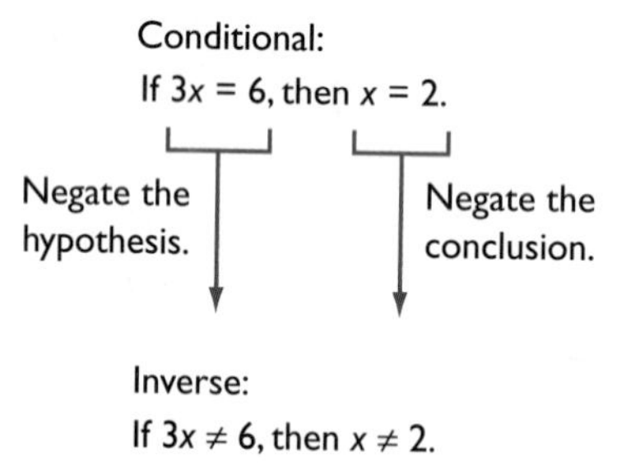

The inverse of a conditional may or may not have the same truth value as the conditional.

Check It Out

Write the negation of each statement.

5. A rectangle has four sides.
6. The donuts were eaten before noon.

Write the inverse of each conditional.

7. If an integer ends with 0 or 5, then it is a multiple of 5.
8. If I am in Seattle, then I am in the state of Washington.

Contrapositive of a Conditional

You form the **contrapositive** of a conditional when you negate the hypothesis and conclusion, and then interchange them.

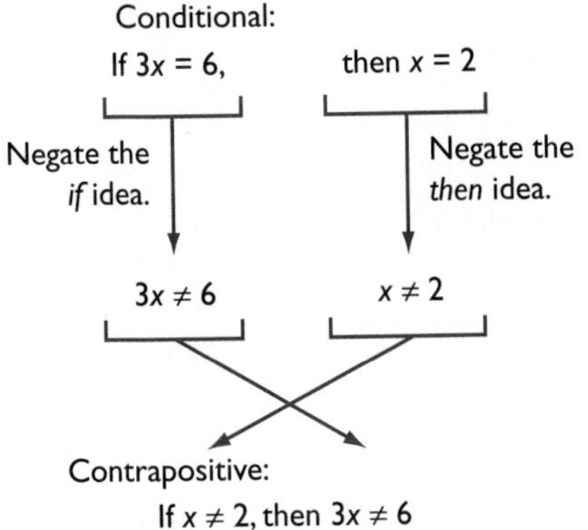

The contrapositive of a conditional statement has the same truth value as the conditional.

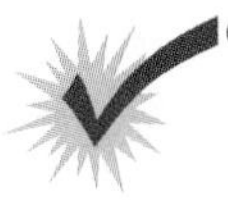

Check It Out

Write the contrapositive of each conditional.

9. If an angle has a measure of 90°, then the angle is a right angle.
10. If $x \neq 3$, then $2x \neq 6$.

Rapunzel, Rapunzel, Let Down Your Hair

Do you think there are two people in the United States with exactly the same number of hairs on their head? In fact, you can prove that there are. Not by counting the hairs on everyone's head, but by logic.

Consider these statements:

A. At the maximum, there are approximately 150,000 hairs on the human scalp.

B. The population of the United States is greater than 150,000.

Because statements a and b are both true, then there are two people in the United States with exactly the same number of hairs on their head.

Here's how to think about this. If you did count hairs, the first 150,001 people could each have a different number. Person 1 could have 1 hair; person 2, 2 hairs, and so on to 150,000. But person 150,001 would have to have a number of hairs between 1 and 150,000 and this would be a duplicate of one of the heads you already counted.

Can you prove that there are two people in your town with the same number of hairs on their head? See Hot Solutions for answer.

5•1 EXERCISES

Write each conditional in if/then form.

1. Perpendicular lines form right angles.
2. Positive integers are greater than zero.
3. Everyone in that town voted in the last election.
4. Equilateral triangles have three equal sides.
5. Numbers that end with 0, 2, 4, 6, or 8 are even numbers.
6. Elena visits her aunt every Friday.

Write the converse of each conditional.

7. If a triangle is equilateral, then it is isosceles.
8. If Chenelle is over 21, then she can vote.
9. If a number is a factor of 8, then it is a factor of 24.
10. If $x = 4$, then $3x = 12$.

Write the negation of each statement.

11. All the buildings are three stories tall.
12. x is a multiple of y.
13. The lines in the diagram intersect at point P.
14. A triangle has three sides.

Write the inverse of each conditional.

15. If $5x = 15$, then $x = 3$.
16. If the weather is good, then I will drive to work.

Write the contrapositive of each conditional.

17. If $x = 6$, then $x^2 = 36$.
18. If the perimeter of a square is 8 in., then each side is 2 in.

For each conditional, write the converse, inverse, and contrapositive.

19. If a rectangle has a length of 4 ft and a width of 2 ft, then its perimeter is 12 ft.
20. If a triangle has three sides of different lengths, then it is scalene.

5•2 Counterexamples

Counterexamples

In the fields of logic and mathematics, if/then statements are either true or false. One way to show that a statement is false is to find one example that agrees with the hypothesis but not with the conclusion. Such an example is a **counterexample.**

When reading the conditional below, you may be tempted to think that it is true.

If a polygon has four equal sides, then it is a square.

The statement is false, however, because there is a counterexample—the rhombus. A rhombus agrees with the hypothesis (it has four equal sides), but it does not agree with the conclusion (a rhombus is not a square).

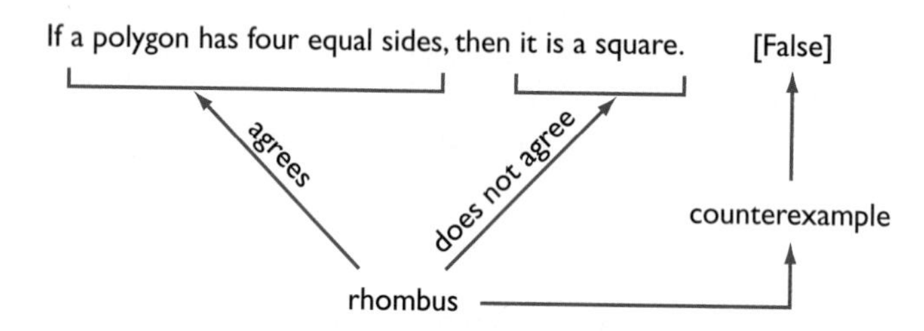

Check It Out

Tell if each statement and its converse are true or false. If a statement is false, give a counterexample.

1. Statement: If two lines in the same plane are parallel, then they do not intersect.
   Converse: If two lines in the same plane do not intersect, then they are parallel.

2. Statement: If an angle has a measure of 90°, then it is a right angle.
   Converse: If an angle is a right angle, then it has a measure of 90°.

5•2 EXERCISES

Find a counterexample that shows that each statement is false.

1. If a number is a factor of 18, then it is a factor of 24.
2. If a figure is made of three segments, then it is a triangle.
3. If a figure is a quadrilateral, then it is a parallelogram.
4. If $x + y$ is an even number, then x and y are even numbers.

Tell whether each conditional is true or false. If false, give a counterexample.

5. If a number is prime, then it is an odd number.
6. If xy is an odd number, then x and y are both odd.
7. If the temperature is 32°F , then pure water will freeze.
8. If you draw a line through a square, then you form two triangles.

Tell if each statement and its converse are true or false. If false, give a counterexample.

9. Statement: If two angles have measures of 30°, then the angles are congruent.
   Converse: If two angles are congruent, then they have measures of 30°.
10. Statement: If $6x = 54$, then $x = 9$.
    Converse: If $x = 9$, then $6x = 54$.

Tell if each statement and its inverse are true or false. If false, give a counterexample.

11. Statement: If $n = 8$, then $n + 9 = 17$.
    Inverse: If $n \neq 8$, then $n + 9 \neq 17$.
12. Statement: If an angle has a measure of 120°, then it is an obtuse angle.
    Inverse: If an angle does not have a measure of 120°, then it is not an obtuse angle.

Tell if each statement and its contrapositive are both true or false. If false, give counterexamples.

13. Statement: If a triangle is isosceles, then it is equilateral.
    Contrapositive: If a triangle is not equilateral, then it is not isosceles.
14. Statement: If a triangle is equilateral, then it is isosceles.
    Contrapositive: If a triangle is not isosceles, then it is not equilateral.

15. Write your own false conditional, then give a counterexample that shows it is false.

5·3 Sets

Sets and Subsets

A *set* is a collection of objects. Each object is called a *member* or *element* of the set. Sets are often named with capital letters.

$A = \{1, 2, 3, 4\}$ $\qquad B = \{a, b, c, d\}$

When a set has no elements, it is the *empty set*. You write { } or Ø to indicate the empty set.

When all the elements of a set are also elements of another set, the first set is a *subset* of the other set.

$\{2, 4\}$ is a subset of $\{1, 2, 3, 4\}$.

$\{2, 4\} \subset \{1, 2, 3, 4\}$ $\quad$ ($\subset$ is the subset symbol)

Remember that every set is a subset of itself and that the empty set is a subset of every set.

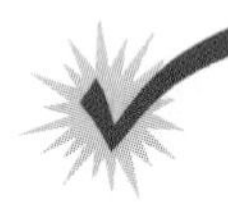

Check It Out

Tell whether each statement is true or false.

1. $\{5\} \subset \{\text{even numbers}\}$
2. $\varnothing \subset \{3, 5\}$
3. $\{2\} \subset \{2\}$

Find all the subsets of each set.

4. $\{1, 4\}$
5. $\{m\}$
6. $\{a, b, c\}$

Union of Sets

You find the **union** of two sets by creating a new set with all the elements from the two sets.

$J = \{1, 3, 5, 7\}$ $\quad L = \{2, 4, 6, 8\}$

$J \cup L = \{1, 2, 3, 4, 5, 6, 7, 8\}$ $\quad$ ($\cup$ is the union symbol)

When the sets have elements in common, list the common elements only once in the intersection.

$P = \{r, s, t, v\}$ $\quad Q = \{a, k, r, t, w\}$

$P \cup Q = \{a, k, r, s, t, v, w\}$

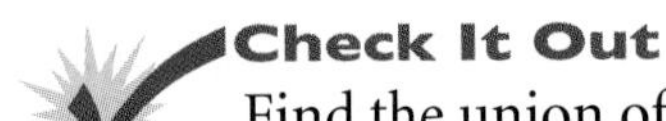

Check It Out

Find the union of each pair of sets.

7. $\{1, 2\} \cup \{9, 10\}$
8. $\{m, a, t, h\} \cup \{m, a, p\}$

Intersection of Sets

You find the **intersection** of two sets by creating a new set that contains all the elements that are common to both sets.

$A = \{8, 12, 16, 20\}$

$B = \{4, 8, 12\}$

$A \cap B = \{8, 12\}$

If the sets have no elements in common, the intersection is the empty set $\{\emptyset\}$.

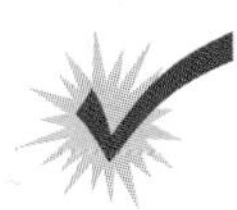

Check It Out

Find the intersection of each pair of sets.

9. $\{9\} \cap \{9, 18\}$
10. $\{a, c, t\} \cap \{b, d, u\}$

Venn Diagrams

A **Venn diagram** shows you how the elements of two or more sets are related.

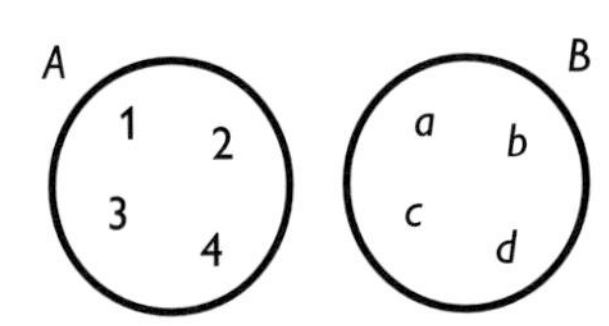

$$A = \{1, 2, 3, 4\}$$
$$B = \{a, b, c, d\}$$
$$A \cup B = \{1, 2, 3, 4, a, b, c, d\}$$

The separate circles for A and B tell you that the sets have no elements in common. That means that $A \cap B = \emptyset$.

When the circles in a Venn diagram overlap, the overlapping part contains the elements that are common to both sets. This diagram shows some sets of attribute shapes.

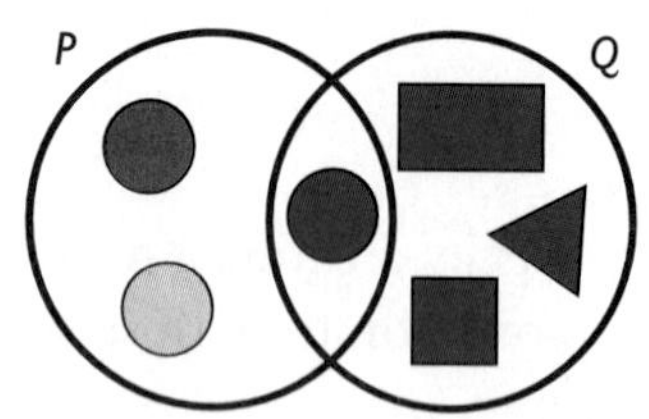

$P = \{\text{large circles}\}$
$Q = \{\text{blue shapes}\}$

The overlapping parts of P and Q contain a shape that has the attributes of both sets, or $P \cap Q = \{\text{large blue circles}\}$.

With more complex Venn diagrams you have to look carefully to identify the overlapping parts and see which elements of the sets are in those parts. The shaded part of the diagram shows where all three sets overlap one another.

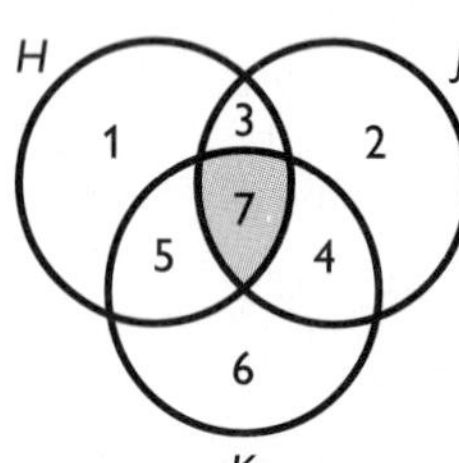

$H = \{1, 3, 5, 7\}$ $H \cup J = \{1, 2, 3, 4, 5, 7\}$
$J = \{2, 3, 4, 7\}$ $H \cup K = \{1, 3, 4, 5, 6, 7\}$
$K = \{4, 5, 6, 7\}$ $J \cup K = \{2, 3, 4, 5, 6, 7\}$
$H \cup J \cup K = \{1, 2, 3, 4, 5, 6, 7\}$
$H \cap J = \{3, 7\}$ $H \cap K = \{5, 7\}$ $J \cap K = \{7, 4\}$

Where all three sets overlap, you can see that $H \cap J \cap K = \{7\}$.

Check It Out

Use Venn diagram for the following items.

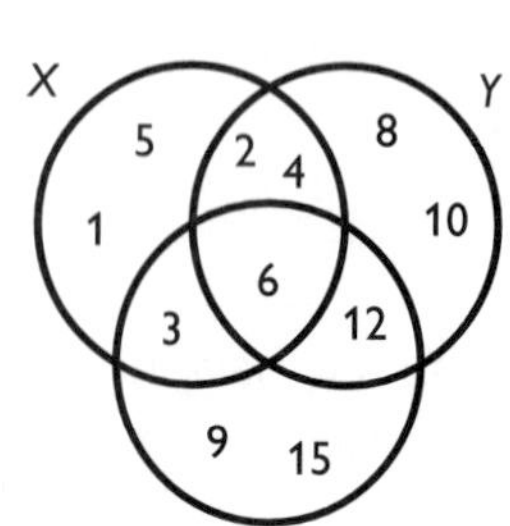

List the elements in:

11. set X.
12. $X \cup Z$.
13. $Y \cap Z$.
14. $X \cap Y \cap Z$.

5·3 EXERCISES

Tell whether each statement is true or false.

1. $\{1, 2, 3\} \subset$ {counting numbers}
2. $\{1, 2, 3\} \subset \{1, 2\}$
3. $\{1, 2, 3\} \subset$ {even numbers}
4. $\emptyset \subset \{1, 2, 3\}$

Find the union of each pair of sets.

5. $\{2, 3\} \cup \{4, 5\}$
6. $\{x, y\} \cup \{y, z\}$
7. $\{r, o, y, a, l\} \cup \{m, o, a, t\}$
8. $\{2, 5, 7, 10\} \cup \{2, 7\}$

Find the intersection of each pair of sets.

9. $\{1, 3, 5, 7\} \cap \{6, 7, 8\}$
10. $\{6, 8, 10\} \cap \{7, 9, 11\}$
11. $\{r, o, y, a, l\} \cap \{m, o, a, t\}$
12. $\emptyset \cap \{4, 5\}$

Use the Venn diagram at the right for items 13–16.

13. List the elements of set T.
14. List the elements of set R.
15. Find $T \cup R$.
16. Find $T \cap R$.

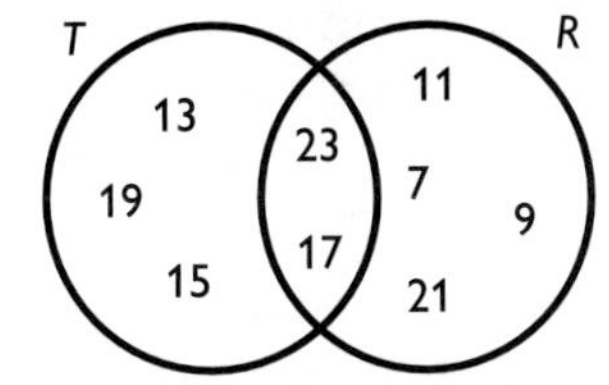

Use the Venn diagram at the right for items 17–25.

17. List the elements of set M.
18. List the elements of set N.
19. Find P.
20. Find $M \cup N$.
21. Find $N \cup P$.
22. Find $M \cup P$.
23. Find $M \cap N$.
24. Find $P \cap N$.
25. Find $M \cap N \cap P$.

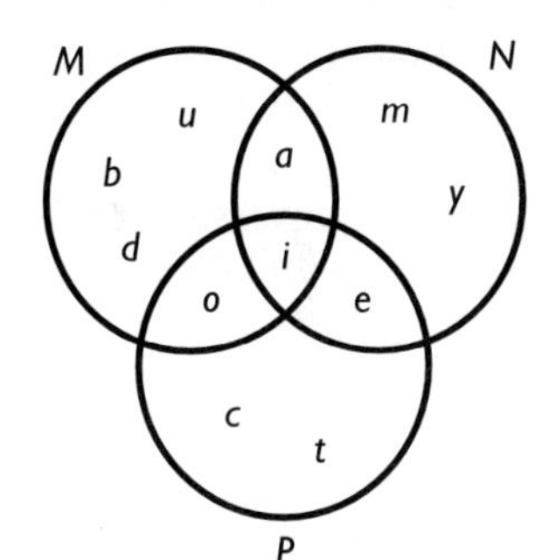

What have you learned?

You can use the problems and the list of words that follow to see what you have learned in this chapter. You can find out more about a particular problem or word by referring to the boldfaced topic number (for example, **5•2**).

Problem Set

Tell whether each statement is true or false.

1. A conditional statement is always true. **5•1**
2. You form the converse of a conditional by interchanging the hypothesis and the conclusion. **5•1**
3. If a conditional statement is true, then its related inverse is always true. **5•2**
4. A counterexample of a conditional agrees with the hypothesis but not with the conclusion. **5•2**
5. The empty set is a subset of every set. **5•3**
6. One counterexample is enough to show that a statement is false. **5•2**

Write each conditional in if/then form. **5•1**

7. A square is a quadrilateral with four equal sides and four equal angles.
8. A right angle has a measure of 90°.

Write the converse of each conditional statement. **5•1**

9. If $y = 9$, then $y^2 = 81$.
10. If an angle has a measure greater than 90° and less than 180°, then the angle is obtuse.

Write the negation of each statement. **5•1**

11. I am glad it's Friday!
12. These two lines are perpendicular.

Write the inverse of each conditional statement. **5•1**

13. If the weather is warm, then we will go for a walk.
14. If the lines do not intersect, then they are parallel.

Write the contrapositive of each conditional statement. **5•1**

15. If a quadrilateral has two pairs of parallel sides, then it is a parallelogram.
16. If you bought your tickets in advance, then you paid less.

Find a counterexample that shows that each of these statements is false. **5•2**

17. The number 24 has only even factors.
18. Find all the subsets of {7, 8, 9}.

Find the union of each pair of sets. **5•3**

19. $\{p, l, o, t\} \cup \{m\}$
20. $\{2, 4, 6, 8\} \cup \{6, 7, 8, 9\}$

Use the Venn diagram for items 21–25. **5•3**

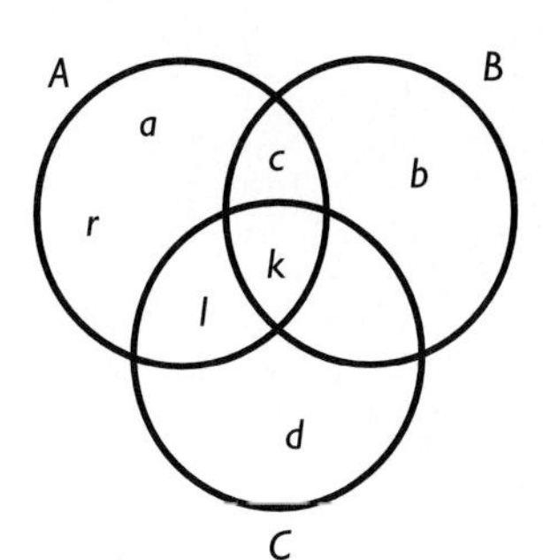

21. List the elements in set *A*.
22. List the elements in set *C*.
23. Find $A \cup B$.
24. Find $B \cap C$.
25. Find $A \cap B \cap C$.

WRITE DEFINITIONS FOR THE FOLLOWING WORDS.

contrapositive **5•1**
converse **5•1**
counterexample **5•2**
intersection **5•3**
inverse **5•1**
union **5•3**
Venn diagram **5•3**

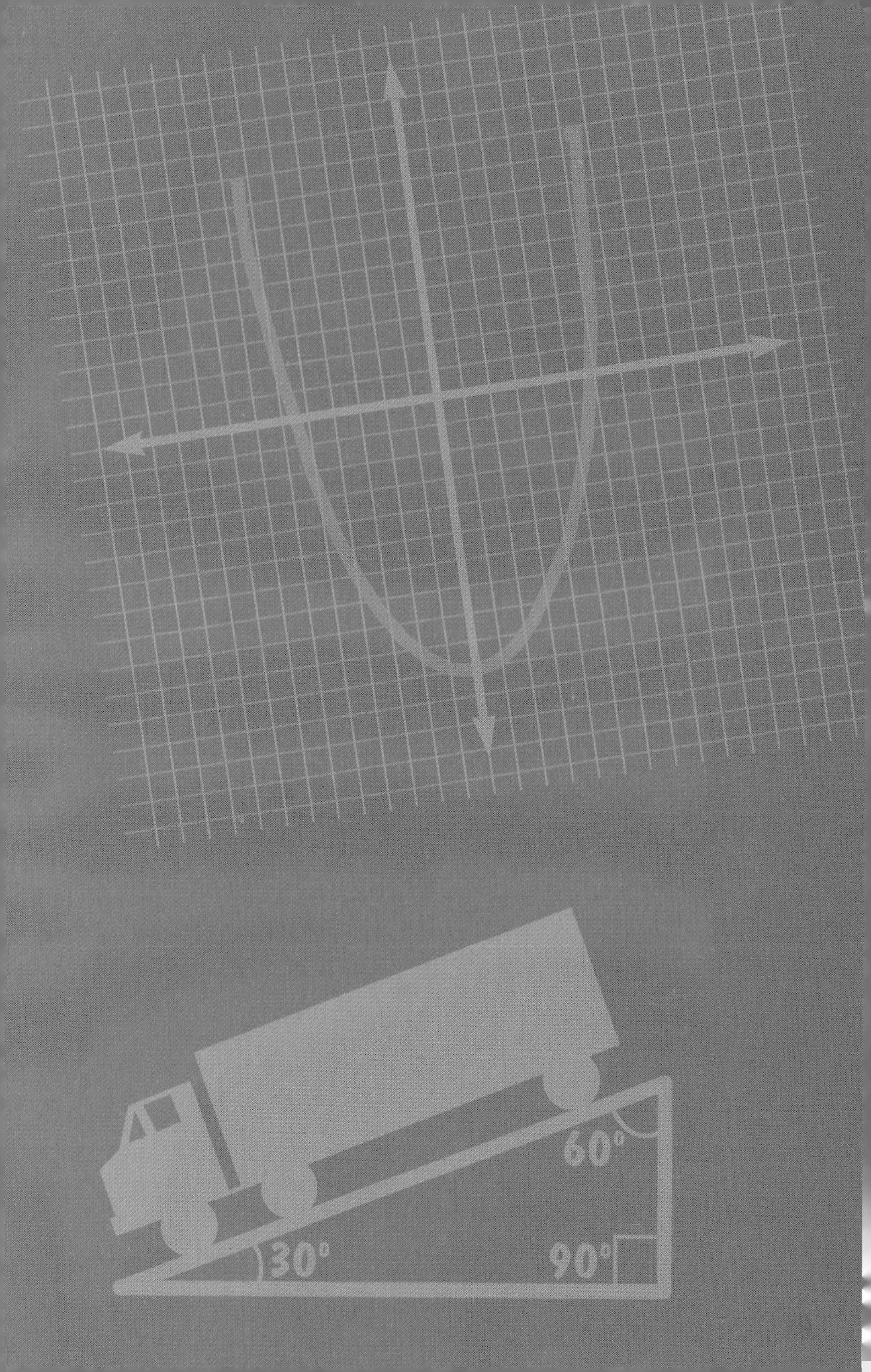

60°
30°
90°

Algebra

6•1	Writing Expressions and Equations	276
6•2	Simplifying Expressions	284
6•3	Evaluating Expressions and Formulas	292
6•4	Solving Linear Equations	296
6•5	Ratio and Proportion	308
6•6	Inequalities	312
6•7	Graphing on the Coordinate Plane	316
6•8	Slope and Intercept	324

What do you already know?

You can use the problems and the list of words that follow to see what you already know about this chapter. The answers to the problems are in Hot Solutions at the back of the book, and the definitions of the words are in Hot Words at the front of the book. You can find out more about a particular problem or word by referring to the boldfaced topic number (for example, **6•2**).

Problem Set

Write an equation for each sentence. **6•1**

1. If 3 is subtracted from twice a number, the result is 9 more than the number.
2. 4 times the sum of a number and 2 is 4 less than twice the number.

Factor out the common factor in each expression. **6•2**

3. $6x + 30$
4. $12n - 15$

Simplify each expression. **6•2**

5. $2a + 5b - a - 2b$
6. $4(3n - 1) - (n + 6)$
7. Find the distance traveled by a jogger who jogs at 8 mi/hr for $2\frac{1}{2}$ hr. Use the formula $d = rt$. **6•3**

Solve each equation. Check your solution. **6•4**

8. $x + 4 = 11$
9. $\frac{y}{4} = -5$
10. $3x - 5 = 22$
11. $\frac{y}{6} - 1 = 8$
12. $9n - 7 = 4n + 8$
13. $y - 8 = 5y + 4$
14. $5(2n - 3) = 4n + 9$
15. $12x - 2(3x - 2) = 5(x + 2)$

Use a proportion to solve each problem. **6•5**

16. In a class, the ratio of boys to girls is $\frac{2}{3}$. If there are 12 boys in the class, how many girls are there?
17. A map is drawn using a scale of 200 mi to 1 cm. The distance between two cities is 1,300 mi. How far apart are the two cities on the map?

Solve each inequality. Graph the solution. **6•6**

18. $x + 7 < 5$
19. $3x + 5 \geq 17$
20. $-4n + 11 < 3$

Locate each point on the coordinate plane and tell in which quadrant or on which axis it lies. **6•7**

21. $A(4, -3)$ 22. $B(-2, -1)$ 23. $C(0, 1)$ 24. $D(-3, 0)$

25. Find the slope of the line that contains the points $(-2, 4)$ and $(8, -2)$. **6•8**

Determine the slope and the y-intercept from the equation of each line. Graph the line. **6•8**

26. $y = -\frac{3}{2}x + 2$ 27. $y = 4$
28. $x + 3y = -6$ 29. $4x - 2y = 0$

Write the equation of the line that contains the given points. **6•8**

30. $(-2, -6)$ and $(5, 1)$
31. $(6, 5)$ and $(-3, -1)$
32. $(-4, 2)$ and $(-4, 7)$
33. $(0, 7)$ and $(5, 7)$
34. $(0, 2)$ and $(9, 1)$
35. $(1, 1)$ and $(3, 2)$

CHAPTER 6

hot words

additive inverse **6•4**
associative property **6•2**
axes **6•7**
commutative property **6•2**
cross product **6•5**
difference **6•1**
distributive property **6•2**
equation **6•1**
equivalent **6•1**
equivalent expression **6•2**
expression **6•1**
formula **6•3**
horizontal **6•7**
inequality **6•6**
like terms **6•2**
order of operations **6•3**
ordered pair **6•7**
origin **6•7**
perimeter **6•3**
point **6•7**
product **6•1**
proportion **6•5**
quadrant **6•7**
quotient **6•1**
rate **6•5**
ratio **6•5**
rise **6•8**
run **6•8**
slope **6•8**
solution **6•4**
sum **6•1**
term **6•1**
variable **6•1**
vertical **6•7**
x-axis **6•7**
y-axis **6•7**
y-intercept **6•8**

6•1 Writing Expressions and Equations

Expressions

In mathematics often the value of a particular number may be unknown. A **variable** is a symbol, usually a letter, that is used to represent an unknown number. Some commonly used variables are

$x \quad n \quad y \quad a \quad ?$

A **term** can be a number, a variable, or a number and variable combined by multiplication or division. Some examples of terms are

$w \quad 5 \quad 3x \quad \frac{y}{8}$

An **expression** can be a term or a collection of terms separated by addition or subtraction signs. Some expressions, with the number of terms, are listed in the chart below.

Expression	Number of Terms	Description
$5y$	1	a number multiplied by a variable
$6z + 4$	2	terms separated by a +
$3x + 7a - 5$	3	terms separated by a + or –
$\frac{9xz}{y}$	1	all multiplication and division; no + symbol

Check It Out

Count the number of terms in each expression.

1. $5x + 12$
2. $3abc$
3. $9xy - 3c - 8$
4. $3a^2b + 2ab$

Writing Expressions Involving Addition

To write an expression, you will often have to interpret a written phrase. For example, the phrase "4 added to some number" could be written as the expression $x + 4$, where the variable x represents the unknown number.

Notice that the words "added to" indicate that the operation between 4 and the number is to be addition. Other words and phrases that indicate addition are "more than," "plus," and "increased by." One other word that indicates addition is **sum.** The sum of two terms is the result of adding them together.

Some common phrases and their corresponding expressions are listed.

Phrase	Expression
3 more than some number	$n + 3$
a number increased by 7	$x + 7$
9 plus some number	$9 + y$
the sum of a number and 6	$n + 6$

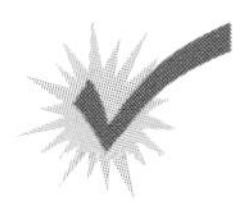

Check It Out

Write an expression for each phrase.

5. a number added to 5
6. the sum of a number and 10
7. some number increased by 8
8. 1 more than some number

Writing Expressions Involving Subtraction

The phrase "4 subtracted from some number" could be written as the expression $x - 4$, where the variable x represents the unknown number. Notice that the words "subtracted from" indicate that the operation between the number and 4 is to be subtraction.

Some other words and phrases that indicate subtraction are "less than," "minus," and "decreased by." One other word that indicates subtraction is **difference.** The difference between two terms is the result of subtracting them.

In a subtraction expression, the order of the terms is very important. You have to know which term is being subtracted and which is being subtracted from. To help interpret the phrase "6 less than a number," replace "a number" with 10. What is 6 less than 10? The answer is 4, which is $10 - 6$, not $6 - 10$. The phrase translates to the expression $x - 6$, not $6 - x$.

Below are some common phrases and their corresponding expressions.

Phrase	Expression
5 less than some number	$n - 5$
a number decreased by 8	$x - 8$
7 minus some number	$7 - y$
the difference between a number and 2	$n - 2$

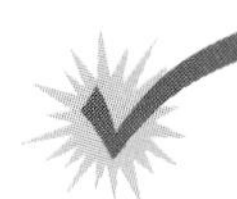

Check It Out

Write an expression for each phrase.

9. a number subtracted from 14
10. the difference between a number and 2
11. some number decreased by 6
12. 4 less than some number

Writing Expressions Involving Multiplication

The phrase "4 multiplied by some number" could be written as the expression $4x$, where the variable x represents the unknown number. Notice that the words "multiplied by" indicate that the operation between the number and 4 is to be multiplication.

Some other words and phrases that indicate multiplication are "times," "twice," and "of." "Twice" is used to mean "2 times." "Of" is used primarily with fractions and percents. One other word that indicates multiplication is **product.** The product of two terms is the result of multiplying them.

Here are some common phrases and their corresponding expressions.

Phrase	Expression
5 times some number	$5a$
twice a number	$2x$
one-fourth of some number	$\frac{1}{4}y$
the product of a number and 8	$8n$

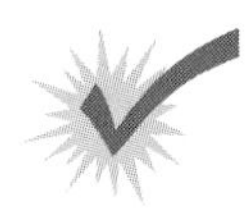

Check It Out

Write an expression for each phrase.

13. a number multiplied by 3
14. the product of a number and 7
15. 25% of some number
16. 12 times some number

Writing Expressions Involving Division

The phrase "4 divided by some number" could be written as the expression $\frac{4}{x}$, where the variable x represents the unknown number. Notice that the words "divided by" indicate that the operation between the number and 4 is to be division.

Some other words and phrases that indicate division are "ratio of" and "divide." One other word that indicates division is **quotient.** The quotient of two terms is the result of dividing them.

Some common phrases and their corresponding expressions are listed below.

Phrase	Expression
the quotient of 20 and some number	$\frac{20}{n}$
a number divided by 6	$\frac{x}{6}$
the ratio of 10 and some number	$\frac{10}{y}$
the quotient of a number and 5	$\frac{n}{5}$

Check It Out

Write an expression for each phrase.

17. a number divided by 7
18. the quotient of 16 and a number
19. the ratio of 40 and some number
20. the quotient of some number and 11

Writing Expressions Involving Two Operations

To translate the phrase "4 added to the product of 5 and some number" to an expression, first realize that "4 added to" means "something" + 4. That "something" is "the product of 5 and some number," which is $5x$, since "product" indicates multiplication. Therefore you can write the expression as $5x + 4$

Phrase	Expression	Think
2 less than the quotient of a number and 5	$\frac{x}{5} - 2$	"2 less than" means "something" – 2; "quotient" indicates division.
5 times the sum of a number and 3	$5(x + 3)$	Write the sum inside parentheses so that the entire sum is multiplied by 5.
3 more than 7 times a number	$7x + 3$	"3 more than" means "something" + 3; "times" indicates multiplication.

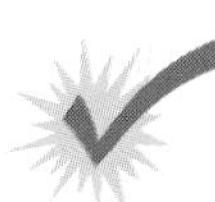

Check It Out

Translate each phrase to an expression.

21. 12 less than the product of 8 and a number
22. 1 subtracted from the quotient of 4 and a number
23. twice the difference between a number and 6

Orphaned Whale Rescued

On January 11, 1997, an orphaned baby gray whale arrived at an aquarium in California. Rescue workers named her J.J. She was three days old, weighed 1,600 lb, and was desperately ill.

Soon her caretakers had her sucking from a tube attached to a thermos. By February 7, on a diet of whale milk formula, she weighed 2,378 pounds. J.J. was gaining 20 to 30 lb a day!

An adult gray whale weighs approximately 35 tons, but the people at the aquarium knew they could release J.J. once she had a solid layer of blubber—when she weighed about 9,000 lb.

Write an equation that tells J.J. is 2,378 lb now and needs to gain 25 lb per day for some number of days until she weighs 9,000 lb. See Hot Solutions for answer.

Equations

An expression is a phrase; an **equation** is a sentence. An equation indicates that two expressions are **equivalent,** or equal. The symbol used in an equation is the equal sign, =.

To translate the sentence "2 less than the product of a number and 5 is the same as 6 more than the number" to an equation, first identify the words that indicate "equals." In this sentence, "equals" is indicated by "is the same as." In other sentences "equals" may be "is," "the result is," "you get," or just "equals."

Once you have identified the =, you can then translate the phrase that comes before the = and write the expression on the left side. Then translate the phrase that comes after the = and write the expression on the right side.

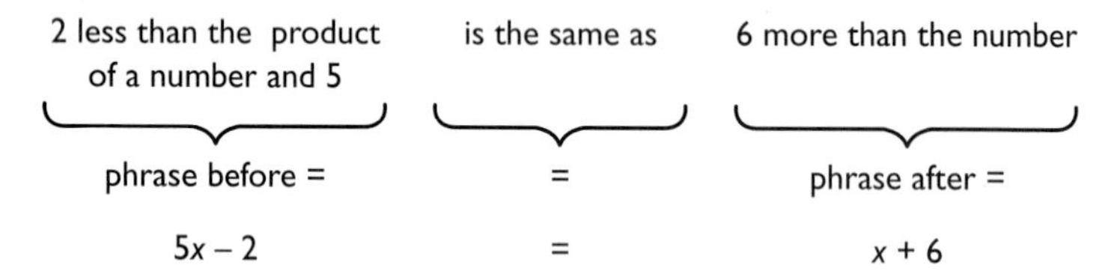

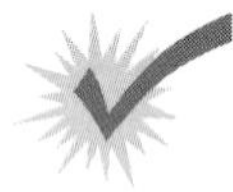

Check It Out

Write an equation for each sentence.

24. 8 subtracted from a number is the same as the product of 5 and the number.
25. 5 less than 4 times a number is 4 more than twice the number.
26. When 1 is added to the quotient of a number and 6, the result is 9 less than the number.

6·1 EXERCISES

Count the number of terms in each expression.

1. $2x + 7$
2. 9
3. $4x + 2y - 3z$
4. $4n - 20$

Write an expression for each phrase.

5. 8 more than a number
6. a number added to 5
7. the sum of a number and 9
8. 5 less than a number
9. 12 decreased by some number
10. the difference between a number and 4
11. one-half of some number
12. twice a number
13. the product of a number and 6
14. a number divided by 8
15. the ratio of 10 and some number
16. the quotient of a number and 3

Write an expression for each phrase.

17. 4 more than the product of a number and 3
18. 5 less than twice a number
19. twice the sum of 8 and a number

Write an equation for each sentence.

20. 8 more than the quotient of a number and 6 is the same as 2 less than the number.
21. If 9 is subtracted from twice a number, the result is 11.
22. 3 times the sum of a number and 5 is 4 more than twice the number.

23. Which of the following words is used to indicate multiplication?
    A. Sum B. Difference C. Product D. Quotient

24. Which of the following does not indicate subtraction?
    A. Less than B. Difference C. Decreased by D. Ratio of

25. Which of the following shows "twice the sum of a number and 8"?
    A. $2(x + 8)$ B. $2x + 8$ C. $2(x - 8)$ D. $2 + (x + 8)$

6•2 Simplifying Expressions

Terms

As you may remember, terms can be numbers, variables, or numbers and variables combined by multiplication or division. Some examples of terms are

$$n \qquad 7 \qquad 5x \qquad x^2$$

Compare the terms 7 and $5x$. The value of $5x$ will change as the value of x changes. If $x = 2$, then $5x = 5(2) = 10$, and if $x = 3$, then $5x = 5(3) = 15$. Notice, though, that the value of 7 never changes—it remains constant. When a term contains a number only, it is called a *constant* term.

Check It Out

Decide whether each term is a constant term.

1. $6x$
2. 9
3. $8(n + 2)$
4. 5

The Commutative Property of Addition and Multiplication

The **commutative property** of addition states that the order of two terms being added may be switched without changing the result: $3 + 4 = 4 + 3$ and $x + 8 = 8 + x$. The commutative property of multiplication states that the order of two terms being multiplied may be switched without changing the result: $3(4) = 4(3)$ and $x \cdot 8 = 8x$.

The commutative property does not hold for subtraction or division. The order of the terms does affect the result: $5 - 3 = 2$, but $3 - 5 = -2$; $8 \div 4 = 2$, but $4 \div 8 = \frac{1}{2}$.

Check It Out

Rewrite each expression using the commutative property of addition or multiplication.

5. $2x + 5$
6. $n \cdot 7$
7. $9 + 4y$
8. $5 \cdot 6$

The Associative Property of Addition and Multiplication

The **associative property** of addition states that the grouping of three terms being added does not affect the result: $(3 + 4) + 5 = 3 + (4 + 5)$ and $(x + 6) + 10 = x + (6 + 10)$. The associative property of multiplication states that the grouping of three terms being multiplied does not affect the result: $(2 \cdot 3) \cdot 4 = 2 \cdot (3 \cdot 4)$ and $5 \cdot 3x = (5 \cdot 3)x$.

The associative property does not hold for subtraction or division. The grouping of the numbers does affect the result: $(8 - 6) - 4 = -2$, but $8 - (6 - 4) = 6$; $(16 \div 8) \div 2 = 1$, but $16 \div (8 \div 2) = 4$.

Prime Time

One week the top five prime-time TV shows were rated like this:

Rating	Program
23.3	Drama
22.6	Comedy
20.9	Movie
19.0	Cartoon
18.6	Sitcom

Let *a* equal the number of homes each rating point stands for. Then write an expression to show the number of homes tuned in to the drama. See Hot Solutions for answer.

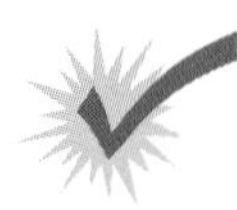

Check It Out

Rewrite each expression using the associative property of addition or multiplication.

9. $(4 + 8) + 11$
10. $(5 \cdot 2) \cdot 9$
11. $(2x + 5y) + 4$
12. $7 \cdot 8n$

The Distributive Property

The **distributive property** of addition and multiplication states that multiplying a sum by a number is the same as multiplying each addend by that number and then adding the two products. So $3(2 + 3) = (3 \cdot 2) + (3 \cdot 3)$.

How would you multiply $7 \cdot 99$ in your head? You might think, $700 - 7 = 693$. If you did, you used the distributive property.

$7(100 - 1)$	Distribute the factor of 7 to each term inside the parentheses.
$= 7 \cdot 100 - 7 \cdot 1$ $= 700 - 7$ $= 693$	Simplify, using order of operations.

The distributive property does not hold for division.

$$3 \div (2 + 3) \neq (3 \div 2) + (3 \div 3)$$

Check It Out

Use the distributive property to find each product.

13. $6 \cdot 98$
14. $3 \cdot 105$
15. $9 \cdot 199$
16. $4 \cdot 318$

Equivalent Expressions

The distributive property can be used to write an **equivalent expression** with two terms. Equivalent expressions are two different ways of writing one expression.

WRITING AN EQUIVALENT EXPRESSION

Write an equivalent expression for $5(9x - 7)$.

$5(9x - 7)$ $5 \cdot 9x - 5 \cdot 7$	• Distribute the factor of 5 to each term inside the parentheses.
$45x - 35$	• Simplify.
$5(9x - 7) = 45x - 35$	• Write the equivalent expressions.

Distributing When the Factor Is Negative

The distributive property is applied in the very same way if the factor to be distributed is negative.

Write an equivalent expression for $-3(5x - 6)$.

$-3(5x - 6)$ $-3 \cdot 5x - (-3) \cdot 6$	• Distribute the factor to each term inside the parentheses.
$15x + 18$ Remember: $(-3) \cdot 6 = -18$ and $-(-18) = +18$.	• Simplify
$-3(5x - 6) = -15x + 18$	• Write the equivalent expressions.

Check It Out

Write an equivalent expression.

17. $2(7x + 4)$
18. $8(3n - 2)$
19. $-1(7y - 4)$
20. $-3(-3x + 5)$

The Distributive Property with Common Factors

Given the expression $10x + 15$, you can use the distributive property to write an equivalent expression. Recognize that each of the two terms has a factor of 5.

Rewrite the expression as $5 \cdot 2x + 5 \cdot 3$. Then write the common factor 5 in front of the parentheses and the remaining factors inside the parentheses: $5(2x + 3)$. You have used the distributive property to *factor out the common factor.*

FACTORING OUT THE COMMON FACTOR

Factor out the common factor from the expression $12n - 30$.

$12n - 30$	• Find a common factor.
$6 \cdot 2n - 6 \cdot 5$	• Rewrite the expression.
$6 \cdot (2n - 5)$	• Use the distributive property.
$12n - 30 = 6 \cdot (2n - 5)$	

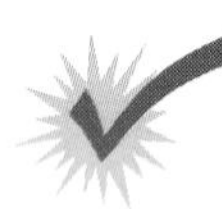

Check It Out

Factor out the common factor in each expression.

21. $7x + 35$
22. $18n - 15$
23. $15c + 60$
24. $40a - 100$

Like Terms

Like terms are terms that contain the same variable with the same exponent. Constant terms are like terms because they do not have any variables. Here are some examples of like terms.

Like Terms	Reason
$3x$ and $4x$	Both contain the same variables.
3 and 11	Both are constant terms.
$2n^2$ and $6n^2$	Both contain the same variable with the same exponent.

Below are some examples of terms that are not like terms.

Not Like Terms	Reason
$3x$ and $5y$	Variables are different.
$4n$ and 12	One term has a variable; the other is constant.
$2x^2$ and $2x$	The variables are the same, but the exponents are different.

Two terms that are like terms may be combined into one term by addition or subtraction. Consider the expression $3x + 4x$. Notice that the two terms have a common factor, x. Use the distributive property to write $x(3 + 4)$. This simplifies to $7x$, so $3x + 4x = 7x$.

COMBINING LIKE TERMS

Simplify $6n - 8n$.

- Recognize that the variable is a common factor. Rewrite the expression using the distributive property.
  $n(6 - 8)$
- Simplify. $n(-2)$
- Use the commutative property of multiplication.
  $-2n$

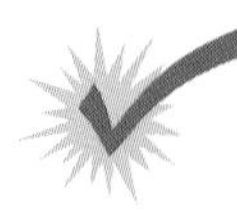

Check It Out

Combine like terms.

25. $4x + 9x$
26. $10y - 6y$
27. $5n + 4n + n$
28. $3a - 7a$

Simplifying Expressions

Expressions are simplified when all of the like terms have been combined. Terms that are not like terms cannot be combined. In the expression $3x - 5y + 6x$, there are three terms. Two of them are like terms, $3x$ and $6x$, which combine to be $9x$. The expression can be written as $9x - 5y$, which is simplified since the two terms are not like terms.

SIMPLIFYING EXPRESSIONS

Simplify the expression $4(2n - 3) - 10n + 17$.

$4(2n - 3) - 10n + 17$	• Combine like terms (if any). • Use the distributive property.
$4 \cdot 2n - 4 \cdot 3 - 10n + 17$	• Simplify.
$8n - 12 - 10n + 17$	• Combine like terms.
$-2n + 5$	• If remaining terms are not like terms, the expression is simplified.

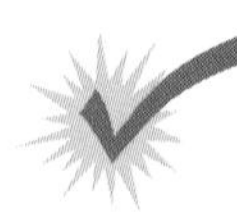

Check It Out

Simplify each expression.

29. $4y + 5z - y + 3z$
30. $x + 4(3x - 5)$
31. $15a + 8 - 2(3a + 2)$
32. $2(5n - 3) - (n - 2)$

6•2 EXERCISES

Decide whether each term is a constant term.

1. $4n$
2. -9

Rewrite each expression using the commutative property of addition or multiplication.

3. $9 + 5$
4. $n \cdot 4$
5. $8x + 11$

Rewrite each expression using the associative property of addition or multiplication.

6. $2 + (7 + 14)$
7. $(8 \cdot 5) \cdot 3$
8. $3 \cdot 6n$

Use the distributive property to find each product.

9. $8 \cdot 99$
10. $6 \cdot 108$

Write an equivalent expression.

11. $4(9x + 5)$
12. $-7(2n + 8)$
13. $12(3a - 10)$
14. $-(-5y - 8)$

Factor out the common factor in each expression.

15. $8x + 32$
16. $6n - 9$
17. $30a - 50$

Combine like terms.

18. $14x - 8x$
19. $7n + 8n - n$
20. $2a - 11a$

Simplify each expression.

21. $8a + b - 3a - 5b$
22. $3x + 2(6x - 5) + 8$
23. $-2(-7n - 3) - (n + 5)$
24. Which property is illustrated by $5(2x + 1) = 10x + 5$?
    A. commutative property of multiplication
    B. distributive property
    C. associative property of multiplication
    D. The example does not illustrate a property.
25. Which of the following shows the expression $24x - 36$ with the greatest common factor factored out?
    A. $2(12x - 18)$
    B. $3(8x - 12)$
    C. $6(4x - 6)$
    D. $12(2x - 3)$

6•3 Evaluating Expressions and Formulas

Evaluating Expressions

Once an expression has been written, you can *evaluate* it for different values of the variable. To evaluate $5x - 3$ for $x = 4$, *substitute* 4 in place of the x: $5(4) - 3$. Use **order of operations** to evaluate: multiply first, then subtract. So $5(4) - 3 = 20 - 3 = 17$.

EVALUATING AN EXPRESSION	
$3x^2 - \frac{4}{x} + 5$, when $x = 2$	• Substitute numeric value for x.
$3(2^2) - \frac{4}{(2)} + 5$	• Use order of operations to simplify. Simplify within parentheses, then evaluate powers.
$3 \cdot 4 - \frac{4}{2} + 5$	• Multiply and divide, in order from left to right.
$12 - 2 + 5$	• Add and subtract, in order from left to right.

When $x = 2$, then $3x^2 - \frac{4}{x} + 5 = 15$.

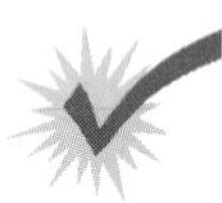

Check It Out

Evaluate each expression for the given value.

1. $9x - 14$ for $x = 4$
2. $5a + 7 + a^2$ for $a = -3$
3. $\frac{n}{3} + 2n - 5$ for $n = 12$
4. $2(y^2 - 2y + 1) + 4y$ for $y = 3$

Evaluating Formulas

The Formula for Perimeter of a Rectangle

The **perimeter** of a rectangle is the distance around the rectangle. The **formula** $P = 2w + 2l$ can be used to find the perimeter, P, if the width, w, and the length, l, are known.

FINDING THE PERIMETER OF A RECTANGLE

Find the perimeter of a rectangle whose width is 5 ft and length is 9 ft.

$P = 2(5) + 2(9)$ • Substitute values into formula for perimeter of rectangle ($P = 2w + 2l$).

$= 10 + 18$ • Simplify, using order of operations.

$= 28$

The perimeter of the rectangle is 28 ft.

Check It Out

Find the perimeter of each rectangle described.

5. $w = 6$ cm, $l = 11$ cm
6. $w = 4.5$ ft, $l = 9.5$ ft

Maglev

Maglev (short for *magnetic levitation*) trains fly above the tracks. Magnetic forces lift and propel the trains. Without the friction of the tracks, the maglevs run at speeds of 150 to 300 mi/hr. Are they the trains of the future? At a speed of 200 mi/hr with no stops, how long would it take to travel the distance between these cities? Round to the nearest quarter of an hour. See Hot Solutions for answers.

235 mi from Boston, MA to New York, NY
440 mi from Los Angeles, CA to San Francisco, CA
750 mi from Mobile, AL to Miami, FL

The Formula for Distance Traveled

The distance traveled by a person, vehicle, or object depends on its rate and the amount of time spent traveling. The formula $d = rt$ can be used to find the distance traveled, d, if the rate, r, and the amount of time, t, are known.

FINDING THE DISTANCE TRAVELED

Find the distance traveled by a train that averages 50 mi/hr for 4 hr.

$d = (50) \cdot (4)$ • Substitute values into the distance formula ($d = rt$).

$d = 200$ mi • Multiply.

The train traveled 200 miles.

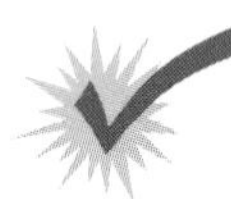

Check It Out

Find the distance traveled.

7. A person rides 12 mi/hr for 3 hr.
8. A plane flies 750 km/hr for 2 hr.
9. A person drives a car 55 mi/hr for 8 hr.
10. A snail moves 2 ft/hr for 4 hr.

6•3 EXERCISES

Evaluate each expression for the given value.

1. $6x - 11$ for $x = 5$
2. $5a^2 + 7 - 3a$ for $a = 4$
3. $\frac{n}{6} - 3n + 10$ for $n = -6$
4. $3(4y - 1) - \frac{12}{y} + 8$ for $y = 2$

Use the formula $P = 2w + 2l$.

5. Find the perimeter of a rectangle that is 60 ft long and 25 ft wide.
6. Find the perimeter of the rectangle.

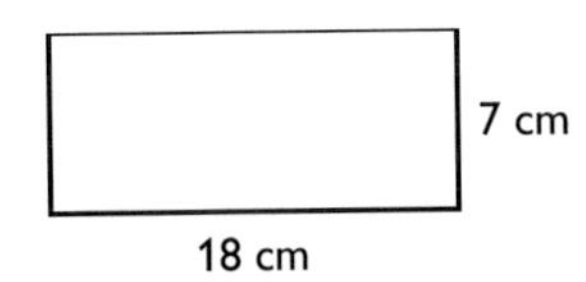

7. Susan had a 20-in. × 30-in. enlargement made of a photograph. She wanted to have it framed. Susan decided that the picture would look better if there was a 3-in. matte all the way around the photo. How many inches of frame would it take to enclose the photo and matte?

Use the formula $d = rt$.

8. Find the distance traveled by a jogger who jogs at 6 mi/hr for $1\frac{1}{2}$ hr.
9. A race car driver averaged 180 mi/hr. If the driver completed the race in $2\frac{1}{2}$ hr, how many miles was the race?
10. The speed of light is approximately 186,000 miles per second. About how far does light travel in 5 seconds?

6•4 Solving Linear Equations

Additive Inverses

Two terms are **additive inverses** if their sum is 0. Some examples are -3 and 3, $5x$ and $-5x$, and $12y$ and $-12y$. The additive inverse of 7 is -7, since $7 + (-7) = 0$, and the additive inverse of $-8n$ is $8n$, since $-8n + 8n = 0$.

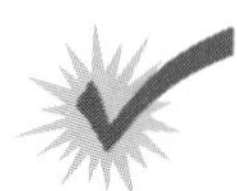

Check It Out

Give the additive inverse of each term.

1. 4
2. $-x$
3. -35
4. $10y$

True or False Equations

The equation $3 + 4 = 7$ represents a true statement. The equation $1 + 4 = 7$ represents a false statement. What about the equation $x + 4 = 7$? You cannot determine whether it is true or false until a value for x is known.

To determine whether the equation $2x + 5 = 11$ is true or false for $x = 1$, $x = 3$, and $x = 5$, simplify the equations.

$2x + 5 = 11$	$2x + 5 = 11$	$2x + 5 = 11$
$2(1) + 5 \stackrel{?}{=} 11$	$2(3) + 5 \stackrel{?}{=} 11$	$2(5) + 5 \stackrel{?}{=} 11$
$2 + 5 \stackrel{?}{=} 11$	$6 + 5 \stackrel{?}{=} 11$	$10 + 5 \stackrel{?}{=} 11$
$7 \stackrel{?}{=} 11$	$11 \stackrel{?}{=} 11$	$15 \stackrel{?}{=} 11$
False	True	False

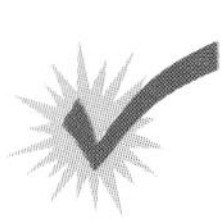

Check It Out

Determine whether each equation is true or false for $x = 2$, $x = 5$, and $x = 8$.

5. $7x - 3 = 11$
6. $3x + 1 = 16$
7. $6x - 8 = 22$
8. $2x - 7 = 9$

The Solution of an Equation

If you look back over the past equations, you will notice that each equation had only one value for the variable that made the equation true. This value is called the **solution** of the equation. If you were to try other values for x in the equations, they would all give false statements.

DETERMINING A SOLUTION

Determine whether 6 is the solution of the equation $4x - 5 = 2x + 6$.

$4x - 5 = 2x + 6$	• Substitute possible solution for x.
$4(6) - 5 \stackrel{?}{=} 2(6) + 6$	• Simplify, using order of operations.
$24 - 5 \stackrel{?}{=} 12 + 6$	
$19 \stackrel{?}{=} 18$	

Check It Out

Determine whether the given value is the solution of the equation.

9. 4; $3x - 5 = 7$
10. 9; $2n + 5 = 3n - 5$
11. 5; $7(y - 3) = 10$
12. 1; $8x + 4 = 15x - 3$

Equivalent Equations

An *equivalent equation* can be obtained from an existing equation by one of four ways.

- Add the same term to both sides of the equation.
- Subtract the same term from both sides.
- Multiply the same term on both sides.
- Divide by the same term on both sides.

Operation	Equation Equivalent to $x = 8$
Add 4 to both sides.	$x + 4 = 12$
Subtract 4 from both sides.	$x - 4 = 4$
Multiply by 4 on both sides.	$4x = 32$
Divide by 4 on both sides.	$\frac{x}{4} = 2$

Check It Out

Write equations equivalent to $x = 9$.

13. Add 3 to both sides.
14. Subtract 3 from both sides.
15. Multiply by 3 on both sides.
16. Divide by 3 on both sides.

Solving Equations

You can use equivalent equations to *solve* an equation. The solution is obtained when the variable is by itself on one side of the equation. The objective, then, is to use equivalent equations to isolate the variable on one side of the equation.

Consider the equation $x + 5 = 9$. In order to solve it, the x has to be on a side by itself. How can you get rid of the $+5$ that is also on that side? Remember that a term and its additive inverse add up to 0. The additive inverse of 5 is -5. To write an equivalent equation, subtract 5 from both sides.

$x + 5 = 9$	• Subtract 5 from both sides.
$x + 5 - 5 = 9 - 5$	• Simplify.
$x = 4$	

You can check the solution to be sure it is correct.

$x + 5 = 9$	• Substitute the possible solution for x.
$4 + 5 \stackrel{?}{=} 9$	• Simplify.
$9 \stackrel{?}{=} 9$	• Since this is a true statement, 4 is the solution.

Solve the equation $n - 8 = 7$. Notice that there is a -8 on the same side as n.

$n - 8 = 7$	• Add 8 to both sides.
$n - 8 + 8 = 7 + 8$	• Simplify.
$n = 15$	

Check the solution.

$n - 8 = 7$	• Substitute the possible solution for n.
$(15) - 8 \stackrel{?}{=} 7$	• Simplify.
$7 \stackrel{?}{=} 7$	• Since this is a true statement, 15 is the solution.

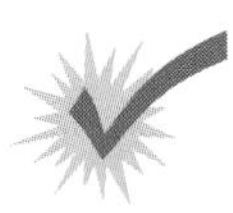

Check It Out

Solve each equation. Check your solution.

17. $x + 4 = 13$
18. $n - 5 = 11$
19. $y + 10 = 3$
20. $a - 8 = 1$

More Solving Equations

Consider the equation $3x = 15$. Notice that there isn't a term being added to or subtracted from the term with the variable. However, the variable still is not isolated. The variable is being multiplied by 3. To write an equivalent equation with the variable isolated, divide by 3 on both sides.

$3x = 15$ • Divide by 3 on both sides.

$\frac{3x}{3} = \frac{15}{3}$ • Simplify.

$x = 5$

Check the solution.

$3x = 15$ • Substitute the possible solution for x.

$3(5) \stackrel{?}{=} 15$ • Simplify.

$15 \stackrel{?}{=} 15$ • Since this is a true statement, 5 is the solution.

Solve the equation $\frac{n}{6} = 3$. The variable is being divided by 6. To write an equivalent equation with the variable isolated, multiply by 6 on both sides.

$\frac{n}{6} = 3$ • Multiply by 6 on both sides.

$\frac{n}{6} \cdot 6 = 3 \cdot 6$ • Simplify.

$n = 18$

Check the solution.

$\frac{n}{6} = 3$ • Substitute the possible solution for n.

$\frac{(18)}{6} \stackrel{?}{=} 3$ • Simplify.

$3 \stackrel{?}{=} 3$ • Since this is a true statement, 18 is the solution.

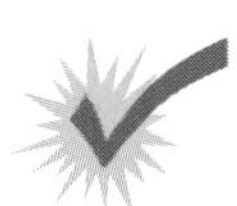

Check It Out

Solve each equation. Check your solution.

21. $5x = 35$
22. $\frac{y}{8} = 4$
23. $9n = -27$
24. $\frac{a}{3} = 12$

Solving Equations Requiring Two Operations

In the equation $4x - 7 = 13$, notice that the variable is being multiplied and has a term being subtracted. Still the objective is to use equivalent equations to isolate the variable. To do this, first isolate the term that contains the variable. Then isolate the variable.

$4x - 7 = 13$	• Add 7 to both sides to isolate the term that contains the variable.
$4x - 7 + 7 = 13 + 7$	• Simplify.
$4x = 20$	• Divide by 4 on both sides to isolate the variable.
$\frac{4x}{4} = \frac{20}{4}$	• Simplify.
$x = 5$	

Check the solution.

$4x - 7 = 13$	• Substitute the possible solution for x.
$4(5) - 7 \stackrel{?}{=} 13$ $20 - 7 \stackrel{?}{=} 13$	• Simplify, using order of operations.
$13 \stackrel{?}{=} 13$	• Since this is a true statement, 5 is the solution

SOLVING EQUATIONS REQUIRING TWO OPERATIONS

Solve the equation $\frac{n}{4} + 8 = 2$.

$\frac{n}{4} + 8 = 2$	• Add or subtract on both sides to isolate the term containing the variable.
$\frac{n}{4} + 8 - 8 = 2 - 8$	• Simplify.
$\frac{n}{4} = -6$	• Multiply or divide on both sides to isolate the variable.
$\frac{n}{4} \cdot 4 = -6 \cdot 4$	• Simplify.
$n = -24$	• Check solution by substituting into original equation.
$\frac{(-24)}{4} + 8 \stackrel{?}{=} 2$	• Simplify, using order of operations.
$-6 + 8 \stackrel{?}{=} 2$	
$2 \stackrel{?}{=} 2$ $n = -24$	• If the statement is true, you substituted the correct solution.

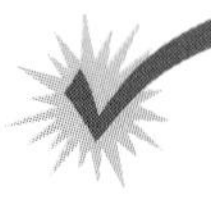

Check It Out

Solve each equation. Check your solution.

25. $6x + 11 = 29$
26. $\frac{y}{5} - 3 = 7$
27. $2n + 15 = 1$
28. $\frac{a}{3} + 11 = 9$

Solving Equations with the Variable on Both Sides

Consider the equation $5x + 4 = 8x - 5$. Notice that both sides of the equation have a term with the variable. To solve this equation, you still have to use equivalent equations to isolate the variable.

To isolate the variable, first use the additive inverse of one of the terms that contain the variable to collect these terms on one side of the equation. (Generally, they should be collected on the side of the equation where the coefficient of the variable is higher—this allows you to work with positive numbers whenever possible.) Then use the additive inverse to collect the constant terms on the other side. Then multiply or divide to isolate the variable.

SOLVING AN EQUATION WITH VARIABLES ON BOTH SIDES

Solve the equation $5x + 4 = 8x - 5$.

$5x + 4 - 5x = 8x - 5 - 5x$	• Add or subtract on both sides to collect terms with the variable on one side.
$4 = 3x - 5$	• Simplify. Combine like terms.
$4 + 5 = 3x - 5 + 5$	• Add or subtract on both sides to collect constant terms on the side opposite the variable.
$9 = 3x$	• Simplify.
$\frac{9}{3} = \frac{3x}{3}$	• Multiply or divide on both sides to isolate the variable.
$3 = x$	• Simplify.
$5(3) + 4 \stackrel{?}{=} 8(3) - 5$ $15 + 4 \stackrel{?}{=} 24 - 5$	• Check by substituting possible solution into original equation.
$19 \stackrel{?}{=} 19$	• Simplify, using order of operations.
$x = 3$	• If the statement is true, you substituted the correct solution.

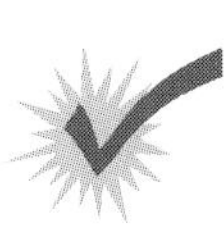

Check It Out

Solve each equation. Check your solution.

29. $9n - 4 = 6n + 8$

30. $12x + 9 = 2x - 11$

Equations Involving the Distributive Property

To solve the equation $3x - 4(2x + 5) = 3(x - 2) + 10$, notice that the terms are not yet ready to be collected on one side of the equation. First you have to use the distributive property.

$3x - 4(2x + 5) = 3(x - 2) + 10$	• Simplify, using distributive property.
$3x - 8x - 20 = 3x - 6 + 10$	• Combine like terms.
$-5x - 20 = 3x + 4$	• Add or subtract on both sides to collect terms with variable on one side.
$-5x - 20 + 5x = 3x + 4 + 5x$	• Combine like terms.
$-20 = 8x + 4$	• Add or subtract on both sides to collect constant terms on the side opposite from the variable.
$-20 - 4 = 8x + 4 - 4$	• Combine like terms.
$-24 = 8x$	• Multiply or divide on both sides to isolate the variable.
$\frac{-24}{8} = \frac{8x}{8}$	• Simplify.
$-3 = x$	• Substitute the possible solution into the original equation.
$3(-3) - 4[2(-3)+5] \stackrel{?}{=} 3[(-3)-2] + 10$ $3(-3) - 4(-6+5) \stackrel{?}{=} 3(-5) + 10$ $-9 - (-4) \stackrel{?}{=} -15 + 10$	• Simplify, using order of operations.
$-5 \stackrel{?}{=} -5$ $x = -3$	• If the statement is true, you substituted the correct solution.

Check It Out

Solve each equation. Check your solution.

31. $5(n - 3) = 10$
32. $7x - (2x + 3) = 9(x - 1) - 5x$

Solving for a Variable in a Formula

Recall the formula $d = rt$, in which the distance traveled, d, was found by multiplying the average rate, r, by the amount of time traveled, t. Could you solve for t in the formula?

$d = rt$	• Divide by r on both sides.
$\frac{d}{r} = \frac{rt}{r}$	• Simplify.
$\frac{d}{r} = t$	

You can solve for w in the formula for the perimeter of a rectangle, $P = 2w + 2l$.

$P = 2w + 2\ell$	• To isolate the term that contains w, subtract $2l$ from both sides.
$P - 2\ell = 2w + 2\ell - 2\ell$	• Combine like terms.
$P - 2\ell = 2w$	• To isolate w, divide both sides by 2.
$\frac{P - 2\ell}{2} = \frac{2w}{2}$	• Simplify.
$\frac{P - 2\ell}{2} = w$	

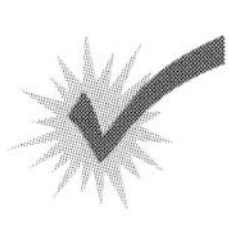

Check It Out

Solve for the indicated variable in each formula.

33. $A = lw$, for w
34. $2y - 3x = 8$, for y

How Risky Is It?

We are bombarded with statistics about risk. We are told we are more likely to die as a result of Earth's collision with an asteroid than as a result of a tornado, more likely to come in contact with germs by handling paper money than by visiting someone in the hospital. We know the odds of finding radon in our houses (1 in 15) and how much one bad sunburn increases the risk of skin cancer (up to 50 percent).

How risky is modern life? Consider these statistics on life expectancy.

Year	Life Expectancy
1900	47.3
1920	54.1
1940	62.9
1960	69.7
1980	73.7
1990	75.4

Graph these points on a coordinate plane. Do the points seem to show a straight line? What is the trend? Has life become more or less risky since the year 1900? See Hot Solutions for answers.

6•4 EXERCISES

Give the additive inverse of each term.

1. 8
2. $-6x$

Determine whether the given value is the solution of the equation.

3. $8;\ 3(y - 3) = 15$
4. $7;\ 6n - 5 = 3n + 11$

Solve each equation. Check your solution.

5. $x + 8 = 15$
6. $n - 3 = 9$
7. $\frac{y}{5} = 9$
8. $4a = -28$
9. $x + 14 = 9$
10. $n - 12 = 4$
11. $7x = 63$
12. $\frac{a}{6} = -2$
13. $3x + 7 = 25$
14. $\frac{y}{9} - 2 = 5$
15. $4n + 11 = 7$
16. $\frac{a}{5} + 8 = 5$
17. $13n - 5 = 10n + 7$
18. $y + 8 = 3y - 6$
19. $7x + 9 = 2x - 1$
20. $6a + 4 = 7a - 3$
21. $8(2n - 5) = 4n + 8$
22. $9y - 5 - 3y = 4(y + 1) - 5$
23. $8x - 3(x - 1) = 4(x + 2)$
24. $14 - (6x - 5) = 5(2x - 1) - 4x$

Solve for the indicated variable in each formula.

25. $d = rt$, for r
26. $A = lw$, for l
27. $4y - 5x = 12$, for y
28. $8y + 3x = 11$, for y

29. Which of the following equations can be solved by adding 6 to both sides and dividing by 5 on both sides?
    A. $5x + 6 = 16$
    B. $\frac{x}{5} + 6 = 16$
    C. $5x - 6 = 14$
    D. $\frac{x}{5} - 6 = 14$

30. Which equation does not have $x = 4$ as its solution?
    A. $3x + 5 = 17$
    B. $2(x + 2) = 10$
    C. $\frac{x}{2} + 5 = 7$
    D. $x + 2 = 2x - 2$

6•5 Ratio and Proportion

Ratio

A **ratio** is a comparison of two quantities. If there are 10 boys and 15 girls in a class, the ratio of the number of boys to the number of girls is 10 to 15, which can be expressed as the fraction $\frac{10}{15}$, which reduces to $\frac{2}{3}$. You can write some other ratios.

Comparison	Ratio	As a Fraction
Number of girls to number of boys	15 to 10	$\frac{15}{10} = \frac{3}{2}$
Number of boys to number of students	10 to 25	$\frac{10}{25} = \frac{2}{5}$
Number of students to number of girls	25 to 15	$\frac{25}{15} = \frac{5}{3}$

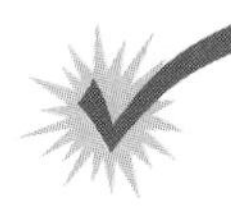

Check It Out

A coin bank contains 3 nickels and 9 dimes. Write each ratio.

1. number of nickels to number of dimes
2. number of dimes to number of coins
3. number of coins to number of nickels

Proportions

A **rate** is a ratio that compares a quantity to 1 unit. Some examples of rates are listed.

$\frac{55 \text{ mi}}{1 \text{ hr}}$ $\frac{5 \text{ apples}}{\$1}$ $\frac{18 \text{ mi}}{1 \text{ gal}}$ $\frac{\$400}{1 \text{ wk}}$ $\frac{60 \text{ sec}}{1 \text{ min}}$

If a car gets 18 mi to 1 gal, then the car can get $\frac{36\text{mi}}{2\text{ gal}}$, $\frac{54\text{ mi}}{3\text{ gal}}$, and so on. The ratios are all equal—they can be reduced to $\frac{18}{1}$.

When two ratios are equal, they form a **proportion.** One way to determine whether two ratios form a proportion is to check their **cross products.** Every proportion has two cross products: the numerator of one ratio multiplied by the denominator of the other ratio. If the cross products are equal, a proportion is formed.

DETERMINING A PROPORTION

Determine whether a proportion is formed.

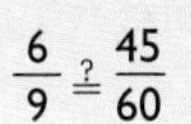

$\frac{6}{9} \stackrel{?}{=} \frac{45}{60}$ $\quad$ $\frac{15}{9} \stackrel{?}{=} \frac{70}{42}$

- Find the cross products.

$6 \cdot 60 \stackrel{?}{=} 45 \cdot 9$ $\quad$ $15 \cdot 42 \stackrel{?}{=} 70 \cdot 9$

$360 \stackrel{?}{=} 405$ $\quad$ $630 \stackrel{?}{=} 630$

- If the sides are equal, the ratios are proportional.

$\frac{6}{9} \stackrel{?}{=} \frac{45}{60}$ is not a proportion. $\quad$ $\frac{15}{9} \stackrel{?}{=} \frac{70}{42}$ is a proportion.

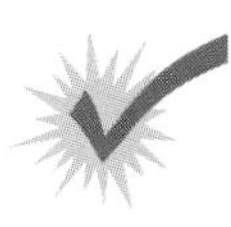

Check It Out

Determine whether a proportion is formed.

4. $\frac{4}{7} = \frac{12}{21}$
5. $\frac{6}{5} = \frac{50}{42}$

Using Proportions to Solve Problems

To use proportions to solve problems, set up two ratios which relate what you know to what you are solving for.

Suppose that you can buy 5 apples for $2. How much would it cost to buy 17 apples? Let c represent the cost of the 17 apples.

If you express each ratio as $\frac{\text{apples}}{\$}$, then one ratio is $\frac{5}{2}$ and another is $\frac{17}{c}$. The two ratios must be equal.

$$\frac{5}{2} = \frac{17}{c}$$

To solve for c, you can use the cross products. Since you have written a proportion, the cross products are equal.

$$5c = 34$$

To isolate the variable, divide by 5 on both sides and simplify.

$$\frac{5c}{5} = \frac{34}{5} \qquad c = 6.8$$

So 17 apples would cost $6.80.

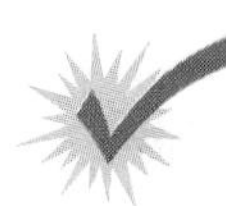

Check It Out

Use proportions to solve items 6–9.

6. A car gets 22 mi/gal. How many gallons would the car need to travel 121 mi?
7. A worker earns $100 every 8 hr. How much would the worker earn in 36 hr?

Use the chart below to answer items 8 and 9.

Top-Five TV Shows in Prime-Time

Rating	Program
23.3	Drama
22.6	Comedy
20.9	Movie
19.0	Cartoon
18.6	Sitcom

8. If 18,430,000 homes watched the cartoon and that is a 19.0 rating, how many homes is one rating point worth?
9. If 18,042,000 homes watched the sitcom, how many homes watched the drama?

6•5 EXERCISES

A basketball team has 20 wins and 10 losses. Write each ratio.

1. number of wins to number of losses
2. number of wins to number of games
3. number of losses to number of games

Determine whether a proportion is formed.

4. $\frac{3}{8} = \frac{16}{42}$
5. $\frac{10}{4} = \frac{25}{10}$
6. $\frac{4}{6} = \frac{15}{22}$

Use a proportion to solve each problem.

7. In a class, the ratio of boys to girls is $\frac{4}{5}$. If there are 12 boys in the class, how many girls are there?
8. An overseas telephone call costs $0.36 per minute. How much would a 6-minute call cost?
9. A map is drawn using a scale of 150 mi to 1 cm. The distance between two cities is 1,200 mi. How far apart are the two cities on the map?
10. A blueprint of a house is drawn using a scale of 5 ft to 2 cm. On the blueprint, a room is 5 cm long. How long will the actual room be?

6•6 Inequalities

Showing Inequalities

When comparing the numbers 7 and 4, you might say that "7 is greater than 4" or you might also say "4 is less than 7." When two expressions are not equal or could be equal, you can write an **inequality.** The symbols are shown in the chart.

Symbol	Meaning	Example
$>$	Is greater than	$7 > 4$
$<$	Is less than	$4 < 7$
$\geq$	Is greater than or equal to	$x \geq 3$
$\leq$	Is less than or equal to	$-2 \leq x$

The equation $x = 5$ has one solution, 5. The inequality $x > 5$ has an infinite number of solutions: 5.001, 5.2, 6, 15, 197, and 955 are just some of the solutions. Note that 5 is not a solution—5 is not greater than 5. Since you cannot list all of the solutions, you can show them on a number line.

To show all the values that are greater than 5, but not including 5, use an open circle on 5 and shade the number line to the right.

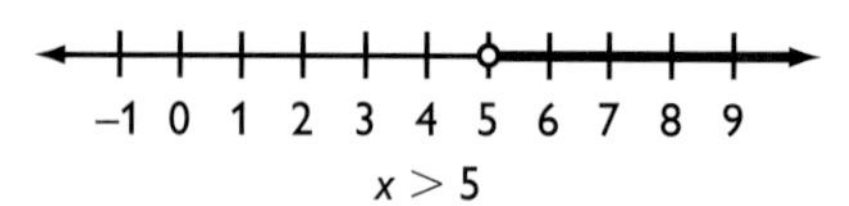

$x > 5$

The inequality $y \leq -1$ also has an infinite number of solutions: -1.01, -1.5, -2, -8, and -54 are just some of the solutions. Note that -1 is also a solution, because -1 is less than *or* equal to -1. On a number line, you want to show all the values that are less than or equal to -1. Since the -1 is to be included, use a closed (filled-in) circle on -1 and shade the number line to the left.

–5 –4 –3 –2 –1 0 1 2

$y \leq -1$

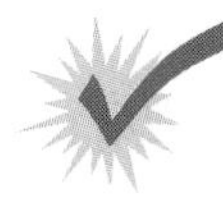

Check It Out

Draw the number line showing the solutions to each inequality.

1. $x \geq 2$
2. $y < -4$
3. $n > -3$
4. $x \leq 1$

Solving Inequalities

Just as you can write equivalent equations, you can write equivalent inequalities. Begin with the inequality $8 > 4$.

$$8 > 4$$

$$-1 \times 8 \; ? \; -1 \times 4$$

$$-8 < -4$$

Notice that when the inequality was multiplied or divided by a negative number on both sides, the inequality sign had to be reversed.

Start with 8 > 4. Perform these operations.	Resulting Inequality
Add 7 to both sides.	$15 > 11$
Subtract 6 from both sides.	$2 > -2$
Multiply by 5 on both sides.	$40 > 20$
Divide by 4 on both sides.	$2 > 1$
Multiply by –3 on both sides.	$-24 < -12$
Divide by –2 on both sides.	$-4 < -2$

To determine the solutions of the inequality $-3x - 5 \geq 10$, use equivalent inequalities to isolate the variable.

$-3x - 5 \geq 10$ • Add or subtract on both sides to isolate the variable term.

$-3x - 5 + 5 \geq 10 + 5$ • Combine like terms.

$-3x \geq 15$ • Multiply or divide on both sides to isolate the variable. If you multiply or divide by a negative number, reverse the inequality sign.

$\frac{-3x}{-3} \leq \frac{15}{-3}$ • Simplify.

$x \geq -5$

Check It Out

Solve each inequality.

5. $x + 8 > 5$
6. $4n \leq -8$
7. $7y + 3 < 24$
8. $-6x + 10 \leq 4$

6·6 EXERCISES

Draw the number line showing the solutions to each inequality.

1. $x < -2$ 2. $y \geq 0$ 3. $n > -1$ 4. $x \leq 7$

Solve each inequality.

5. $x - 3 < 7$
6. $2y \geq 8$
7. $n + 8 > 3$
8. $-6a \leq 18$
9. $2x + 4 \geq 14$
10. $9x - 11 < 16$
11. $-3n + 7 \leq 1$
12. $8 - y > 5$

Write the inequality for each number line.

13.

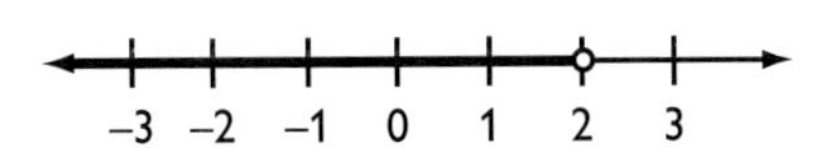

14.

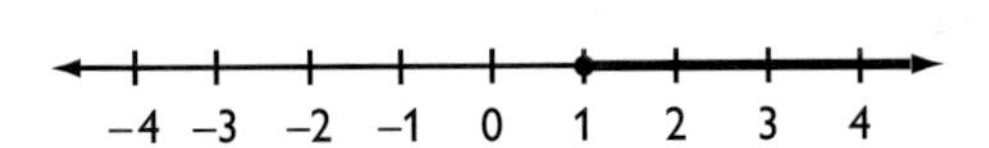

15. Which operation(s) would require that the inequality sign be reversed?
   A. addition of -2 B. subtraction of -2
   C. multiplication by -2 D. division by -2
16. If $x = -3$, is it true that $3(x - 4) \leq 2x$?
17. If $x = 6$, is it true that $2(x - 4) < 8$?
18. Which operation would require that the inequality symbol be reversed?
   A. $\times$ 5 on both sides B. $+ (-5)$ on both sides
   C. $\div$ 5 on both sides D. $\div (-5)$ on both sides
19. Which of the following statements is false?
   A. $-7 \leq 2$ B. $0 \leq -4$
   C. $6 \geq -6$ D. $3 \geq 3$
20. Which of the following inequalities does not have $x < 2$ as its solution?
   A. $-4x < -8$ B. $x + 6 < 8$
   C. $4x - 1 < 7$ D. $-x > -2$

6•7 Graphing on the Coordinate Plane

Axes and Quadrants

When you cross a **horizontal** (left to right) number line with a **vertical** (up and down) number line, the result is a two-dimensional coordinate plane.

The number lines are called **axes.** The horizontal number line is the ***x*-axis**, and the vertical number line is the ***y*-axis**. The plane is divided into four regions, called **quadrants.** Each quadrant is named by a Roman numeral, as shown in the diagram.

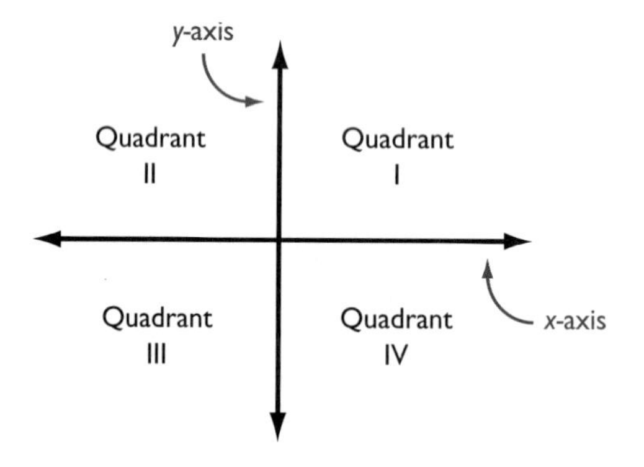

Check It Out

Fill in the blank.

1. The vertical number line is called the ___.
2. The upper-left region of the coordinate plane is called ___.
3. The lower-right region of the coordinate plane is called ___.
4. The horizontal number line is called the ___.

Writing an Ordered Pair

Any location on the coordinate plane can be represented by a **point.** The location of any point is given in relation to where the two axes intersect, called the **origin.**

Two numbers are required to identify the location of a point. The x-coordinate tells how far to the left or right of the origin the point lies. The y-coordinate tells how far up or down from the origin the point lies. Together the x-coordinate and y-coordinate form an **ordered pair,** (x, y).

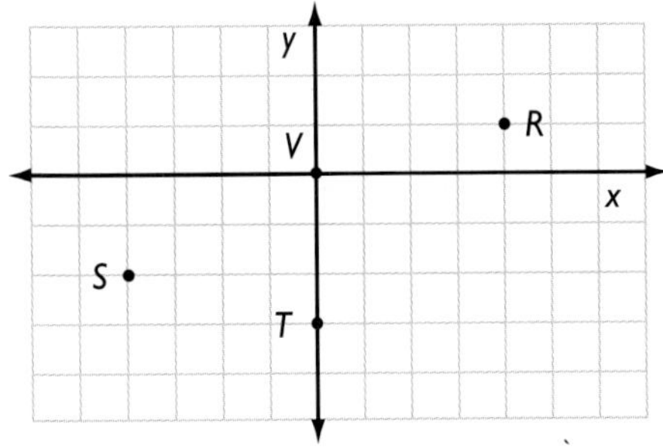

Since point R is 4 units to the right of the origin and 1 unit up, its ordered pair is $(4, 1)$. Point S is 4 units to the left of the origin and 2 units down, so its ordered pair is $(-4, -2)$. Point T is 0 units from the origin and 3 units down, so its ordered pair is $(0, -3)$. Point V is the origin and its ordered pair is $(0, 0)$.

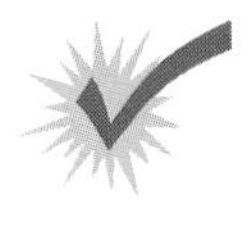

Check It Out

Give the ordered pair for each point.

5. M
6. N
7. P
8. Q

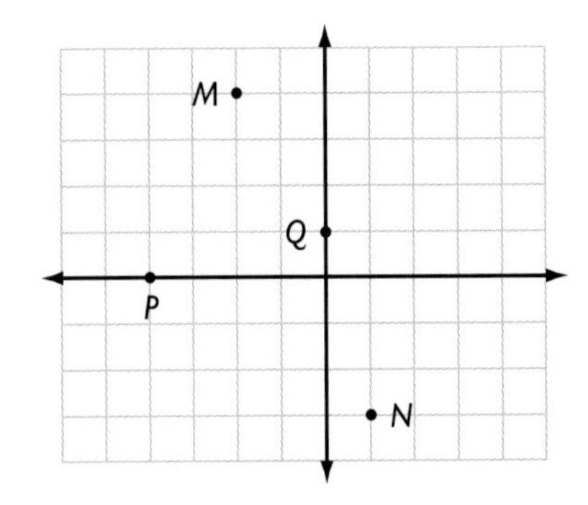

Locating Points on the Coordinate Plane

To locate point $A(3, -4)$ from the origin, move 3 units to the right and 4 units down. Point A lies in Quadrant IV. To locate point $B(-1, 4)$ from the origin, move 1 unit to the left and 4 units up. Point B lies in Quadrant II. Point $C(5, 0)$ is, from the origin, 5 units to the right and 0 units up or down. Point C lies on the x-axis. Point $D(0, -2)$ is, from the origin, 0 units to the left or right and 2 units down. Point D lies on the y-axis.

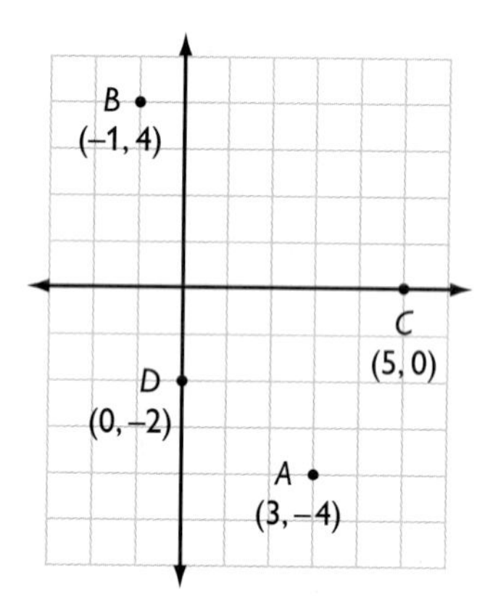

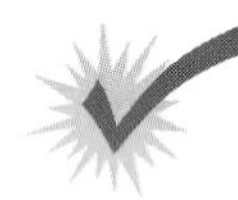

Check It Out

Draw each point on a coordinate plane and tell where it lies.

9. $H(-5, 2)$
10. $J(2, -5)$
11. $K(-3, -4)$
12. $L(-1, 0)$

The Graph of an Equation with Two Variables

Consider the equation $y = 2x - 1$. Notice that it has two variables, x and y. The point (3, 5) is a solution of this equation. If you substitute 3 for x and 5 for y (in the ordered pair, 3 is the x-coordinate and 5 is the y-coordinate), you get the true statement $5 = 5$. The point (2, 4) is not a solution of the equation. Substituting 2 for x and 4 for y results in the false statement $4 = 3$.

Choose a value for x	Substitute the value into the equation $y = 2x - 1$	Solve for y	Ordered Pair
0	$y = 2(0) - 1$	–1	(0, –1)
1	$y = 2(1) - 1$	1	(1, 1)
3	$y = 2(3) - 1$	5	(3, 5)
–1	$y = 2(-1) - 1$	–3	(–1, –3)

If you locate the points on a coordinate plane, you will notice that they all lie along a straight line.

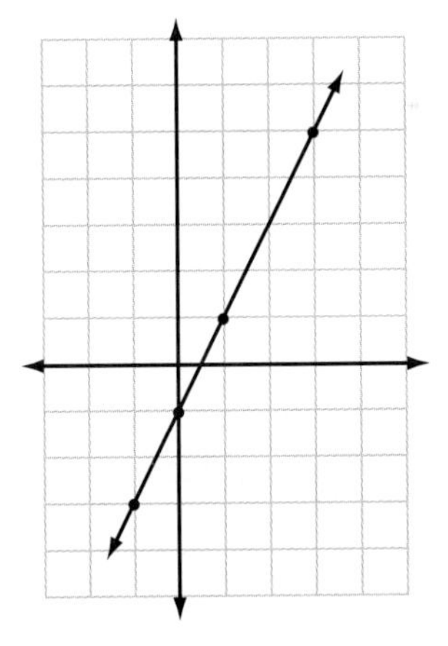

The coordinates of any point on the line will result in a true statement if substituted into the equation.

GRAPHING THE EQUATION OF A LINE

Graph the equation $y = \frac{1}{3}x - 2$.

- Choose five values for x.

  Since the value of x is to be multiplied by $\frac{1}{3}$, choose values that are multiples of 3, such as $-3, 0, 3, 6,$ and 9.

- Calculate the corresponding values for y.

  When $x = -3, y = \frac{1}{3}(-3) - 2 = -3$
  When $x = 0, y = \frac{1}{3}(0) - 2 = -2$
  When $x = 3, y = \frac{1}{3}(3) - 2 = -1$
  When $x = 6, y = \frac{1}{3}(6) - 2 = 0$
  When $x = 9, y = \frac{1}{3}(9) - 2 = 1$

- Write the five solutions as ordered pairs.

  $(-3, -3), (0, -2), (3, -1), (6, 0),$ and $(9, 1)$

- Locate the points on a coordinate plane, and draw the line.

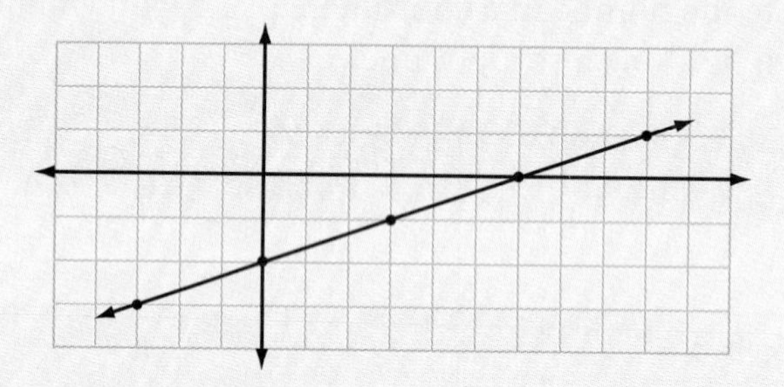

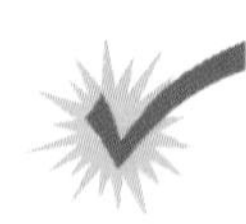

Check It Out

Find five solutions of each equation. Graph each line.

13. $y = 3x - 2$
14. $y = 2x + 1$
15. $y = \frac{1}{2}x - 3$
16. $y = -2x + 3$

More Graphing Equations with Two Variables

How could solutions of the equation $2x - 3y = 6$ be determined? You could use equivalent equations to isolate the y on one side of the equation.

$2x - 3y = 6$	• Add or subtract on both sides to isolate the y term.
$2x - 3y - 2x = 6 - 2x$	• Combine like terms. (Use commutative property to change order of terms.)
$-3y = -2x + 6$	• Multiply or divide on both sides to isolate y.
$\frac{-3y}{-3} = \frac{-2x + 6}{-3}$	• Simplify.
$y = \frac{-2x}{-3} + \frac{6}{-3}$	
$y = \frac{2}{3}x - 2$	

Now you can find five solutions and graph the line.

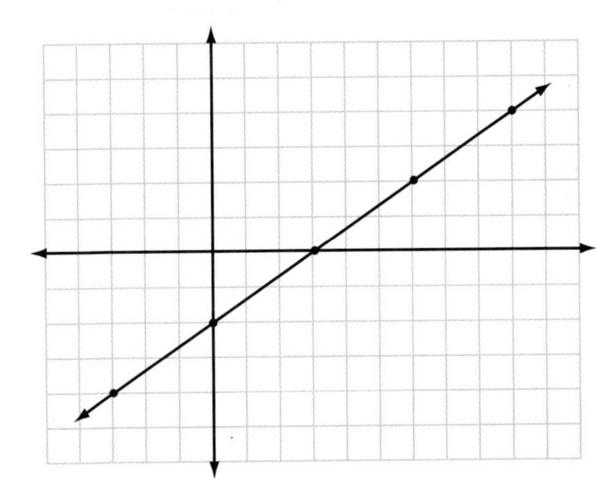

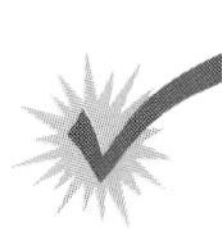

Check It Out

Graph each line.

17. $2x + y = 5$
18. $x + 2y = 4$
19. $4x - 2y = 6$
20. $2x - 5y = 10$

Horizontal and Vertical Lines

Choose several points that lie on a horizontal line.

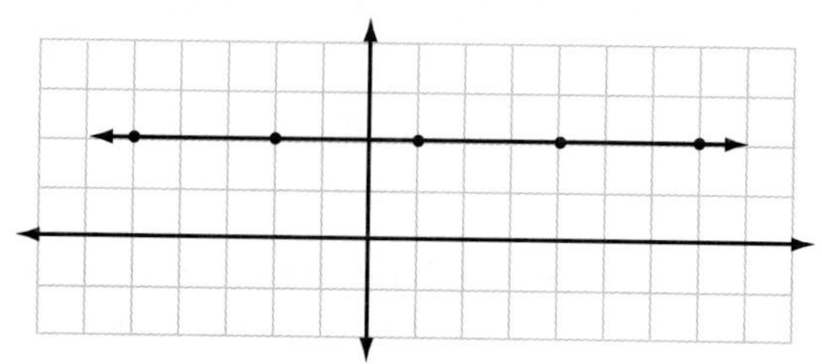

Notice that any point that lies on the line has a y-coordinate of 2. The equation of this line is $y = 2$.

Choose several points that lie on a vertical line.

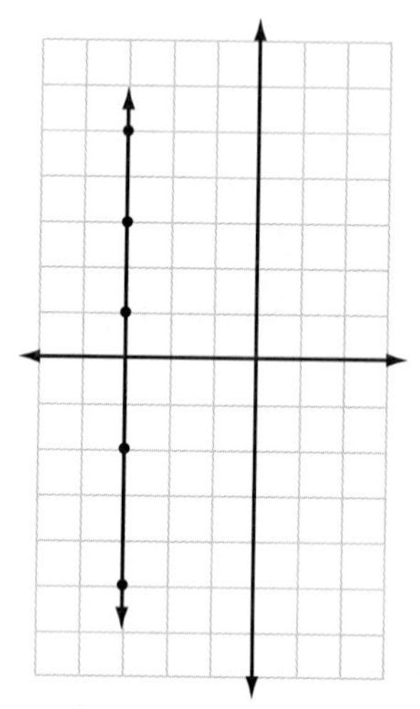

Notice that any point that lies on the line has an x-coordinate of -3. The equation of this line is $x = -3$.

Check It Out

Graph each line.

21. $x = 4$
22. $y = -3$
23. $x = -1$
24. $y = 6$

6•7 EXERCISES

Fill in the blanks.

1. The horizontal number line is called the ___.
2. The lower left region of the coordinate plane is called ___.
3. The upper right region of the coordinate plane is called ___.

Give the ordered pair for each point.

4. *A*
5. *B*
6. *C*
7. *D*

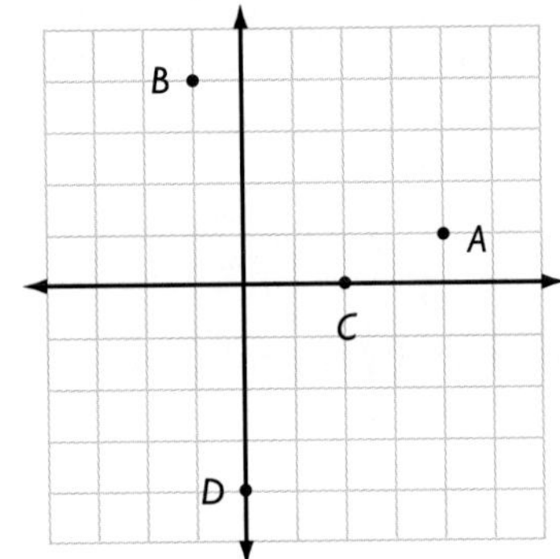

Locate each point on the coordinate plane and tell where it lies.

8. $H(2, 5)$ 9. $J(-1, -2)$ 10. $K(0, 3)$ 11. $L(-4, 0)$

Find five solutions of each equation. Graph each line.

12. $y = 2x - 2$ 13. $y = -3x + 3$ 14. $y = \frac{1}{2}x - 1$

Graph each line.

15. $2x - y = 3$ 16. $x - 3y = 6$ 17. $4x + y = 8$
18. $3x + 5y = 15$ 19. $x = 5$ 20. $y = -2$

6•8 Slope and Intercept

Slope

One characteristic of a line is its **slope.** Slope is a measure of a line's slant. To describe the way a line slants, you state how the coordinates on the line change when you move right. Choose two points along the line. The **run** is the difference in the x-coordinates. The **rise** is the difference in the y-coordinates.

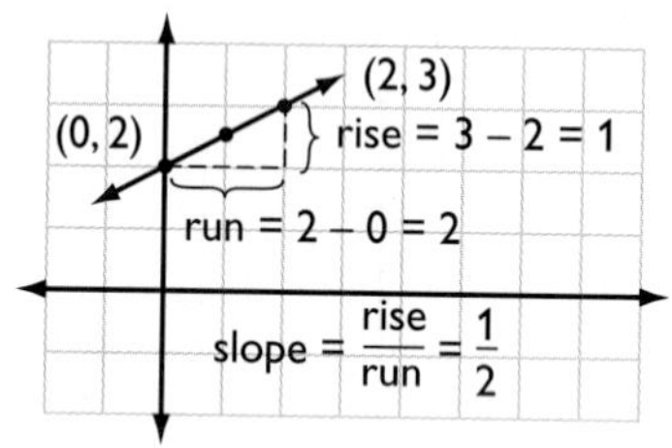

The slope, then, is given by the ratio of rise (vertical movement) to run (horizontal movement).

Slope $= \frac{\text{rise}}{\text{run}}$

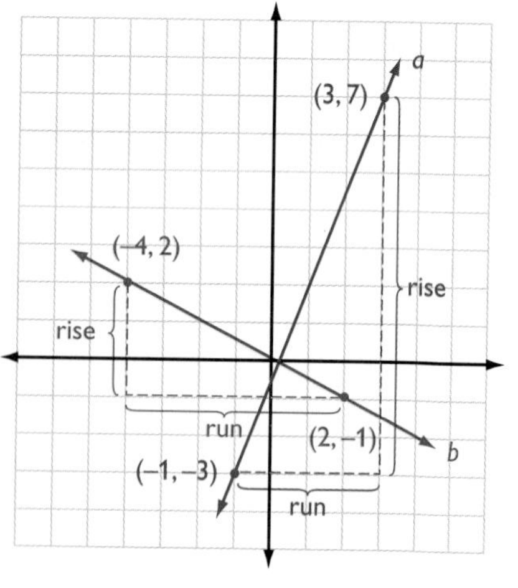

Notice that for line a, the rise between the two marked points is 10 units and the run is 4 units. The slope of the line, then, is $\frac{10}{4} = \frac{5}{2}$. For line b, the rise between the two marked points is -3 and the run is 6, so the slope of the line is $\frac{-3}{6} = -\frac{1}{2}$.

The slope along a straight line is always the same. For line a, regardless of the two points chosen, the slope will always simplify to be $\frac{5}{2}$.

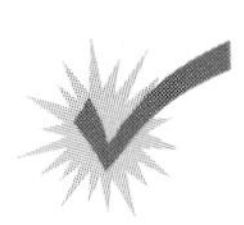

Check It Out

Determine the slope of each line.

1.

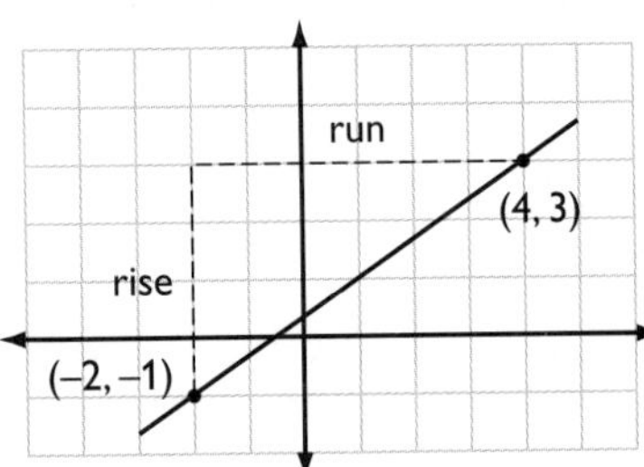

2.

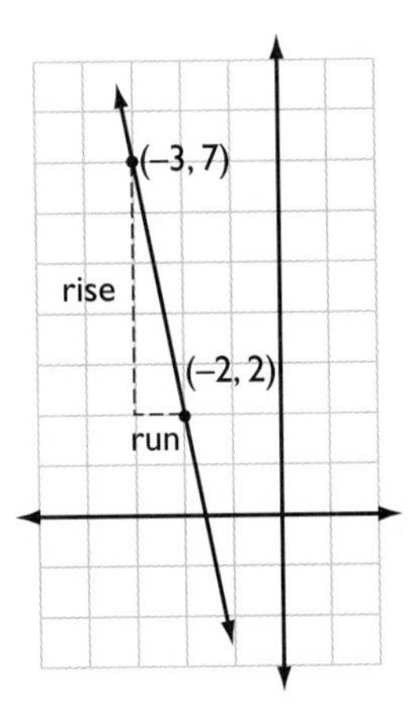

Calculating the Slope of a Line

You can calculate the slope of a line just from knowing two points on the line. The rise is the difference of the y-coordinates and the run is the difference of the x-coordinates. For the line that passes through the points (1, −2) and (4, 5), the slope can be calculated as shown. The variable m is used to represent slope.

$$m = \frac{\text{rise}}{\text{run}} = \frac{5 - (-2)}{4 - 1} = \frac{7}{3}$$

The slope could also have been calculated another way.

$$m = \frac{\text{rise}}{\text{run}} = \frac{-2 - 5}{1 - 4} = \frac{-7}{-3} = \frac{7}{3}$$

The order in which you subtract the coordinates does not matter, as long as you find both differences in the same order.

CALCULATING THE SLOPE OF A LINE

Find the slope of the line that contains the points $(-2, 3)$ and $(4, -1)$.

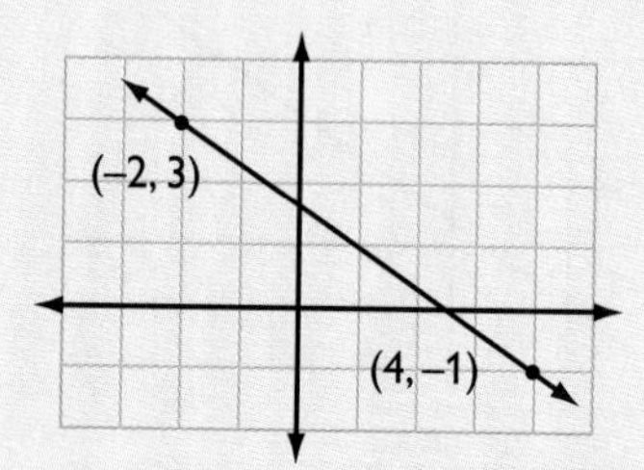

$m = \frac{-1 - 3}{4 - (-2)}$ or $m = \frac{3 - (-1)}{-2 - 4}$

•Use the definition $m = \frac{\text{rise}}{\text{run}} = \frac{\text{difference of } y\text{–coordinates}}{\text{difference of } x\text{–coordinates}}$ to find the slope.

$m = \frac{-4}{6}$ or $m = \frac{4}{-6}$

$m = \frac{-2}{3}$ or $m = \frac{2}{-3}$

•Simplify.

The slope is $-\frac{2}{3}$.

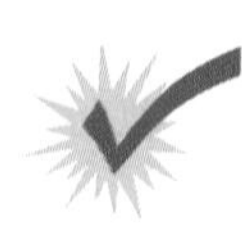

Check It Out

Find the slope of the line that contains the given points.

3. $(-1, 7)$ and $(4, 2)$
4. $(-3, -4)$ and $(1, 2)$
5. $(-2, 0)$ and $(4, -3)$
6. $(0, -3)$ and $(2, 7)$

Slopes of Horizontal and Vertical Lines

Choose two points on a horizontal line, $(-1, 2)$ and $(3, 2)$. Calculate the slope of the line.

$$m = \frac{\text{rise}}{\text{run}} = \frac{2 - 2}{3 - (-1)} = \frac{0}{4} = 0$$

A horizontal line has no rise; its slope is 0.

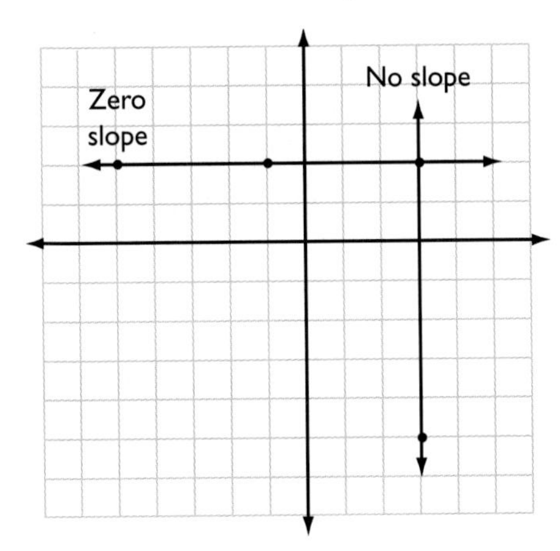

Choose two points on a vertical line, $(3, 2)$ and $(3, -5)$. Calculate the slope of the line.

$$m = \frac{\text{rise}}{\text{run}} = \frac{-5 - 2}{3 - 3} = \frac{-7}{0}, \text{ which is undefined.}$$

A vertical line has no run; it has *no slope.*

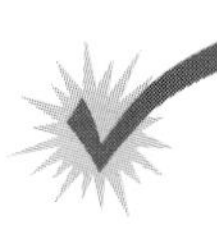

Check It Out

Find the slope of the line that contains the given points.

7. $(-1, 4)$ and $(5, 4)$
8. $(2, -1)$ and $(2, 6)$
9. $(-5, 0)$ and $(-5, 7)$
10. $(4, -4)$ and $(-1, -4)$

The y-Intercept

A second characteristic of a line, after the slope, is the ***y*-intercept**. The *y*-intercept is the location along the *y*-axis where the line crosses, or intercepts, the axis.

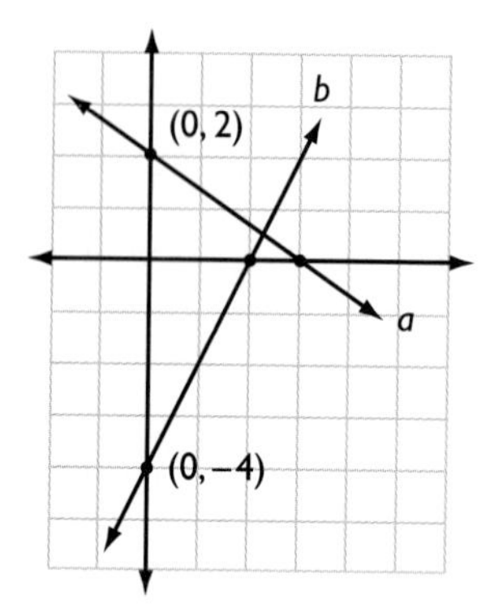

The *y*-intercept of line *a* is 2, and the *y*-intercept of line *b* is −4.

Check It Out

Identify the *y*-intercept of each line.

11. *c*
12. *d*

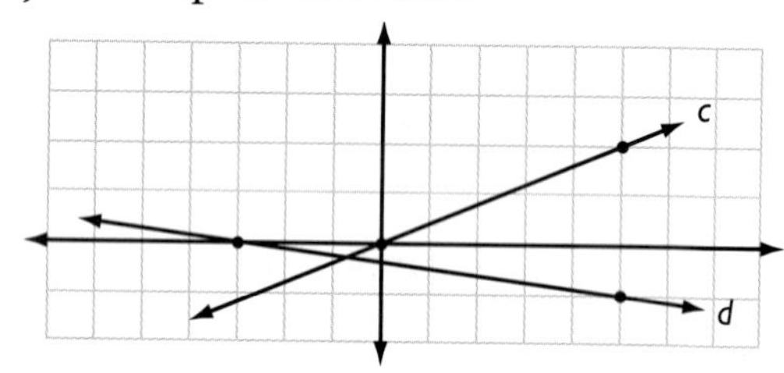

Using the Slope and y-Intercept to Graph a Line

A line can be graphed if the slope and the y-intercept are known. First you locate the y-intercept on the y-axis. Then you use the rise and run of the slope to locate a second point on the line. Connect the two points to plot your line.

GRAPHING A LINE USING THE SLOPE AND y-INTERCEPT

Graph the line with slope -2 and y-intercept 3.

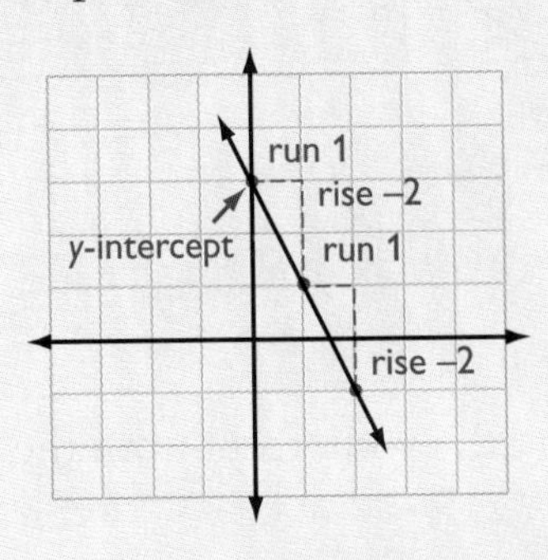

- Locate the y-intercept on the y-axis.
- Use the slope to locate other points on the line. If the slope is a whole number a, remember $a = \frac{a}{1}$, so rise is a and run is 1.
- Draw a line through the points.

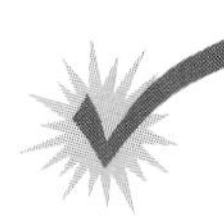

Check It Out

Graph each line.

13. Slope $= \frac{1}{3}$, y-intercept is -2.
14. Slope $= \frac{-2}{5}$, y-intercept is 4.
15. Slope $= 3$, y-intercept is -3.
16. Slope $= -2$, y-intercept is 0.

Slope-Intercept Form

The equation $y = mx + b$ is in the *slope-intercept form* for the equation of a line. When an equation is in this form, the slope of the line is given by m and the y-intercept is located at b. The graph of the equation $y = \frac{2}{3}x - 4$ is a line that has a slope of $\frac{2}{3}$ and a y-intercept at -4. The graph is shown.

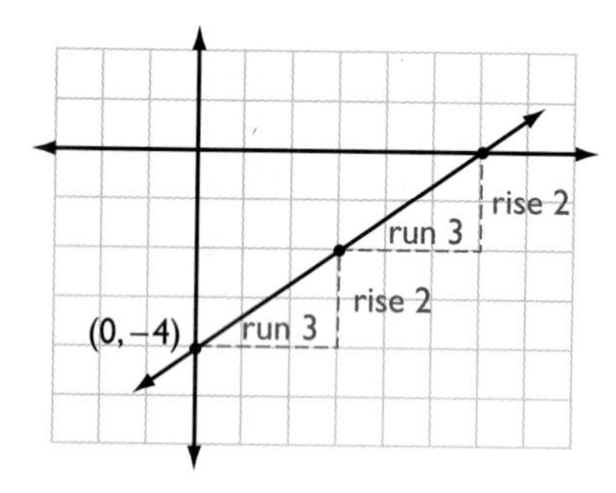

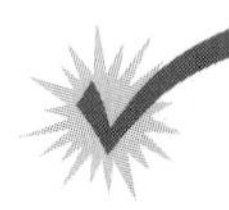

Check It Out

Determine the slope and the y-intercept from the equation of each line.

17. $y = -2x + 3$
18. $y = \frac{1}{5}x - 1$
19. $y = \frac{-3}{4}x$
20. $y = 4x - 3$

Writing Equations in Slope-Intercept Form

To write the equation $4x - 3y = 9$ in slope-intercept form, you isolate the y on one side of the equation. You can use equivalent equations to isolate the y.

$4x - 3y = 9$	• Add or subtract to isolate the term with the y.
$4x - 3y - 4x = 9 - 4x$	• Combine like terms. Use commutative property to reorder.
$-3y = -4x + 9$	• Multiply or divide to isolate the y.
$\frac{-3y}{-3} = \frac{-4x + 9}{-3}$	• Simplify.

$$y = \frac{-4}{-3}x + \frac{9}{-3}$$

$$y = \frac{4}{3}x - 3$$

In slope-intercept form, the equation $4x - 3y = 9$ is $y = \frac{4}{3}x - 3$. The slope is $\frac{4}{3}$ and the y-intercept is located at -3. The graph of the line is shown.

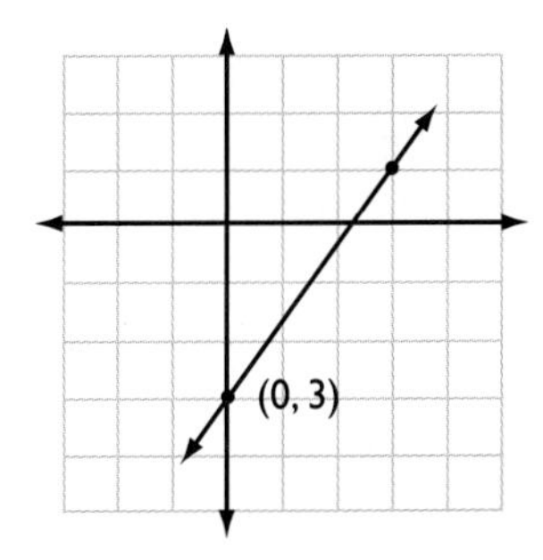

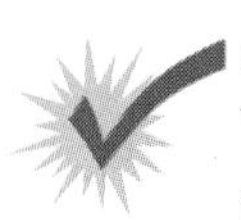

Check It Out

Write each equation in slope-intercept form. Graph the line.

21. $x + 2y = 6$
22. $2x - 3y = 9$
23. $4x - 2y = 4$
24. $7x + y = 8$

Slope-Intercept Form and Horizontal and Vertical Lines

As you will remember, the equation of a horizontal line is in the form $y =$ (number). In the graph below, the horizontal line has the equation $y = 2$. Is this equation in slope-intercept form? Yes, since the equation could be written $y = 0x + 2$. The y is isolated on one side of the equation. The slope, remember, is 0, and the y-intercept is 2.

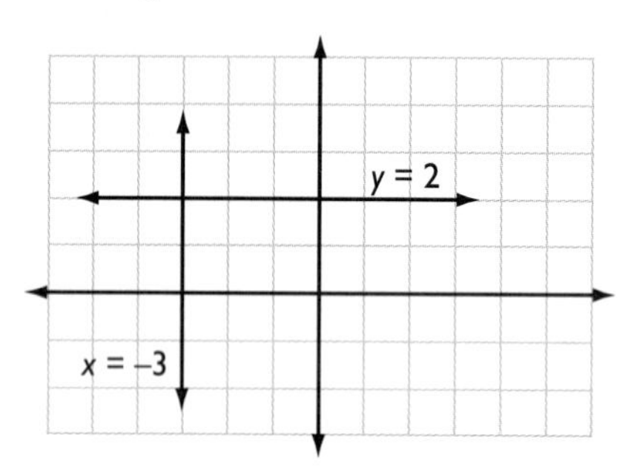

The equation of a vertical line is in the form $x =$ (number). In the graph above, the vertical line has the equation $x = -3$. Is this equation in slope-intercept form? No, since there isn't a y isolated on one side of the equation. A vertical line, remember, has no slope, nor does it have a y-intercept.

Check It Out

Give the slope and y-intercept of each line. Graph the line.

25. $y = -3$
26. $x = 4$
27. $y = 1$
28. $x = -2$

Writing the Equation of a Line

If you know the slope and the y-intercept of a line, you can write the equation of the line. If a line has a slope of 3 and a y-intercept of -2, substitute 3 for m and -2 for b into the slope-intercept form for the equation of a line. The equation of the line is $y = 3x - 2$.

WRITING THE EQUATION OF A LINE

Write the equation of the line.

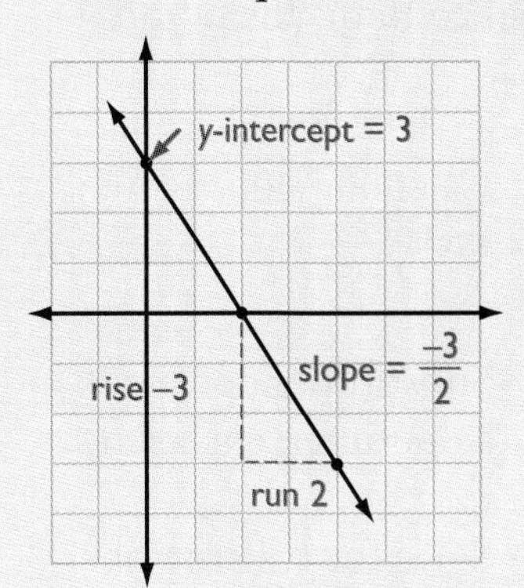

- Identify the y-intercept (b).
- Find the slope ($m = \frac{\text{rise}}{\text{run}}$).
- Substitute y-intercept and slope into the slope-intercept form. ($y = mx + b$)

$y = -\frac{3}{2}x + 3$

Check It Out

Write the equation of each line.

29. slope $= -2$, y-intercept at 4
30. slope $= \frac{2}{3}$, y-intercept at -2
31.

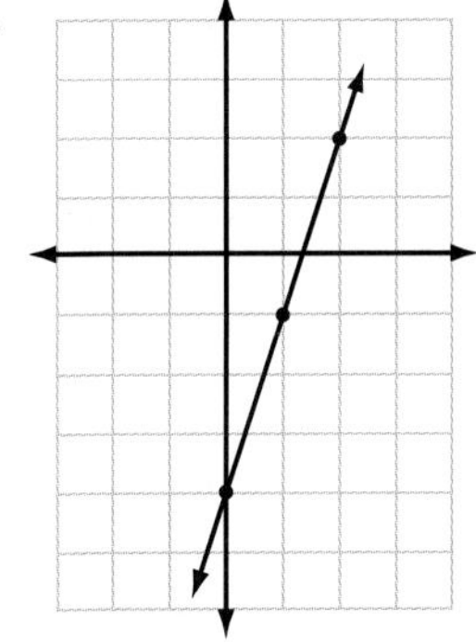

Writing the Equation of a Line from Two Points

If you know only two points that are on a line, you can still write its equation. First you have to find the slope. Then you can find the y-intercept.

WRITING THE EQUATION OF A LINE FROM TWO POINTS

Write the equation of the line that contains the points (6, −1) and (−2, 3).

$\frac{3-(-1)}{-2-6} = \frac{4}{-8} = \frac{1}{-2}$

Slope $= m = -\frac{1}{2}$

• Calculate slope using $m = \frac{\text{rise}}{\text{run}}$.

$y = -\frac{1}{2}x + b$

• Substitute slope for m in slope-intercept form ($y = mx + b$).

$-1 = -\frac{1}{2}(6) + b$ or $3 = \frac{-1}{2}(-2) + b$

• Solve for b. Remember that the 2 given points must be solutions of the equation. Substitute y-coordinates for y and x-coordinates for x.

$-1 = -\frac{6}{2} + b$ or $3 = \frac{2}{2} + b$

$-1 = -3 + b$ or $3 = 1 + b$

• Simplify.

$-1 + 3 = -3 + b + 3$ or $3 - 1 = 1 + b - 1$

• Add or subtract to isolate b.

$2 = b$ or $2 = b$

• Combine like terms.

$y = -\frac{1}{2}x + 2$

• Substitute the values you found for m and b into the slope-intercept form.

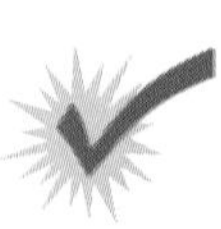

Check It Out

Write the equation of the line with the given points.

32. (1, −1) and (5, 3)
33. (−2, 9) and (3, −1)
34. (8, 3) and (−4, −6)
35. (−1, 2) and (4, 2)

6·8 EXERCISES

Determine the slope of each line.

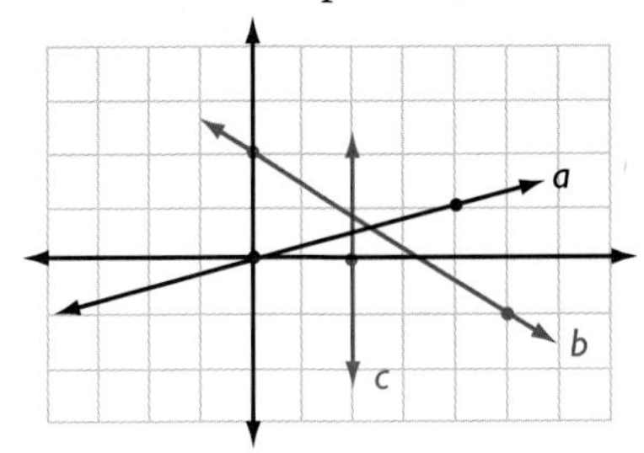

1. slope of a
2. slope of b
3. slope of c
4. contains $(-3, 1)$ and $(5, -3)$
5. contains $(0, -5)$ and $(2, 6)$

Graph each line.

6. Slope $= \frac{-1}{3}$, y-intercept is 2.
7. Slope $= 4$, y-intercept is -3.

Determine slope and y-intercept from the equation of each line.

8. $y = -3x - 2$
9. $y = \frac{-3}{4}x + 3$
10. $y = x + 2$
11. $y = 6$
12. $x = -2$

Write each equation in slope-intercept form. Graph the line.

13. $2x + y = 4$
14. $x - y = 1$

Write the equation of each line.

15. Slope $= 3$, y-intercept at -7
16. Slope $= \frac{-1}{3}$, y-intercept at 2
17. Line a above

Write the equation of the line that contains the given points.

18. $(-4, -5)$ and $(6, 0)$
19. $(-4, 2)$ and $(-3, 1)$
20. $(-6, 4)$ and $(3, -2)$

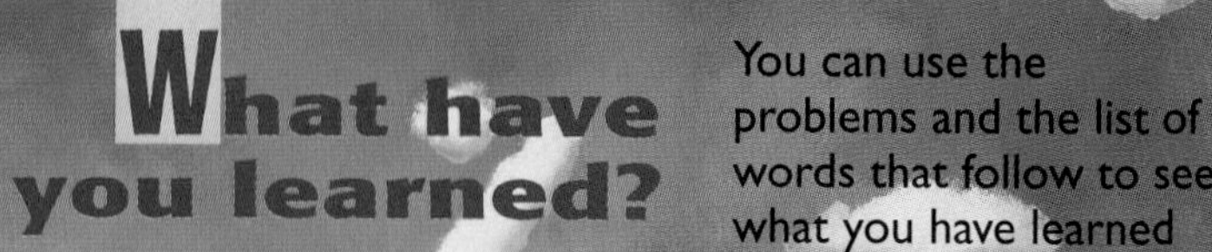

You can use the problems and the list of words that follow to see what you have learned in this chapter. You can find out more about a particular problem or word by referring to the boldfaced topic number (for example, **6•2**).

Problem Set

Write an equation for each sentence. **6•1**

1. If 7 is subtracted from the product of 3 and a number, the result is 5 more than the number.
2. 6 times the sum of a number and 2 is 8 less than twice the number.

Factor out the greatest common factor in each expression. **6•2**

3. $4x + 28$
4. $9n - 6$

Simplify each expression. **6•2**

5. $11a - b - 4a + 7b$
6. $8(2n - 1) - (2n + 5)$

7. Find the distance traveled by an in-line skater who skates at 12 mi/hr for $1\frac{1}{2}$ hr. Use the formula $d = rt$. **6•3**

Solve each equation. Check your solution. **6•4**

8. $x + 9 = 20$
9. $\frac{y}{3} = -8$
10. $6x - 7 = 29$
11. $\frac{y}{2} - 5 = 1$
12. $y - 10 = 7y + 8$
13. $10x - 2(2x - 3) = 3(x + 6)$

Use a proportion to solve items 14–15. **6•5**

14. In a class, the ratio of boys to girls is $\frac{3}{2}$. If there are 12 girls in the class, how many boys are there?
15. A map is drawn using a scale of 80 mi to 1 cm. On the map, the two cities are 7.5 cm apart. What is the actual distance between the two cities?

Solve each inequality. Graph the solution. **6•6**

16. $x + 9 \leq 6$
17. $4x + 10 > 2$

Draw each point on the coordinate plane and tell where it lies. **6•7**

18. $A(1, 5)$
19. $B(4, 0)$
20. $C(0, -2)$
21. $D(-2, 3)$

22. Find the slope of the line that contains the points $(2, -4)$ and $(-8, 2)$. **6•8**

Determine the slope and the y-intercept from the equation of each line. Graph the line. **6•8**

23. $y = \frac{-1}{2}x - 2$
24. $y = -4$
25. $x - 2y = 8$
26. $8x + 4y = 0$

Write the equation of the line with the given points. **6•8**

27. $(3, -2)$ and $(3, 5)$
28. $(5, 2)$ and $(0, 6)$
29. $(2, 2)$ and $(-2, -2)$
30. $(-2, 3)$ and $(2, 3)$

hot words

WRITE DEFINITIONS FOR THE FOLLOWING WORDS.

additive inverse **6•4**
associative property **6•2**
axes **6•7**
commutative property **6•2**
cross product **6•5**
difference **6•1**
distributive property **6•2**
equation **6•1**
equivalent **6•1**
equivalent expression **6•2**
expression **6•1**
formula **6•3**
horizontal **6•7**
inequality **6•6**
like terms **6•2**
order of operations **6•3**
ordered pair **6•7**
origin **6•7**
perimeter **6•3**
point **6•7**
product **6•1**
proportion **6•5**
quadrant **6•7**
quotient **6•1**
rate **6•5**
ratio **6•5**
rise **6•8**
run **6•8**
slope **6•8**
solution **6•4**
sum **6•1**
term **6•1**
variable **6•1**
vertical **6•7**
x-axis **6•7**
y-axis **6•7**
y-intercept **6•8**

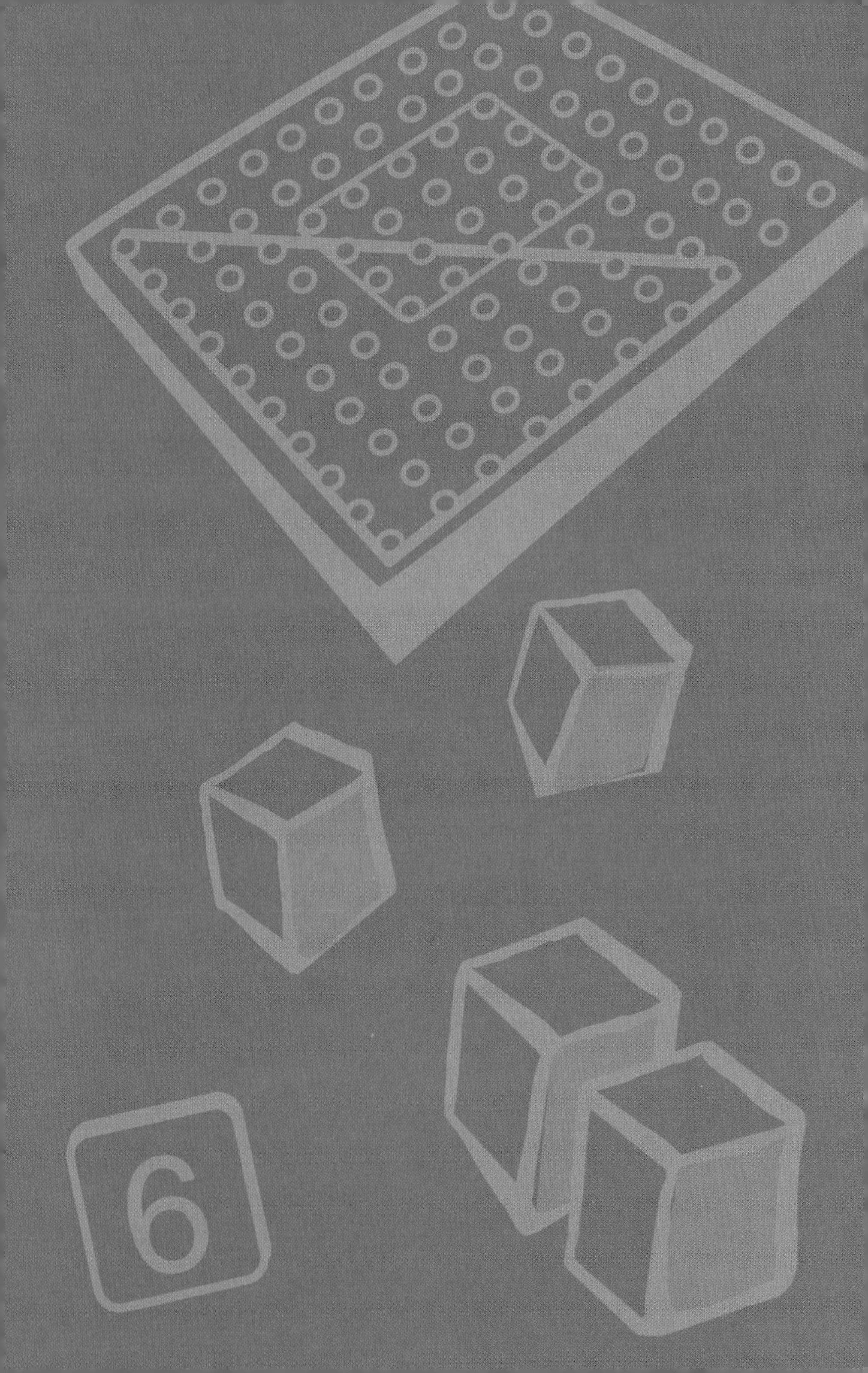
6

Geometry

7•1	Naming and Classifying Angles and Triangles	342
7•2	Naming and Classifying Polygons and Polyhedrons	350
7•3	Symmetry and Transformations	360
7•4	Perimeter	366
7•5	Area	372
7•6	Surface Area	378
7•7	Volume	382
7•8	Circles	388
7•9	Pythagorean Theorem	394
7•10	Tangent Ratio	398

What do you already know?

You can use the problems and the list of words that follow to see what you already know about this chapter. The answers to the problems are in Hot Solutions at the back of the book, and the definitions of the words are in HotWords at the front of the book. You can find out more about a particular problem or word by referring to the boldfaced topic number (for example, **7•2**).

Problem Set

Refer to this figure for items 1–3. **7•1**

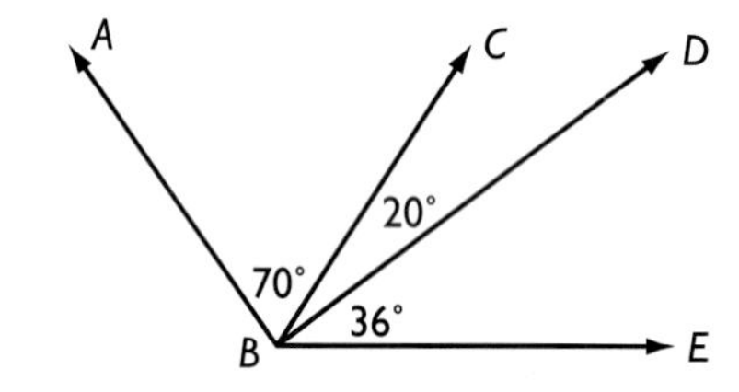

1. Name the angle adjacent to $\angle EBC$.
2. Name the right angle in this figure.
3. Name the angle that is 20°.
4. Two angles of a triangle measure 17.5° and 110.5°. What is the measure of the third angle? **7•1**
5. Find the measure of each angle of a regular pentagon. **7•2**
6. What kind of figure is both a rectangle and a rhombus? **7•2**
7. A right triangle has legs of 5 cm and 12 cm. What is the perimeter of the triangle? **7•4**
8. What is the perimeter of a regular octagon with 6 in. sides? **7•4**
9. Find the area of a right triangle with legs 12 m and 8 m. **7•5**
10. A rectangle measures 8 in. × 12 in. What is its area? **7•5**
11. A trapezoid has bases of 10 ft and 16 ft. The distance between these bases is 5 ft. What is the area of the trapezoid? **7•5**
12. Each face of a triangular prism is a square, 10 cm on each side. If the area of each base is 43.3 cm^2, what is the surface area of the prism? **7•6**
13. Find the surface area of a cylinder with height 9 ft and circumference 8 ft. Round to the nearest square foot. **7•6**

14. Find the volume of a cylinder that has a 5 in. diameter and is 6 in. high. Round to the nearest cubic inch. **7•7**
15. A pyramid, a rectangular prism, and a cylinder all have the same base area and the same height. Which two figures have the same volume? **7•7**
16. What is the circumference and area of a circle with radius 25 ft? Round to the nearest foot or square foot. **7•8**
17. A triangle has sides of 11 ft, 12 ft, and 16 ft. Use the Pythagorean Theorem to find if this is a right triangle. **7•9**
18. Find the length of the unknown leg of a right triangle that has hypotenuse 21 in. and one leg that is 16 in. **7•9**
19. In right triangle *ABC*, $\overline{AC}$ is the hypotenuse and $\angle B$ is the right angle. Which side is opposite $\angle A$, which side is adjacent to $\angle A$, and what is the tangent ratio for $\angle A$? **7•10**
20. In the right triangle of problem 19, if *AB* is 12 m and *BC* is 5.6 m, what is the measure of $\angle A$? **7•10**

CHAPTER 7

hot words

acute angle **7•10**
angle **7•1**
arc **7•8**
base **7•5**
circle **7•6**
circumference **7•6**
congruent **7•1**
cube **7•2**
cylinder **7•6**
degree **7•1**
diagonal **7•2**
diameter **7•8**
face **7•2**
hexagon **7•2**
hypotenuse **7•9**
isosceles triangle **7•4**
legs of a triangle **7•5**
line **7•1**
opposite angle **7•2**
parallel **7•2**
parallelogram **7•2**
pentagon **7•2**
perimeter **7•4**
perpendicular **7•5**
pi **7•7**
point **7•1**
polygon **7•1**
polyhedron **7•2**
prism **7•2**
pyramid **7•2**
Pythagorean Theorem **7•4**
Pythagorean triple **7•9**
quadrilateral **7•2**
radius **7•8**
ray **7•1**
rectangular prism **7•2**
reflection **7•3**
regular shape **7•2**
rhombus **7•2**
right angle **7•1**
right triangle **7•4**
rotation **7•3**
segment **7•8**
surface area **7•6**
symmetry **7•3**
tangent **7•10**
tetrahedron **7•2**
transformation **7•3**
translation **7•3**
trapezoid **7•2**
triangular prism **7•6**
vertex **7•1**
volume **7•7**

7·1 Naming and Classifying Angles and Triangles

Points, Lines, and Rays

In the world of math, it is sometimes necessary to refer to a specific **point** in space. Simply draw a small dot with a pencil tip to represent a point. A point has no size; its only function is to show position.

Every point needs a name, so we name a point by using a single capital letter.

· *A*

Point *A*

If you draw two points on a sheet of paper, a **line** can be used to connect them. Imagine this line as being perfectly straight and continuing without end in opposite directions. It has no thickness.

Lines need names just like points do, so that we can refer to them easily. To name a line, pick any two points on the line.

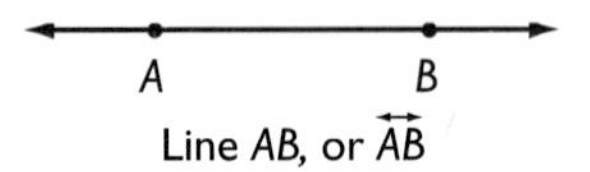

Line *AB*, or $\overleftrightarrow{AB}$

Since the length of any line is infinite, we sometimes use parts of a line. A **ray** is part of a line that extends without end in one direction. In $\overrightarrow{AB}$, which is read as "ray *AB*," *A* is the endpoint. The second point that is used to name the ray can be any point other than the endpoint. You could also name this ray *AC*.

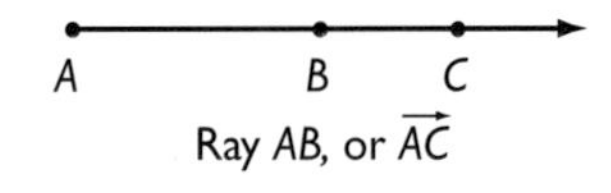

Ray *AB*, or $\overrightarrow{AC}$

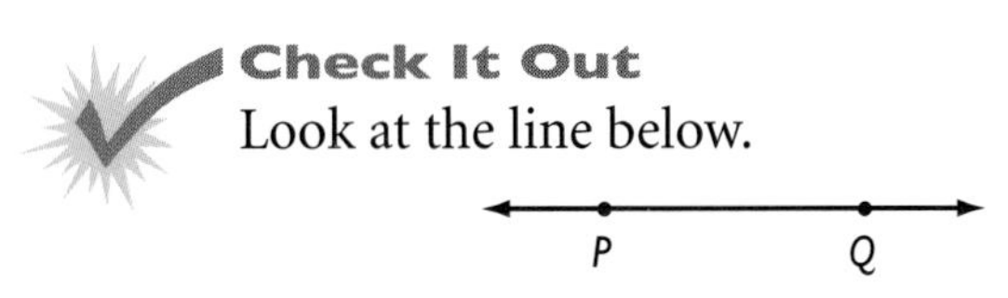

Check It Out

Look at the line below.

1. Name the line in two different ways.
2. What is the endpoint of $\overrightarrow{PQ}$?

Naming Angles

Imagine two different rays with the same endpoint. Together they form what is called an **angle.** The point they have in common is called the **vertex** of the angle. The rays form the **sides** of the angle.

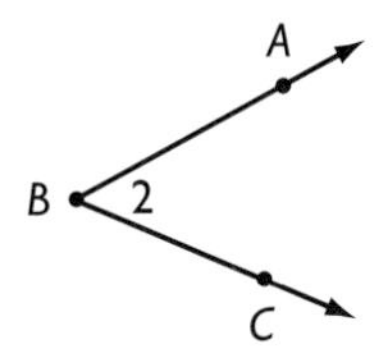

The angle above is made up of $\overrightarrow{BA}$ and $\overrightarrow{BC}$. B is the common endpoint of the two rays. Point B is the vertex of the angle. Instead of writing the word *angle*, you can use the symbol for an angle, which is $\angle$.

There are several ways to name an angle. You can name it using the three letters of the points that make up the two rays with the vertex as the middle letter ($\angle ABC$, or $\angle CBA$). You can also use just the letter of the vertex to name the angle ($\angle B$). Sometimes you might want to name an angle with a number ($\angle 2$).

When more than one angle is formed at a vertex, you use three letters to name each of the angles. Since S is the vertex of three different angles, each angle needs three letters to name it: $\angle PSR$; $\angle PSQ$; $\angle RSQ$.

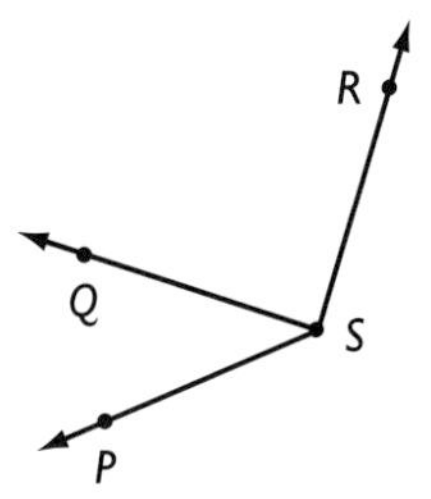

Check It Out

Look at the angles formed by the rays below.

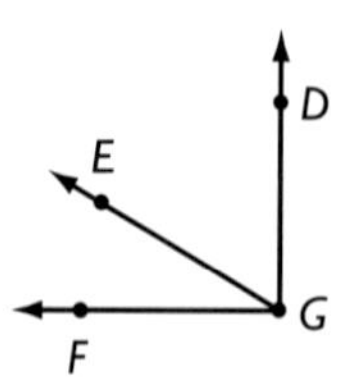

3. Name the vertex.
4. Name all the angles.

Measuring Angles

You measure an angle in **degrees,** using a *protractor* (p. 445). The number of degrees in an angle will be greater than 0 and less than or equal to 180.

MEASURING WITH A PROTRACTOR

Measure $\angle XYZ$.

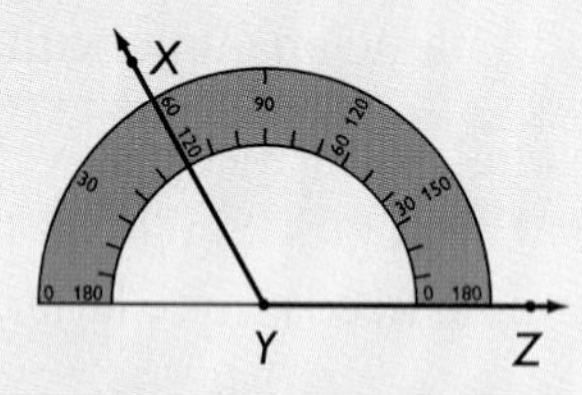

- Place the center point of the protractor on the vertex of the angle. Align the 0° line on the protractor with one side of the angle.
- Read the number of degrees on the scale where it intersects the second side of the angle.

$m\angle XYZ = 120°$

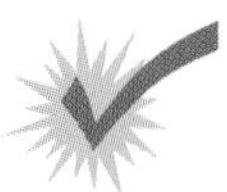

Check It Out

Measure the angles.

5. ∠*GHI*
6. ∠*IHJ*
7. ∠*GHJ*

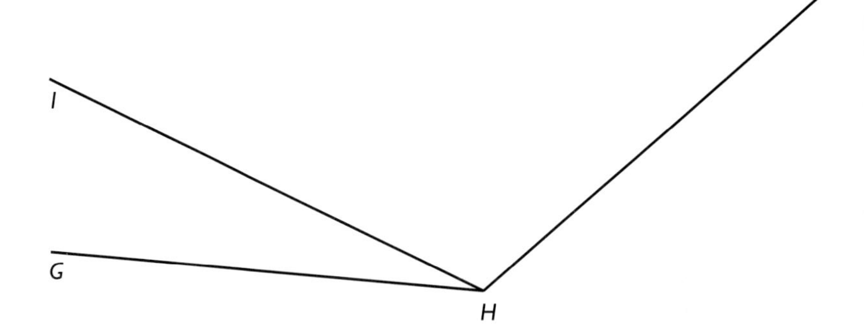

Classifying Angles

You can classify angles by their measures.

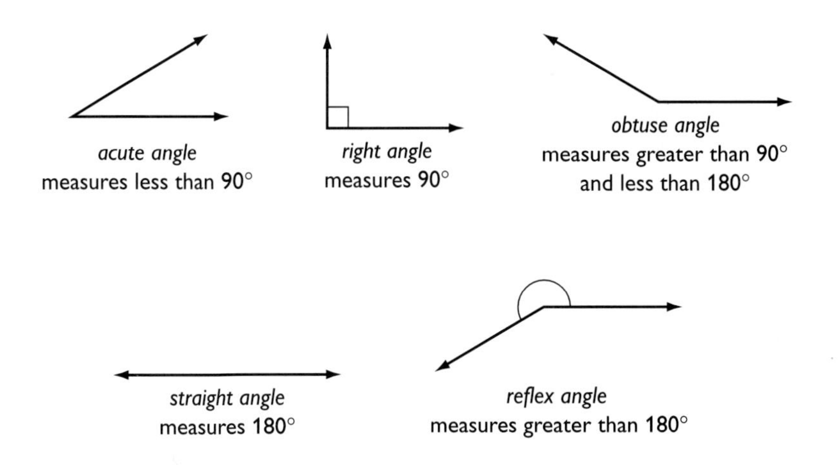

Angles that share a side are called *adjacent angles.* You can add measures if the angles are adjacent.

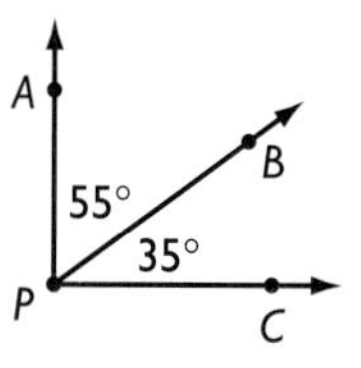

$m\angle APB = 55°$
$m\angle BPC = 35°$
$m\angle APC = 55° + 35° = 90°$
Since the sum is 90°, you know that $\angle APC$ is a right angle.

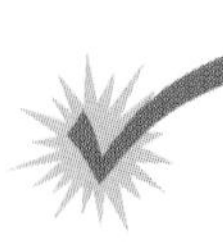

Check It Out

Use a protractor to measure and classify each angle.

8. $\angle DBC$
9. $\angle ABC$
10. $\angle ABD$

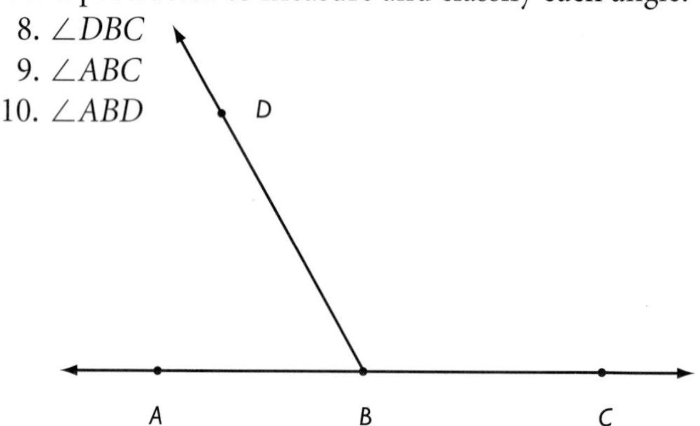

Triangles

Triangles are **polygons** that have three sides, three vertices, and three angles.

You name a triangle using the three vertices, in any order. $\triangle ABC$ is read "triangle *ABC*."

Classifying Triangles

Like angles, triangles are classified by their angle measures. They are also classified by the number of **congruent** sides, which are sides with equal length.

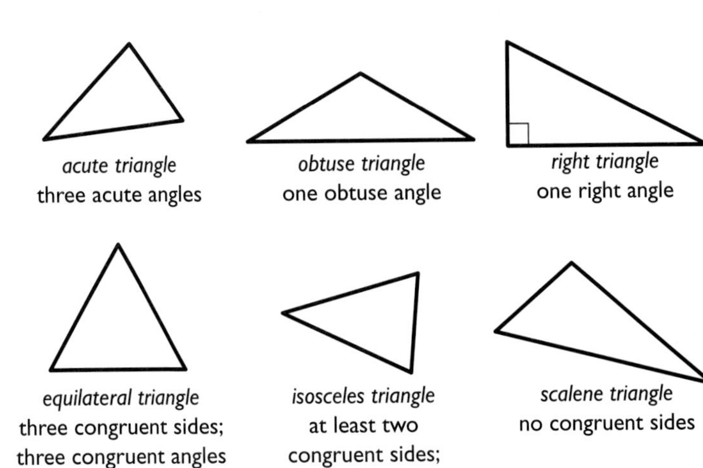

The **sum** of the measures of the three angles in a triangle is always 180°.

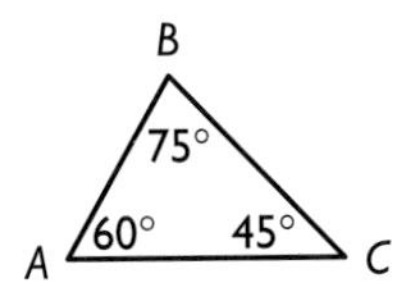

In $\triangle ABC$, $m\angle A = 60°$, $m\angle B = 75°$, and $m\angle C = 45°$.

$$60° + 75° + 45° = 180°$$

So the sum of the angles of $\triangle ABC$ is 180°.

FINDING THE MEASURE OF THE UNKNOWN ANGLE IN A TRIANGLE

$\angle P$ is a right angle, so its measure is 90°. The measure of $\angle Q$ is 40°. Find the measure of $\angle R$.

$90° + 40° = 130°$

$180° - 130° = 50°$

$\angle R = 50°$

- Add the two known angles.
- Subtract the sum from 180°.
- The difference is the measure of the third angle.

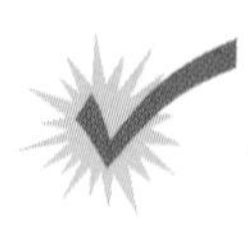

Check It Out

Find the measures of the third angle of each triangle.

11.

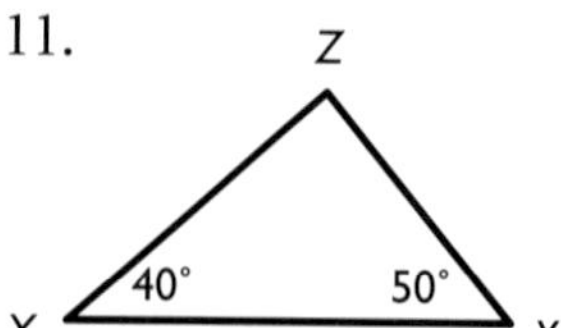

12.

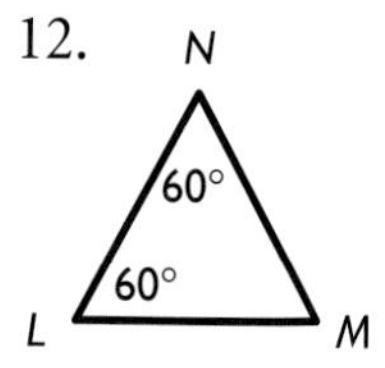

The Triangle Inequality

The length of the third side of a triangle is always less than the sum of the other two sides and greater than their difference. So $(a + b) > c > (a - b)$.

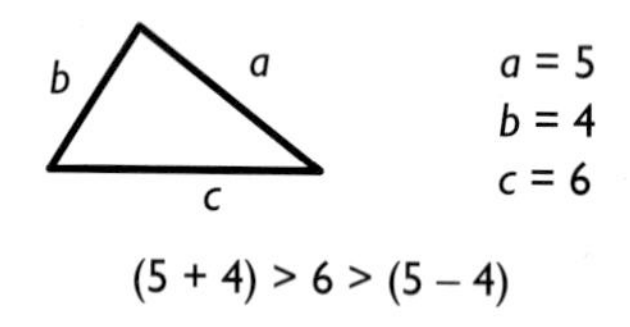

$(5 + 4) > 6 > (5 - 4)$

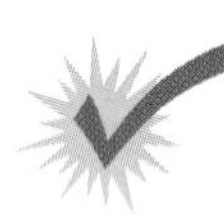

Check It Out

Which of the following cannot be the lengths of the sides of a triangle?

13. A. 8m, 1m, 8m
    B. 15 cm, 4 cm, 12 cm
    C. 5 in., 5 in., 9 in.
    D. 12 ft, 3 ft, 8 ft
14. A. 2m, 3m, 3m
    B. 4 cm, 6 cm, 2 cm
    C. 7 in., 7 in., 7 in.
    D. 1 ft, 6 ft, 9 ft

7·1 EXERCISES

Use the figures to answer the questions.

1. Name the line in two different ways.
2. What is the endpoint of $\overrightarrow{ST}$?

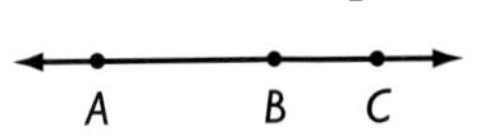

3. Name the line in six different ways.
4. Give two names for the ray that begins at point A and goes to the right.
5. Name the vertex.
6. Name an acute angle.
7. Name two obtuse angles.
8. What is the measure of $\angle UVW$?

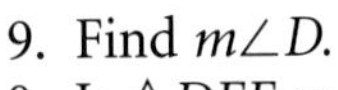

9. Find $m\angle D$.
10. Is $\triangle DEF$ an acute, right, or obtuse triangle?

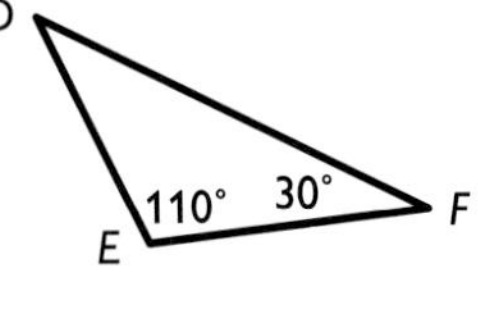

11. Find $m\angle T$.
12. Is $\triangle RST$ an acute, right, or obtuse triangle?

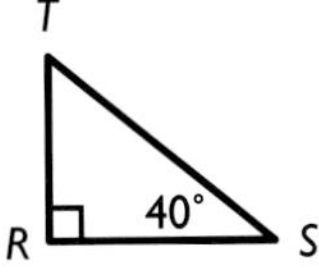

13. Can a triangle have sides that are 5 cm, 8 cm, and 2 cm in length? Why?
14. If two sides of a triangle are 4 cm and 7 cm, then the third side is greater than but less than what distances?
15. Use the map to answer the questions.
    A. What line represents Main Street?
    B. How would you represent Grove Street?
    C. Two of the streets intersect at a right angle. Name the two streets.

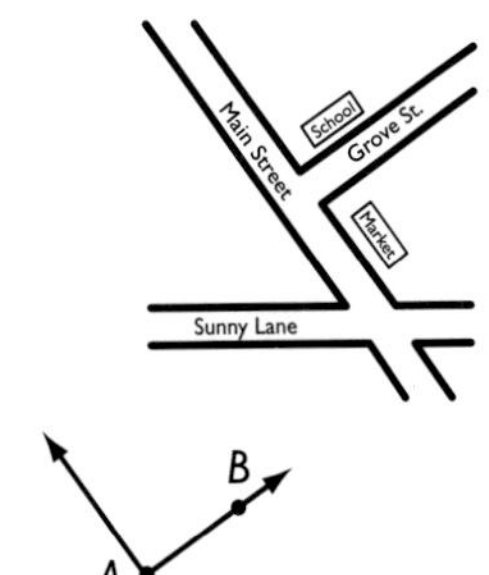

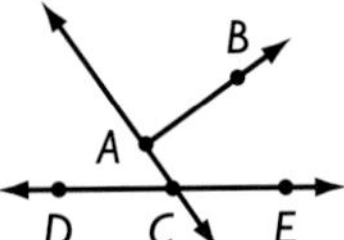

7·2 Naming and Classifying Polygons and Polyhedrons

Quadrilaterals

You may have noticed that there is a wide variety of four-sided figures, or **quadrilaterals,** to work with in geometry. All quadrilaterals have four sides and four angles. The sum of the angles of a quadrilateral is 360°. There are also many different types of quadrilaterals, which are classified by their sides and angles.

To name a quadrilateral, list the four vertices, either clockwise or counterclockwise.

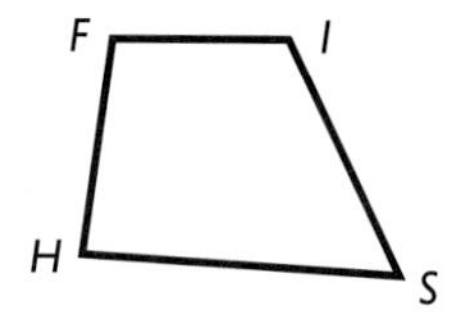

Angles of a Quadrilateral

The sum of the angles of a quadrilateral is 360°.
If you know the measures of three angles in a quadrilateral, you can find the measure of the fourth angle.

FINDING THE MEASURE OF THE UNKNOWN ANGLE IN A QUADRILATERAL

Find $m\angle A$ in quadrilateral *ABCD*.

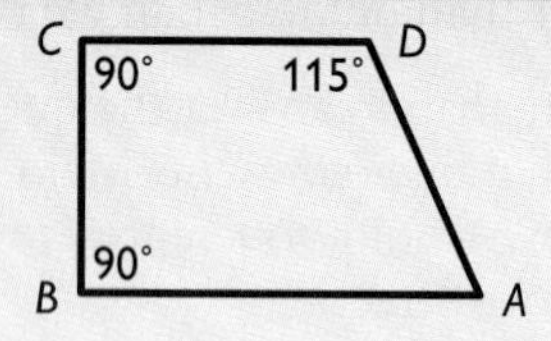

- Add the measures of the three known angles. $90° + 90° + 115° = 295°$
- Subtract the sum from 360°. $360° - 295° = 65°$
- The **difference** is the measure of the fourth angle. $m\angle A = 65°$

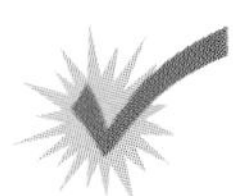

Check It Out

1. Name the quadrilateral in at least two ways.
2. What is the sum of the angles in this quadrilateral?
3. Find $m\angle P$.

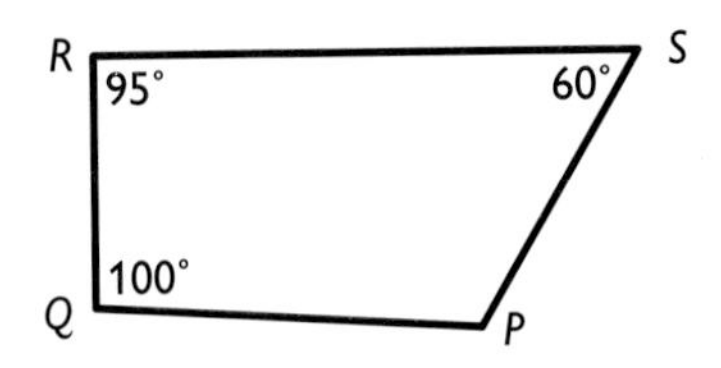

Types of Quadrilaterals

A **rectangle** is a quadrilateral with four right angles. *WXYZ* is a rectangle. Its **length** is 5 cm and its **width** is 3 cm.

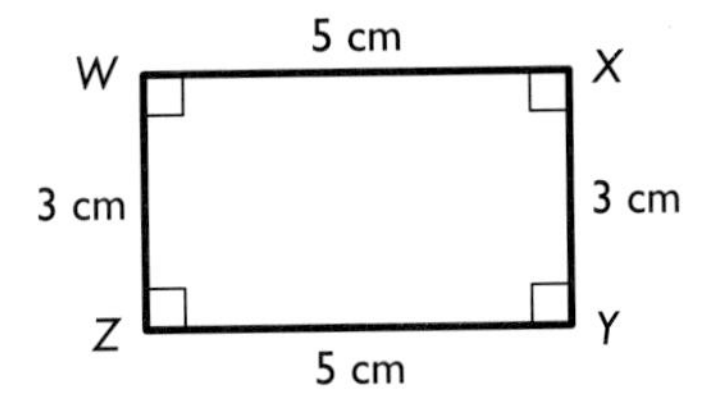

Opposite sides of a rectangle are equal in length. If all four sides of the rectangle are equal, the rectangle is called a **square.** A square is a **regular shape** because all of its sides are of equal length and all of the interior angles are congruent. Some rectangles may be squares, but *all* squares are rectangles.

A **parallelogram** is a quadrilateral with opposite sides that are **parallel.** In a parallelogram, opposite sides are equal, and **opposite angles** are equal. *ABCD* is a parallelogram.

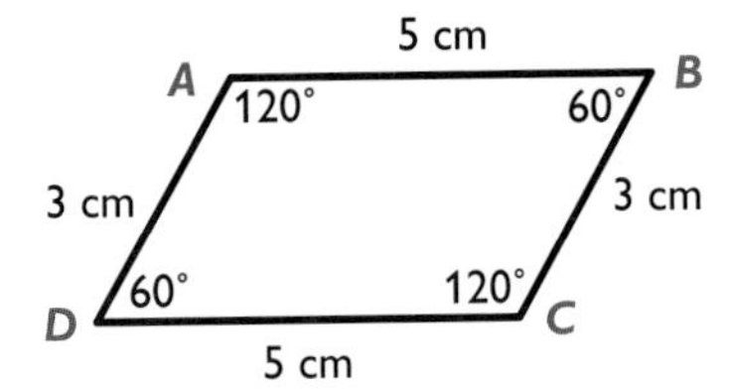

Some parallelograms may be rectangles, but all rectangles are parallelograms. Therefore squares are also parallelograms. If all four sides of a parallelogram are the same length, the parallelogram is called a **rhombus.** *HIJK* is a rhombus.

Every square is a rhombus, although not every rhombus is a square, because a square also has equal angles.

In a **trapezoid,** two sides are parallel and two are not. A trapezoid is a quadrilateral, but it is not a parallelogram. *PARK* is a trapezoid.

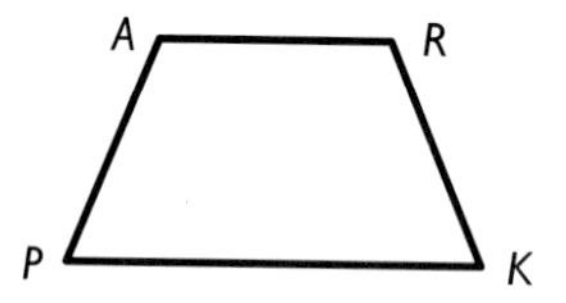

Check It Out

4. Is quadrilateral *RSTU* a rectangle? a parallelogram? a square? a rhombus? a trapezoid?

5. Is a square a rhombus? Why or why not?

Polygons

A polygon is a closed figure that has three or more sides. Each side is a **line segment,** and the sides meet only at the endpoints, or vertices.

This figure is a polygon. These figures are not.

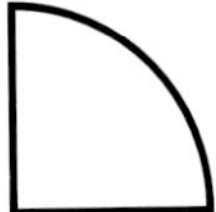

A rectangle, a square, a parallelogram, a rhombus, a trapezoid, and a triangle are all polygons.

There are some aspects of polygons that are always true. For example, a polygon of n sides has n angles and n vertices. A polygon with three sides has three angles and three vertices. A polygon with eight sides has eight angles and eight vertices, and so on.

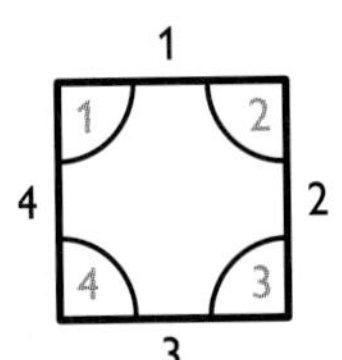

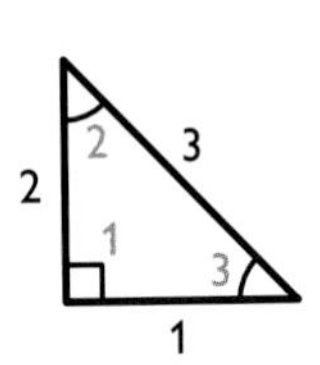

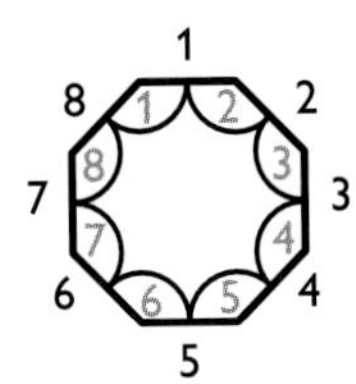

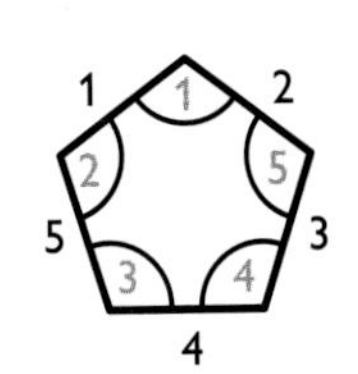

A line segment connecting two vertices of a polygon is either a side or a **diagonal.** $\overline{AE}$ is a side of polygon *ABCDE*. $\overline{AD}$ is a diagonal.

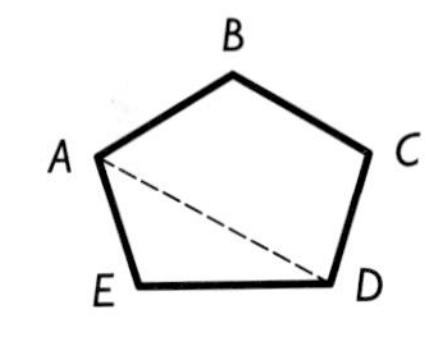

Types of Polygons

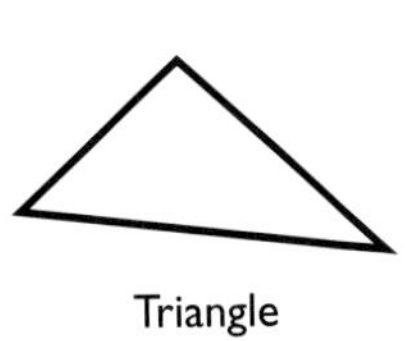

Triangle
3-sides

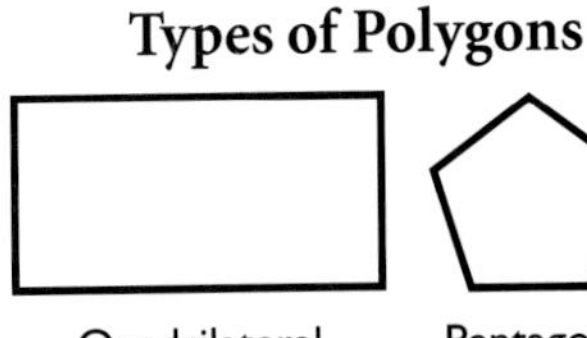

Quadrilateral
4-sides

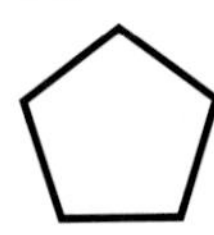

Pentagon
5-sides

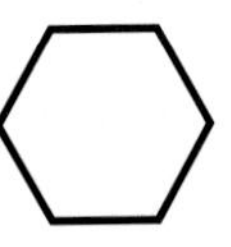

Hexagon
6-sides

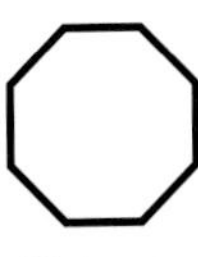

Octagon
8-sides

A seven-sided polygon is called a **heptagon,** a nine-sided polygon is called a **nonagon,** and a ten-sided polygon is called a *decagon*.

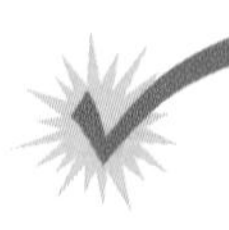

Check It Out

State whether or not the figure is a polygon. If it is a polygon, classify it according to the number of sides it has.

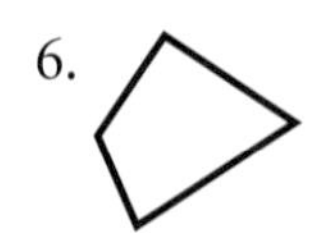

6. 7. 8.

Oh, Obelisk!

Ancient Egyptian obelisks were carved horizontally out of the rock quarry. Exactly how the Egyptians lifted the obelisks into vertical position is a mystery. But clues suggest that the Egyptians slid them down a dirt ramp, inched them higher with levers, and finally pulled them upright with ropes.

A crew from a television station attempted to move a 43-ft-long block of granite using this method. They tilted the 40-ton obelisk down a ramp at a 33° angle. With levers, they inched the obelisk up to about a 40° angle. Then 200 people tried to haul it with ropes to a standing position. They couldn't budge it. Finally, out of time and money, they abandoned the attempt.

How many additional degrees did the crew need to raise the obelisk before it would have stood upright? See Hot Solutions for answer.

Angles of a Polygon

You know that the sum of the angles of a triangle is 180° and that the sum of the angles of a quadrilateral is 360°. The sum of the angles of *any* polygon totals at least 180° (triangle). Each additional side adds 180° to the measure of the first three angles. To see why, look at a **pentagon.**

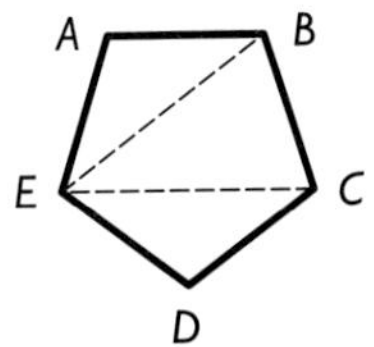

Drawing diagonals $\overline{EB}$ and $\overline{EC}$ shows that the sum of the angles of a pentagon is the sum of the angles in three triangles.

$$3 \times 180° = 540°$$

So the sum of the angles of a pentagon is 540°.

You can use the formula $(n - 2) \times 180°$ to find the sum of the angles of a polygon. Just let n equal the number of sides of a polygon. The answer you get is the sum of the measures of all the angles of the polygon.

FINDING THE SUM OF THE ANGLES OF A POLYGON

$(n - 2) \times 180°$ = sum of polygon with n sides

Find the sum of the angles of an octagon.

Think: An octagon has 8 sides. Subtract 2. Then multiply the difference by 180.

- Use the formula: $(8 - 2) \times 180° = 6 \times 180° = 1{,}080°$

So the sum of the angles of an octagon is 1,080°.

As you know, a **regular polygon** has equal sides and equal angles. You can use what you know about finding the sum of the angles of a polygon to find the measure of each angle of a regular polygon.

Find the measure of each angle in a regular **hexagon.**

Begin by using the formula $(n - 2) \times 180°$. A hexagon has 6 sides, and so you should substitute 6 for n.

$$(6 - 2) \times 180° = 4 \times 180° = 720°$$

Then divide the sum of the angles by the number of angles. Since a hexagon has 6 angles, divide by 6.

$$720° \div 6 = 120°$$

The answer tells you that each angle of a regular hexagon measures 120°.

Check It Out

9. Find the sum of the angles of a decagon.
10. Find the measure of each angle in a regular pentagon.

Polyhedrons

Solid shapes can be curved, like these.

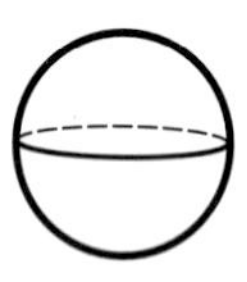

Sphere

Cylinder

Cone

Some solid shapes have flat surfaces. Each of the figures below is a **polyhedron.**

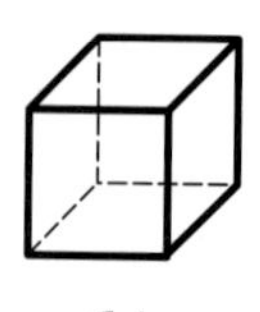

Cube

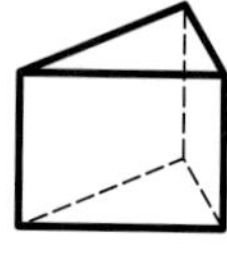

Prism

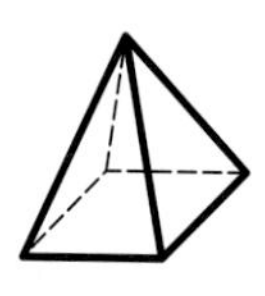

Pyramid

A polyhedron is any solid whose surface is made up of polygons. Triangles, quadrilaterals, and pentagons make up the **faces** of the common polyhedrons below.

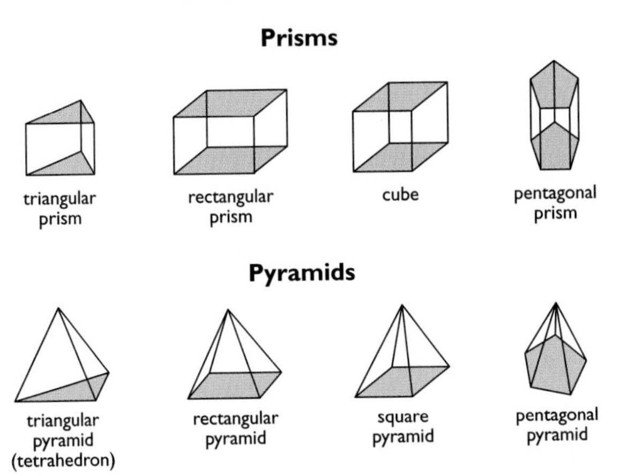

A **prism** has two bases, or "end" faces. The bases of a prism are the same size and shape and are parallel to each other. Its other faces are parallelograms. The bases of the prisms are shaded in the chart. When all six faces of a **rectangular prism** are square, the figure is a **cube.**

A **pyramid** is a structure that has one base in the shape of a polygon. It has triangular faces that meet a point called the *apex.* The base of each pyramid is shaded in the chart.
A triangular pyramid is a **tetrahedron.** A tetrahedron has four faces. Each face is triangular. A triangular prism, however, is *not* a tetrahedron.

Check It Out

Identify each polyhedron.

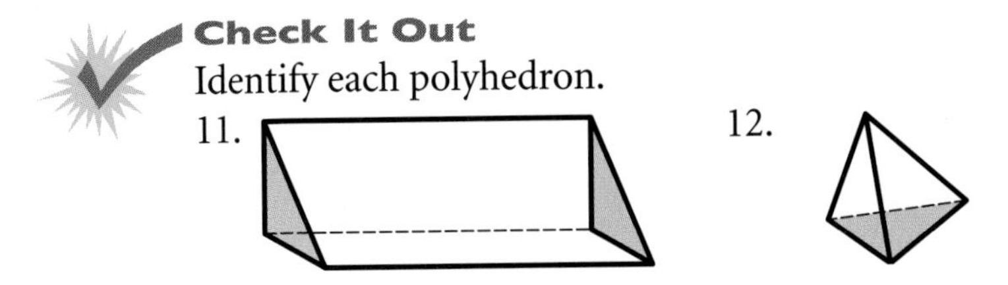

7•2 EXERCISES

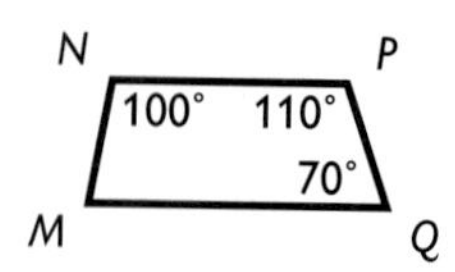

1. Give two other names for quadrilateral *MNPQ*.
2. Find $m\angle M$.

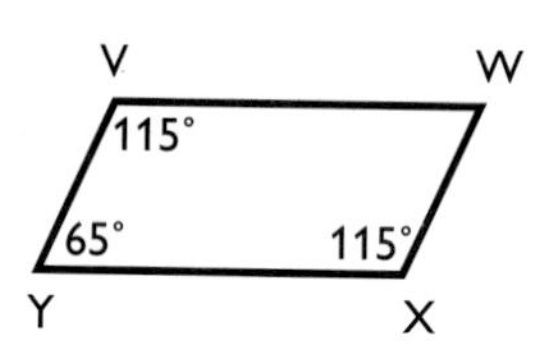

3. Give two other names for quadrilateral *RSTU*.
4. Find $m\angle U$.
5. Give two other names for quadrilateral *VWXY*.
6. Find $m\angle W$.

Tell whether each statement below is true or false.

7. A square is a parallelogram.
8. Every rectangle is a parallelogram.
9. Not all rectangles are squares.
10. Some trapezoids are parallelograms.
11. Every square is a rhombus.
12. All rhombuses are quadrilaterals.
13. A quadrilateral cannot be both a rectangle and a rhombus.

Identify each polygon.

14. 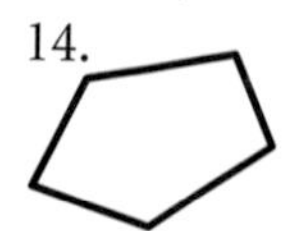

15. 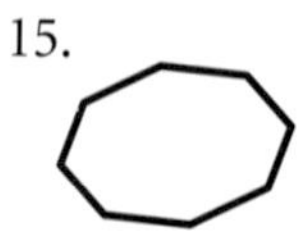

16. 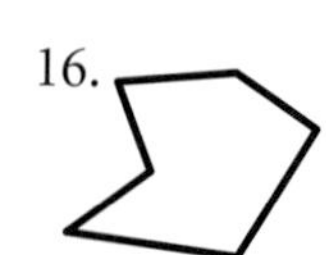

17. 18.

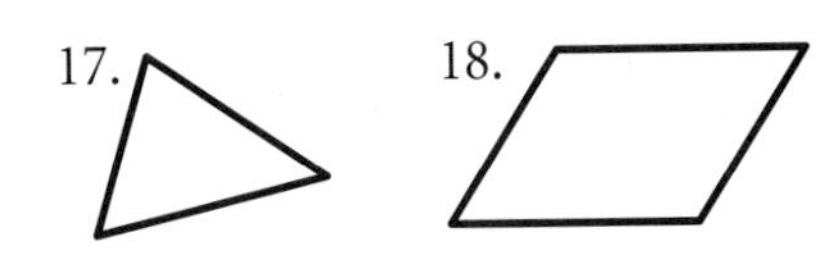

Find the sum of the angles for each type of polygon.

19. pentagon
20. nonagon
21. heptagon
22. What is the measure of each angle in a regular octagon?

Identify each polyhedron.

23.

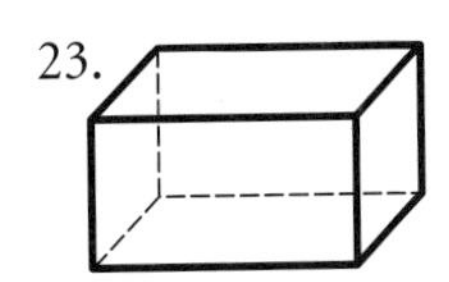

24.

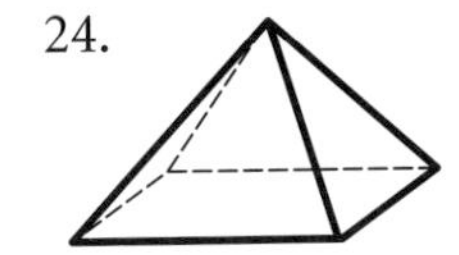

25.

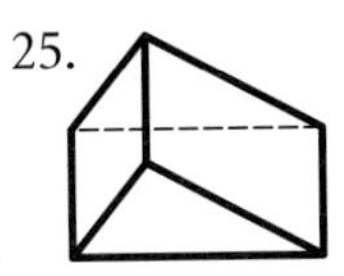

Identify each real-world polygon or polyhedron.

26. The infield of a baseball diamond
27. Home plate on a baseball diamond

28. 29. 30.

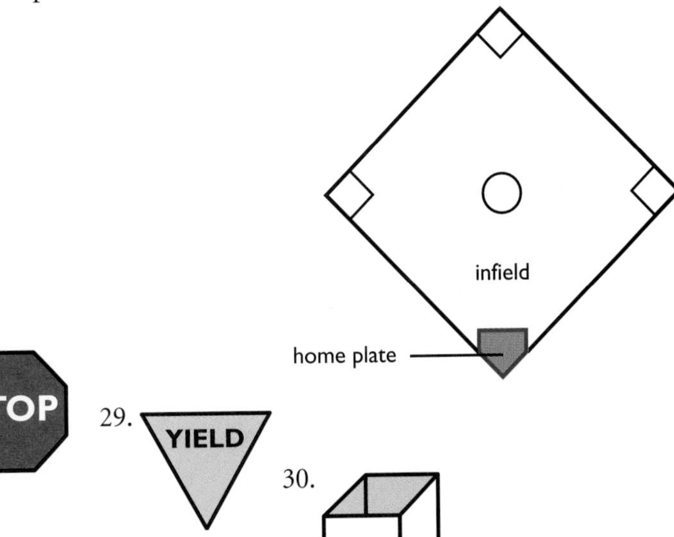

7•3 Symmetry and Transformations

Whenever you move a shape that is in a plane, you are performing a **transformation.**

Reflections

A **reflection** (or **flip**) is one kind of transformation. When you hear the word "reflection," you may think of a mirror. The mirror image, or reverse image, of a point or shape is called a *reflection.*

The reflection of a point is another point on the other side of a line of **symmetry.** Both the point and its reflection are the same distance from the line.

P' reflects point P on the other side of line l. P' is read "P-prime." P' is called the *image* of P.

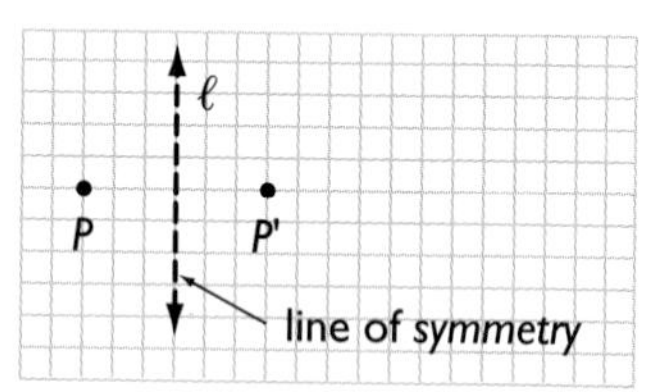

Any point, line, or polygon can be reflected. Quadrilateral $DEFG$ is reflected on the other side of line m. The image of $DEFG$ is $D'E'F'G'$.

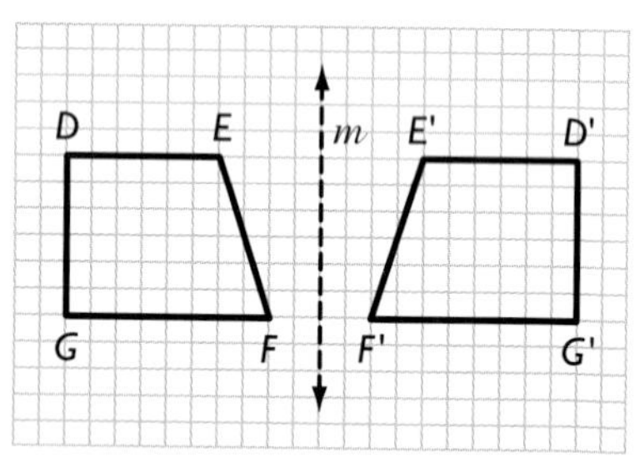

To find an image of a shape, pick several key points in the shape. For a polygon, use the vertices. For each point, measure the distance to the line of symmetry. The image of each point will be the same distance from the line of symmetry on the opposite side.

In the quadrilateral reflection on the opposite page, point D is 10 units from the line of symmetry, and point D' is also 10 units from the line on the opposite side. You can measure the distance from the line for each point, and the corresponding image point will be the same distance.

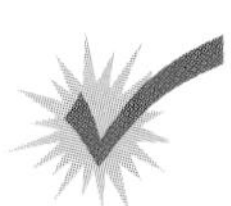

Check It Out

1. Copy the shape on grid paper. Then find the reflection and draw and label the images.

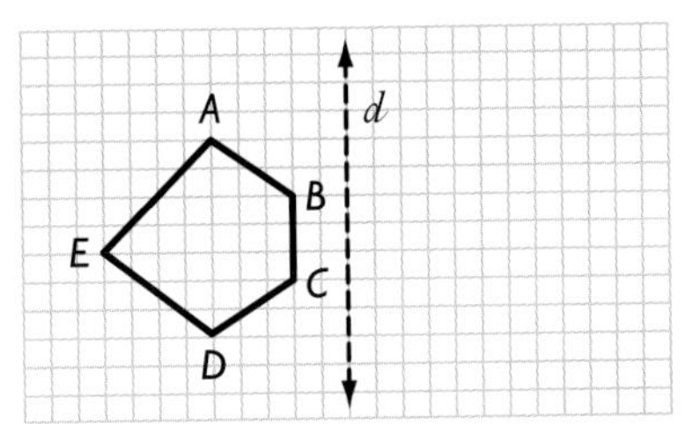

Fish Farming

Fish don't grow on trees.

Fish farming, one of the fastest growing food industries, now supplies about 20 percent of the world's fish and shellfish.

A Japanese oyster farmer builds big floating bamboo rafts in the ocean, then hangs clean shells from them by ropes. Oyster larva attach to the shells and grow in thick masses. The rafts are supported with barrels so they don't sink to the bottom where the oysters' natural predators, starfish, can get them.

An oyster farmer might have 100 rafts, each about 10 m by 15 m. What is the total area of the rafts? **See Hot Solutions for answer.**

Reflection Symmetry

You have seen that a line of symmetry is used to show the reflection symmetry of a point, a line, or a shape. A line of symmetry can also *separate* a shape into two parts, where one part is a reflection of the other. Each of these figures is symmetrical with respect to the line of symmetry.

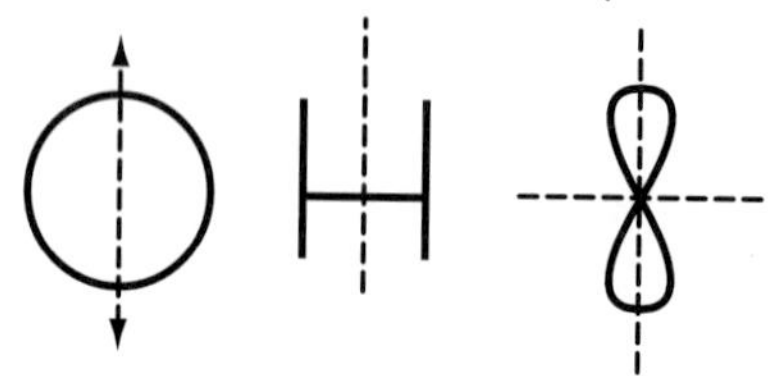

Sometimes a figure has more than one line of symmetry. Here are more shapes that have more than one line of symmetry.

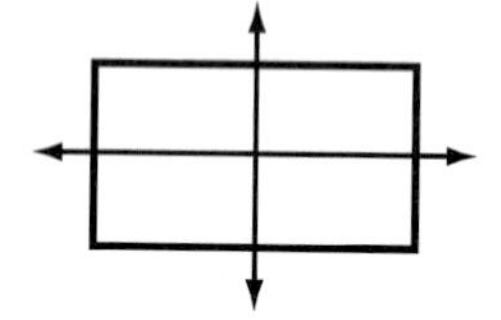

A rectangle has two lines of symmetry.

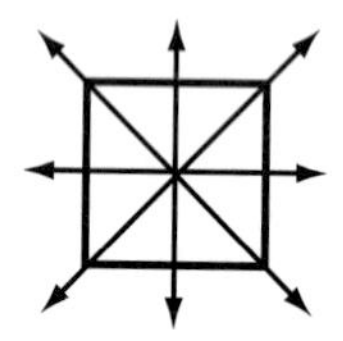

A square has four lines of symmetry.

Any line through the center of a circle is a line of symmetry. So a circle has an infinite number of lines of symmetry.

Check It Out

Tell whether each figure has reflection symmetry. If your answer is yes, tell how many lines of symmetry can be drawn through the figure.

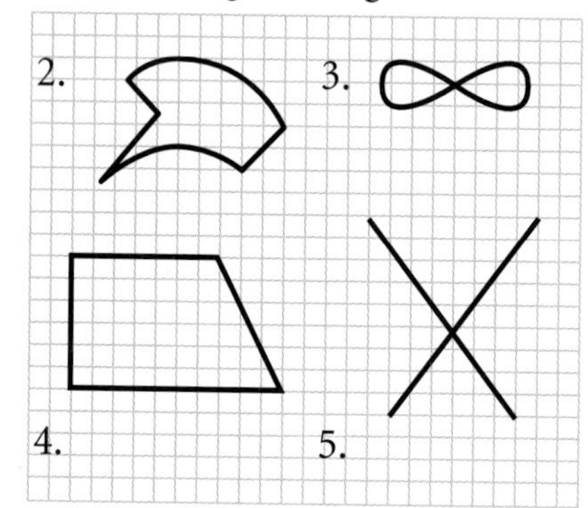

Rotations

A **rotation** (or **turn**) is a transformation that turns a line or a shape around a fixed point. This point is called the *center of rotation.* Rotations are usually measured in the counterclockwise direction.

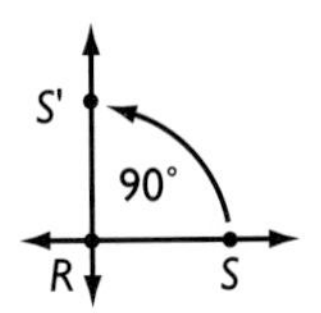

$\overleftrightarrow{RS}$ is rotated 90° around point R.

If you rotate a figure 360°, it comes back to where it started. Despite the rotation, its position is unchanged.

If you rotate $\overrightarrow{AB}$ 360° around point P, $\overrightarrow{AB}$ is still in the same place.

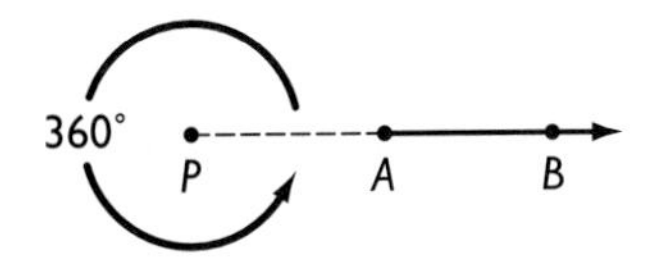

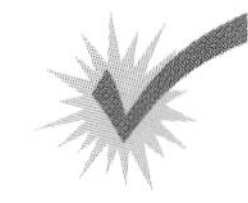

Check It Out

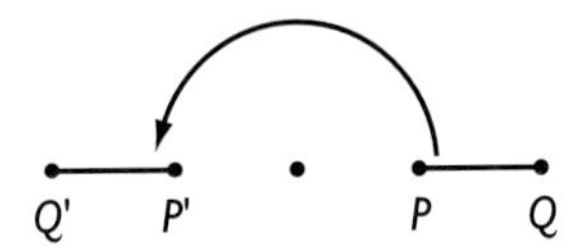

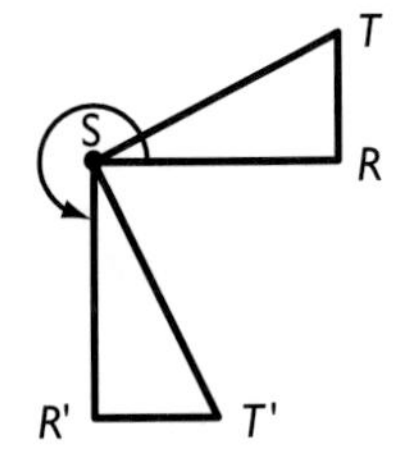

6. How many degrees has $\overrightarrow{PQ}$ been rotated?
7. How many degrees has $\triangle TSR$ been rotated?

Translations

A **translation** (or **slide**) is another kind of transformation. When you slide a figure to a new position without turning it, you are performing a translation.

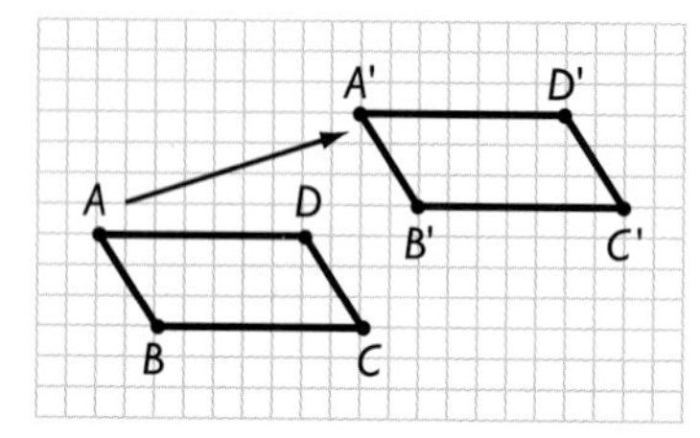

Rectangle *ABCD* moves right and up. *A'B'C'D'* is the image of *ABCD* under a translation. *A'* is 9 units to the right and 4 units up from *A*. All other points on the rectangle moved the same way.

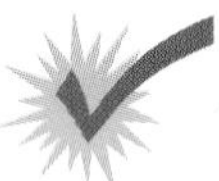

Check It Out

Write whether the figures below represent a translation.

8. 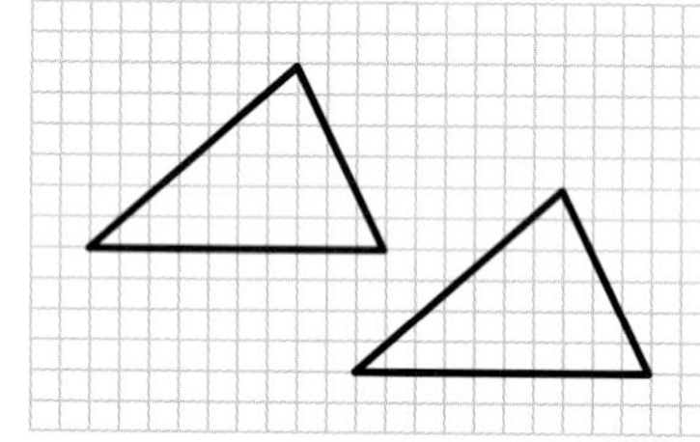

9. 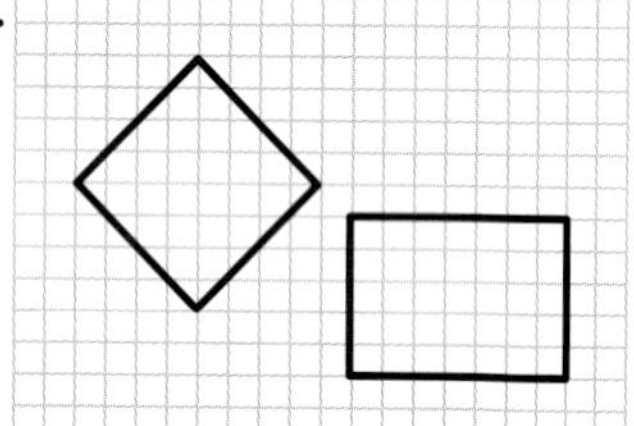

10. 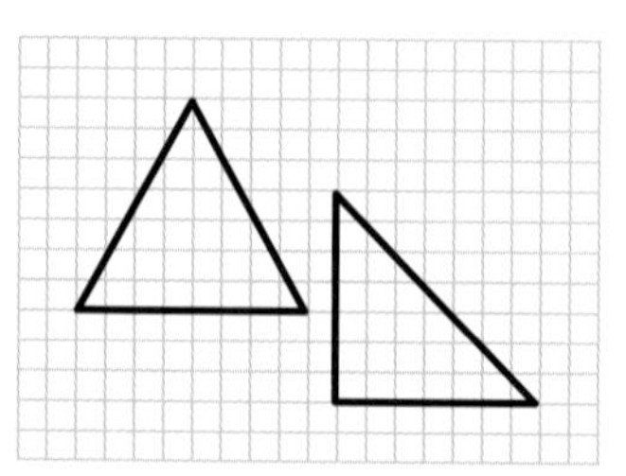

7·3 EXERCISES

What is the reflection across line l of each of the following?

1. Point P 2. $\triangle ABC$ 3. AC

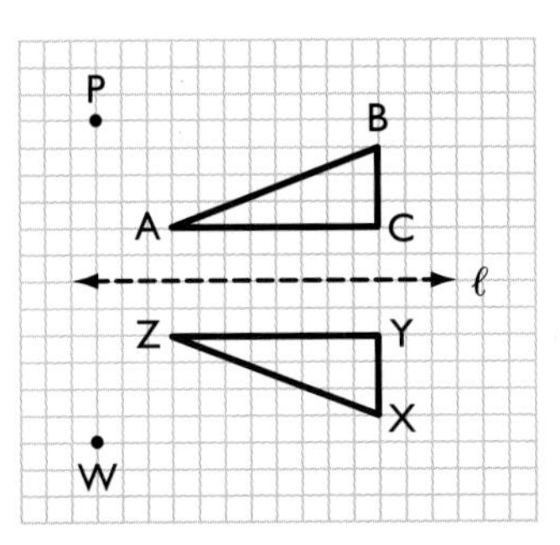

Copy each shape. Then draw all lines of symmetry.

4. 5. 6.

Which type of transformation is illustrated?

7.

A B D C A' B' D' C'

8.

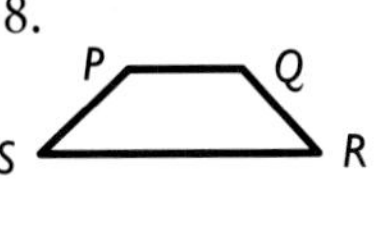

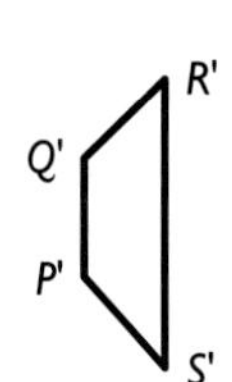

9.

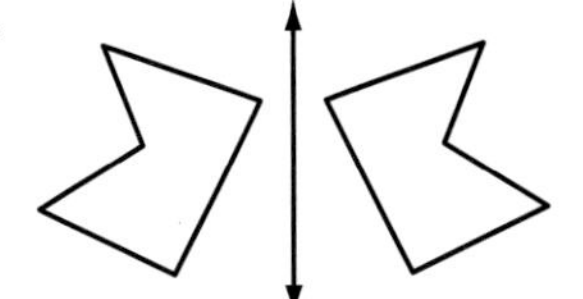

10.

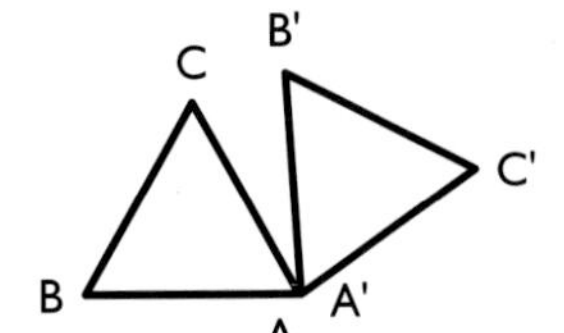

7·4 Perimeter

Perimeter of a Polygon

Ramon is planning to put a fence around his pasture. To determine how much fencing he needs, he must calculate the **perimeter** of, or *distance around*, his pasture.

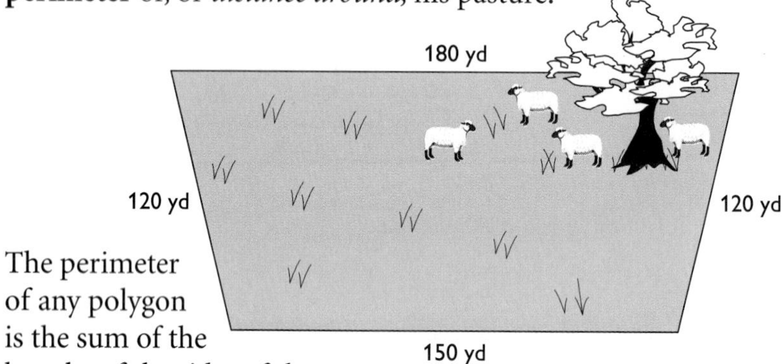

RAMON'S PASTURE

The perimeter of any polygon is the sum of the lengths of the sides of the polygon. So to find the perimeter of his pasture, Ramon needs to measure the length of each side and to add them together. He finds that two of the sides are 120 yd long, one side is 180 yd, and another is 150 yd. How much fencing will Ramon need to enclose his pasture?

$$P = 120 \text{ yd} + 120 \text{ yd} + 150 \text{ yd} + 180 \text{ yd} = 570 \text{ yd}$$

The perimeter of the pasture is 570 yd. Ramon will need 570 yd of fencing to enclose the pasture.

FINDING THE PERIMETER OF A POLYGON

To find the perimeter of any polygon, add up the lengths of all its sides.

Find the perimeter of the hexagon.

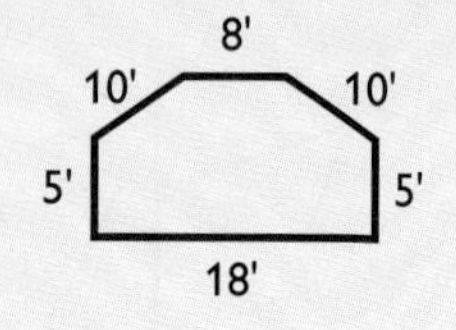

$P = 5 \text{ ft} + 10 \text{ ft} + 8 \text{ ft} + 10 \text{ ft} + 5 \text{ ft} + 18 \text{ ft} = 56 \text{ ft}$
The perimeter of this hexagon is 56 ft.

Regular Polygon Perimeters

The sides of a regular polygon are all the same length. If you know the perimeter of a regular polygon, you can find the length of each side.

To find the length of each side of a regular octagon with a perimeter of 36 cm, let x = length of a side.

$$36 \text{ cm} = 8x$$
$$4.5 \text{ cm} = x$$

Each side is 4.5 cm long.

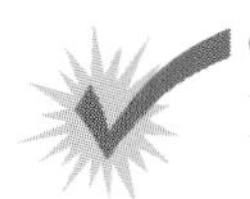

Check It Out

Find the perimeter of each polygon.

1.

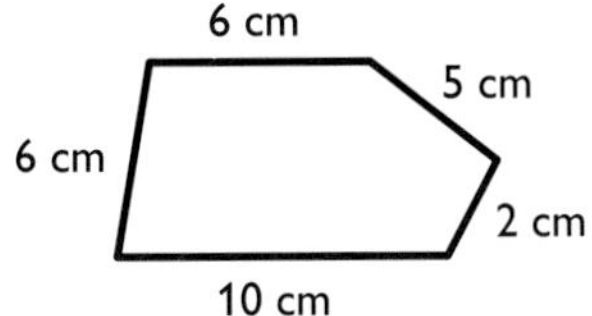

2.

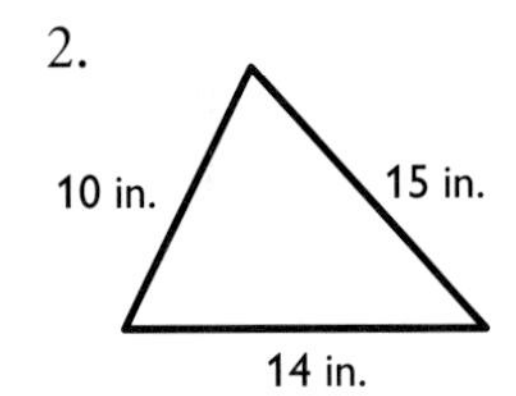

Find the length of each side.

3. a square with a perimeter of 24 m
4. a regular pentagon with a perimeter of 100 feet

Perimeter of a Rectangle

Opposite sides of a rectangle are equal in length. So to find the perimeter of a rectangle, you only need to know its length and width.

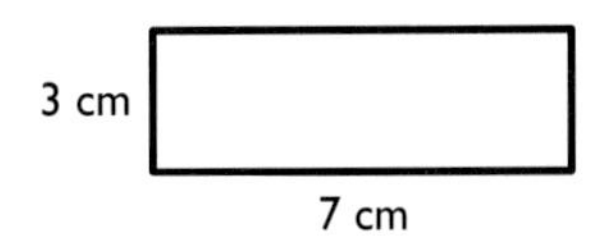

The perimeter is 7 cm + 3 cm + 7 cm + 3 cm = 20 cm.
You can write this as $(2 \times 7 \text{ cm}) + (2 \times 3 \text{ cm}) = 20 \text{ cm}$.

FINDING THE PERIMETER OF A RECTANGLE

For a rectangle with length, l, and width, w, the perimeter, P, can be found by using the formula $P = 2l + 2w$.

$$P = 2l + 2w$$
$$= (2 \times 15 \text{ m}) + (2 \times 9 \text{ m})$$
$$= 30 \text{ m} + 18 \text{ m} = 48 \text{ m}$$

15 m

9 m

The perimeter is 48 m.

A square is a rectangle whose length and width are equal. So the formula for finding the perimeter of a square, whose sides measure s, is $P = 4 \times s$ or $P = 4s$.

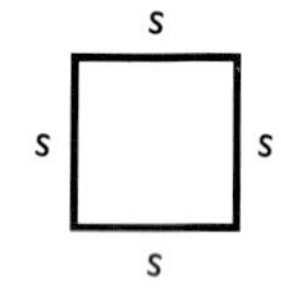

Check It Out

Find the perimeter.

5. rectangle with length 16 cm and width 14 cm
6. square with sides that are 12 cm

The Pentagon

Located near Washington, D.C., the Pentagon is one of the largest office buildings in the world. The United States Army, Navy, and Air Force all have their headquarters there.

The building covers an area of 29 acres and has 3,707,745 ft^2 of usable office space.

The structure consists of five concentric regular pentagons with 10 spokelike corridors connecting them. The outside perimeter of the building is about 4,620 ft. What is the length of an outermost side? See Hot Solutions for answer.

Perimeter of a Right Triangle

If you know the lengths of two sides of a **right triangle,** you can find the length of the third side by using the **Pythagorean Theorem.**

For a review of the *Pythagorean Theorem,* see page 395.

FINDING THE PERIMETER OF A RIGHT TRIANGLE

Use the Pythagorean Theorem to find the perimeter of the right triangle.

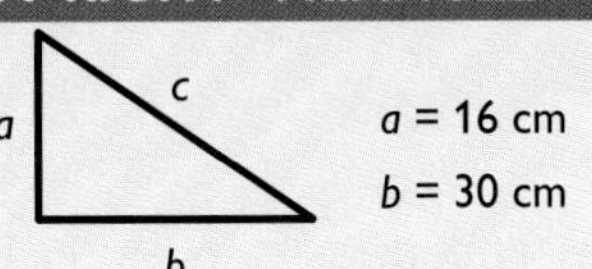

- Use the equation $c^2 = a^2 + b^2$ to find the length of the hypotenuse.

$$c^2 = 16^2 + 30^2$$
$$= 256 + 900$$
$$= 1{,}156$$

- The square root of c^2 is the length of the hypotenuse.

$$c = 34$$

- Add the lengths of the sides. The sum is the perimeter of the triangle.

$$16 \text{ cm} + 30 \text{ cm} + 34 \text{ cm} = 80 \text{ cm}$$

The perimeter is 80 cm.

Check It Out

Use the Pythagorean Theorem to find the perimeter of each triangle.

7.

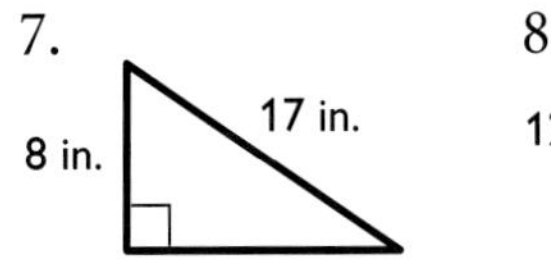

8.

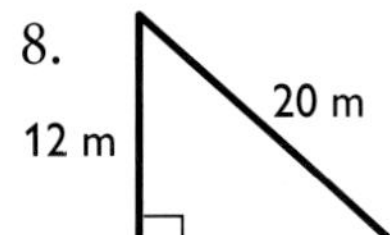

7·4 EXERCISES

Find the perimeter of each polygon.

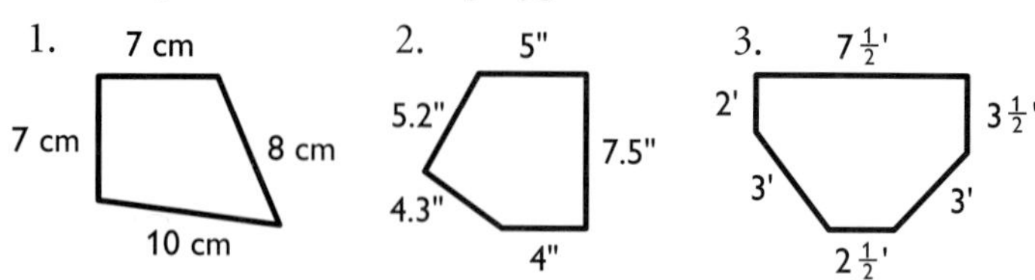

4. Find the perimeter of a regular decagon 4.8 cm on a side.
5. The perimeter of a regular hexagon is 200 in. Find the length of each side.
6. The perimeter of a square is 16 ft. What is the length of each side?

Find the perimeter of each rectangle.

7. $l = 12$ m, $w = 8$ m
8. $l = 35$ ft, $w = 19$ ft
9. $l = 6.1$ m, $w = 4.3$ m
10. $l = 2$ cm, $w = 1.5$ cm
11. The perimeter of a rectangle is 15 m. The length is 6 m. What is the width?
12. Find the perimeter of a square whose sides are 1.5 cm long.
13. Find the perimeter of the triangle below.

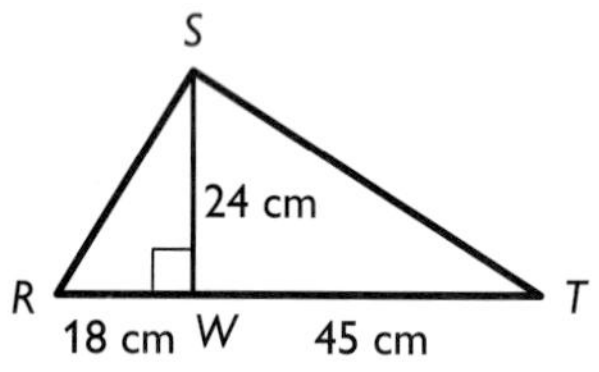

14. Two sides of a triangle are 9 in. and 7 in. If it is an **isosceles triangle,** what are the two different possible perimeters?
15. The perimeter of an equilateral triangle is 27 cm. What does each side measure?
16. If one side of a regular pentagon measures 18 in., what is the perimeter?
17. If a side of a regular nonagon measures 8 cm, what is the perimeter?

In-line skaters have created a race course on a deserted parking lot.

18. How long is the race?
19. If they changed the course to go around the edge of the parking lot, how long would it be?

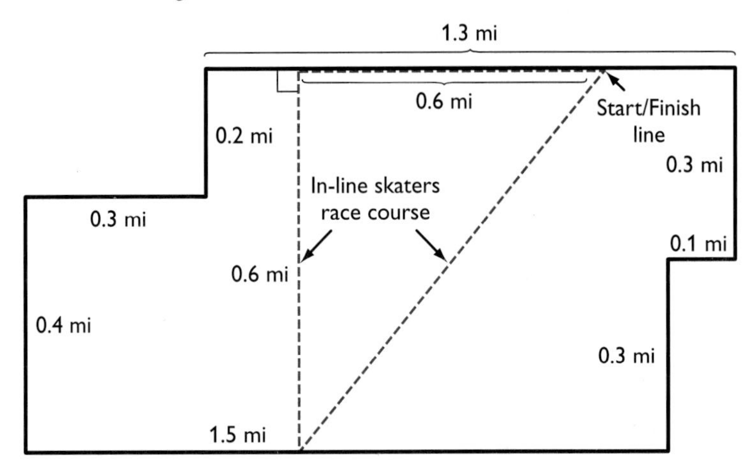

20. Cleve is using chalk lines to create a baseball diamond. He needs to make a square with sides that are 60 ft long. In addition, he wants to mark off two batter's boxes (left-handed and right-handed) with a length of 5 ft and a width of 3 ft. Cleve's chalkbag will make 375 ft of chalk line. How many feet does he need to chalk? Does he have enough chalk to complete the job?

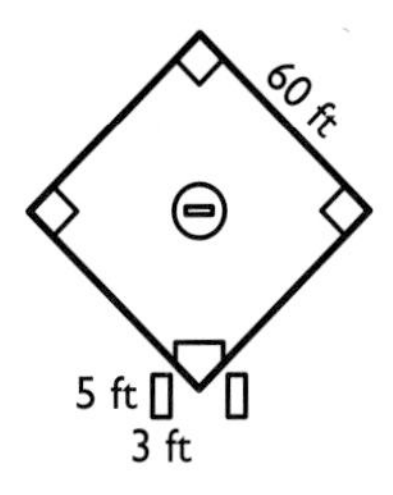

7•5 Area

What Is Area?

Area measures the size of a surface. Your desktop is a surface with area, and so is the state of Montana. Instead of measuring with units of length, such as inches, centimeters, feet, and kilometers, you measure area in square units, such as **square inches (in.^{2})** and **square centimeters (cm^2)**.

This square has an area of one square centimeter. It takes exactly three of these squares to cover this rectangle, which tells you that the area of the rectangle is three square centimeters, or 3 cm^2.

Estimating Area

When an exact answer is not needed or is hard to find, you can **estimate** the area of a surface.

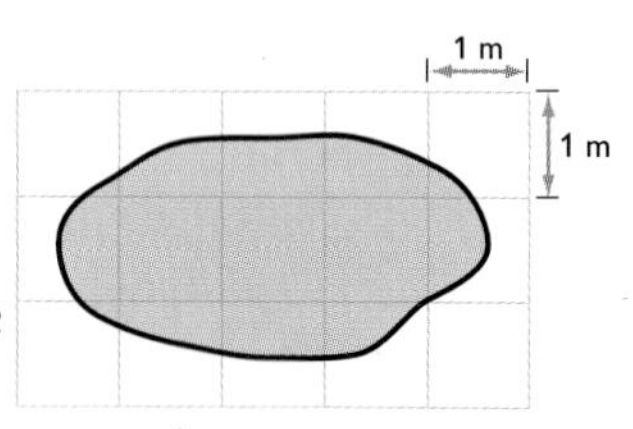

In the shaded figure to the right, three squares are completely shaded, and so you know that the area is greater than 3 m^2. The rectangle around the shape covers 15 m^2, and obviously the shaded area is less than that. So you can estimate the area of the shaded figure by saying that it is greater than 3 m^2 but less than 15 m^2.

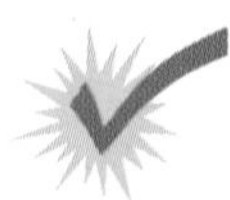

Check It Out

1. Estimate the area of the shaded region. Each square represents 1 cm^2.

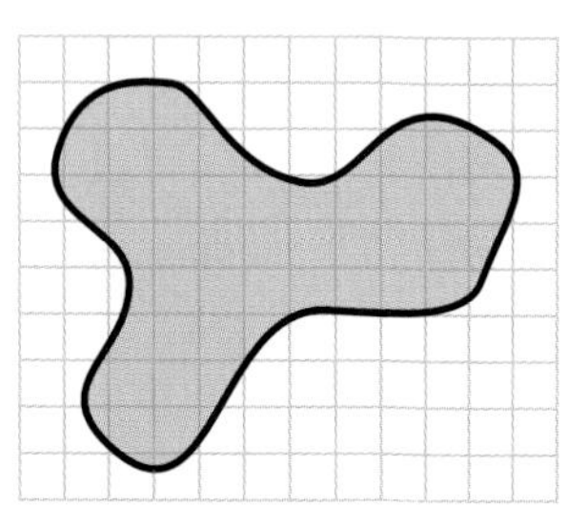

Area of a Rectangle

You can count squares to find the area of this rectangle.

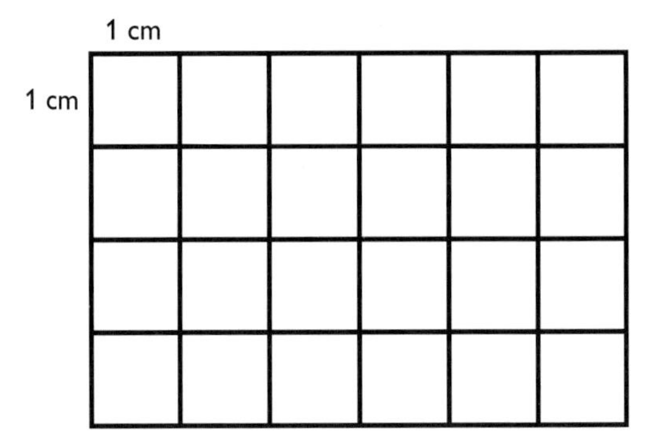

There are 24 squares and each is a square centimeter. So the area of this rectangle is 24 cm^2.

You can also use the formula for finding the area of a rectangle: $A = l \times w$, or $A = lw$. The length of the rectangle above is 6 cm and the width is 4 cm. Using the formula, you find that

$$A = 6 \text{ cm} \times 4 \text{ cm}$$
$$= 24 \text{ cm}^2$$

FINDING THE AREA OF A RECTANGLE

Find the area of this rectangle.

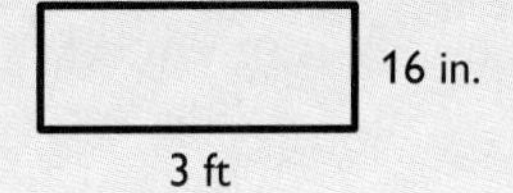

- The length and the width must be in the same units.

  3 ft = 36 in. So l = 36 in. and w = 16 in.

- Use the formula for the area of a rectangle.

$$A = l \times w$$
$$= 36 \text{ in.} \times 16 \text{ in.}$$
$$= 576 \text{ in.}^2$$

The area of the rectangle is 576 in.^2

If the rectangle is a square, the length and the width are the same. So for a square whose sides measure s units, you can use the formula $A = s \times s$, or $A = s^2$.

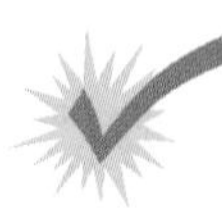

Check It Out

2. Find the area of a rectangle if $l = 40$ in. and $w = 2$ ft.
3. Find the area of a square whose sides are 6 cm.

Area of a Parallelogram

To find the area of a parallelogram, you multiply the **base** by the **height.**

Area = base × height
$A = b \times h$,
or $A = bh$

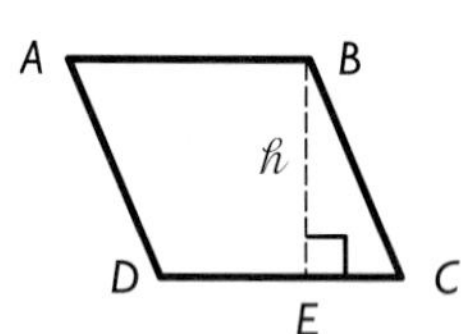

The height of a parallelogram is always **perpendicular** to the base. So in parallelogram *ABCD,* the height, *h,* is equal to *BE,* not *BC.* The base, *b,* is equal to *DC.*

FINDING THE AREA OF A PARALLELOGRAM

Find the area of a parallelogram with a base of 12 in. and a height of 7 in.

$A = b \times h$
$= 12 \text{ in.} \times 7 \text{ in.}$
$= 84 \text{ in.}^2$

The area of the parallelogram is 84 in.^{2} or 84 sq in.

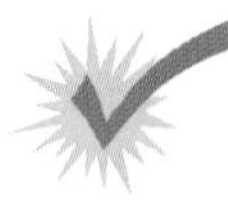

Check It Out

4. Find the area of a parallelogram if $b = 9$m and $h = 6$ m.
5. Find the length of the base of a parallelogram if the area is 32 m^2 and the height is 4 m.

Area of a Triangle

If you were to cut a parallelogram along a diagonal, you would have two triangles with equal bases, *b*, and equal height, *h*.

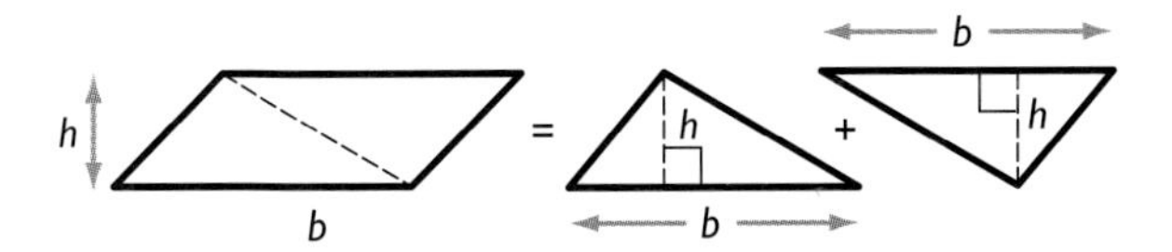

A triangle has half the area of a parallelogram with the same base and height. The area of a triangle equals $\frac{1}{2}$ the base times the height, and so the formula is $A = \frac{1}{2} \times b \times h$, or $A = \frac{1}{2}bh$.

$A = \frac{1}{2} \times b \times h$
$A = \frac{1}{2} \times 13.5 \text{ cm} \times 8.4 \text{ cm}$
$= 0.5 \times 13.5 \text{ cm} \times 8.4 \text{ cm}$
$= 56.7 \text{ cm}^2$

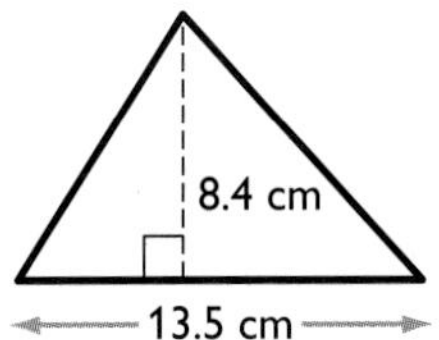

The area of the triangle is 56.7 cm^2.

FINDING THE AREA OF A TRIANGLE

Find the area of $\triangle PQR$. Note that in a right triangle the two **legs** constitute a height and a base.

$A = \frac{1}{2}bh$
$= \frac{1}{2} \times 5 \text{ m} \times 3 \text{ m}$
$= 0.5 \times 5 \text{ m} \times 3 \text{ m}$
$= 7.5 \text{ m}^2$

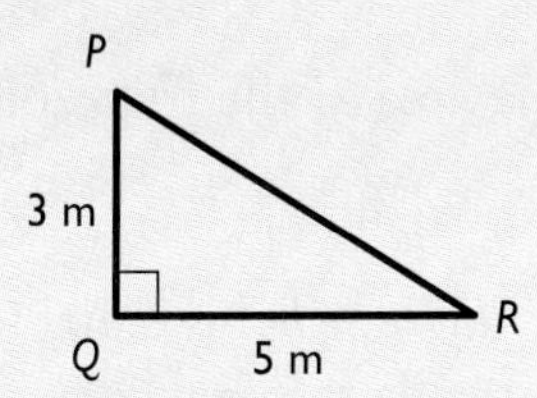

The area of the triangle is 7.5 m^2.

For a review of *right triangles,* see page 394.

Check It Out

6. Find the area of a triangle where $b = 20$ in. and $h = 6$ in.
7. Find the area of a right triangle whose sides are 24 cm, 45 cm, and 51 cm.

Area of a Trapezoid

A trapezoid has two bases, which are labeled b_1 and b_2. You read b_1 as "*b* sub-one." The area of a trapezoid is equal to the area of two triangles.

b_1 h b_2 = b_1 h + h b_2

You know that the formula for the area of a triangle is $A = \frac{1}{2}bh$, and so it makes sense that the formula for finding the area of a trapezoid would be $A = \frac{1}{2}b_1h + \frac{1}{2}b_2h$ or, in simplified form, $A = \frac{1}{2}h(b_1 + b_2)$.

FINDING THE AREA OF A TRAPEZOID

Find the area of trapezoid *WXYZ*.

$$\begin{aligned} A &= \tfrac{1}{2}h(b_1 + b_2) \\ &= \tfrac{1}{2} \times 4\,(5 + 11) \\ &= 2 \times 16 \\ &= 32 \text{ cm}^2 \end{aligned}$$

W X Y Z
b_1 = 5 cm
h = 4 cm
b_2 = 11 cm

The area of the trapezoid is 32 cm^2.

Since $\frac{1}{2}h(b_1 + b_2)$ is equal to $h \times \frac{b_1 + b_2}{2}$, you can remember the formula this way:

A = height times the average of the bases

For a review of how to find an *average* or *mean*, see page 222.

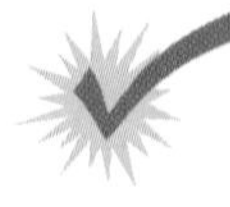

Check It Out

8. The height of a trapezoid is 3 ft. The bases are 2 ft and 6 ft. What is the area?
9. The height of a trapezoid is 4 ft. The bases are 8 ft and 7 ft. What is the area?

7·5 EXERCISES

1. Estimate the area of the figure below.

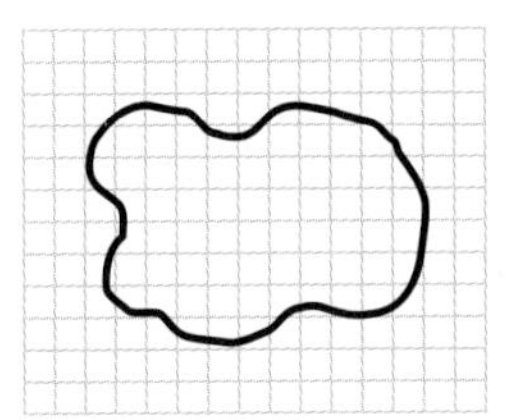

2. If each square unit in the figure above is $\frac{1}{16}$ in., estimate the area in inches.

Find the area of each rectangle given the length, l, and the width, w.

3. $l = 3$ m, $w = 2.5$ m
4. $l = 200$ cm, $w = 1.5$ m

Find the area of each parallelogram.

5.

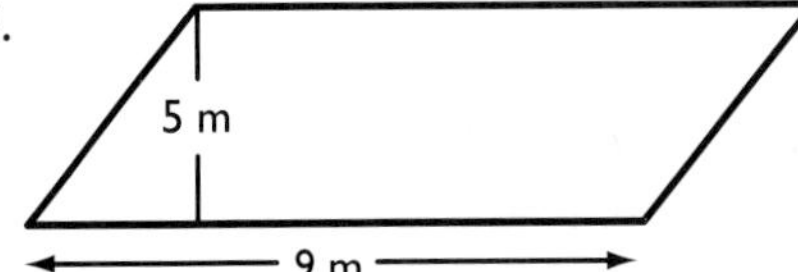

6.

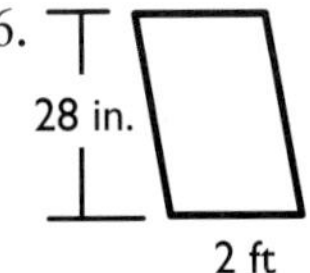

Find the area of each triangle given the base, b, and the height, h.

7. $b = 5$ in., $h = 4$ in.
8. $b = 6.8$ cm, $h = 1.5$ cm

9. Find the area of a trapezoid with bases 7 in. and 9 in. and height 1 ft.
10. Mr. Lopez plans to give the plot of land shown below to his two daughters. How many square yards of land will each daughter receive if the land is divided evenly between them?

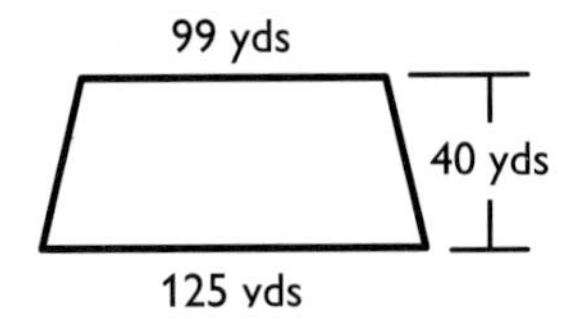

7·6 Surface Area

The **surface area** of a solid is the total area of its exterior surfaces. You can think about surface area in terms of the parts of a solid shape that you would paint. Like area, surface area is expressed in square units. To see why, "unfold" the rectangular prism.

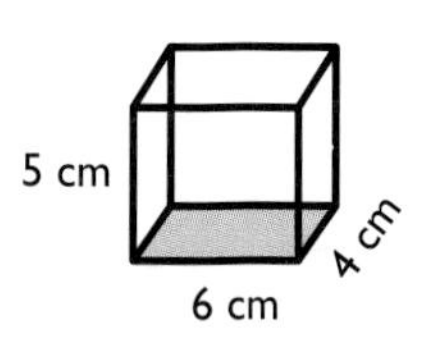

Mathematicians call this unfolded prism a **net.** A net is a **pattern** that can be folded to make a three-dimensional figure.

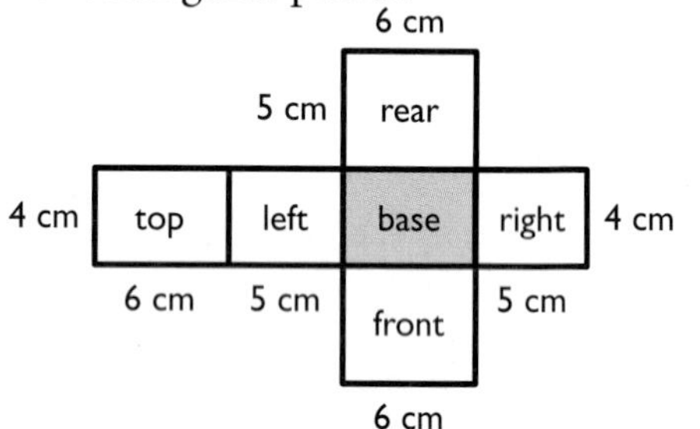

Surface Area of a Rectangular Prism

A rectangular prism has six rectangular faces. To find the surface area of a rectangular prism, find the sum of the areas of the six faces, or rectangles. *Remember:* Opposite faces are equal. For a review of *polyhedrons* and *prisms*, see page 356.

FINDING THE SURFACE AREA OF A RECTANGULAR PRISM

Use the net to find the area of the rectangular prism above.

- Use the formula $A = lw$ to find the area of each face.
- Then add the six areas.
- Express the answer in square units.

$$\begin{aligned}
\text{Area} &= \text{top + base} + \text{left + right} + \text{front + rear} \\
&= 2 \times (6 \times 4) + 2 \times (5 \times 4) + 2 \times (6 \times 5) \\
&= 2 \times 24 + 2 \times 20 + 2 \times 30 \\
&= 48 + 40 + 60 \\
&= 148 \text{ cm}^2
\end{aligned}$$

The surface area of the rectangular prism is 148 cm^2.

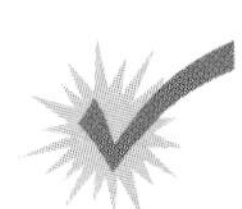

Check It Out

Find the surface area of each shape.

1.

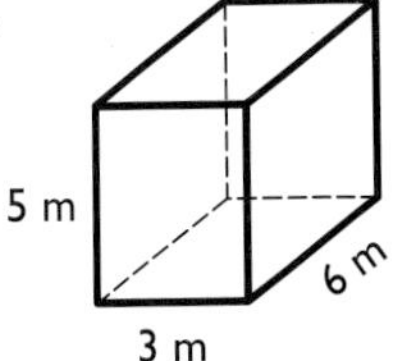

2.

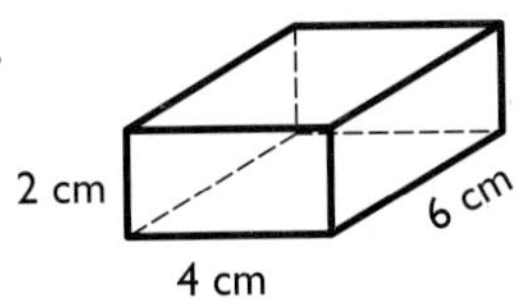

Surface Areas of Other Solids

The unfolding technique can be used to find the surface area of any polyhedron. Look at the **triangular prism** and its net

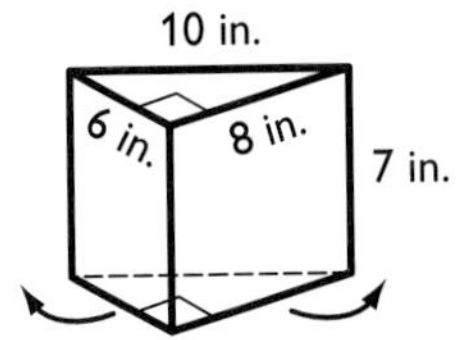

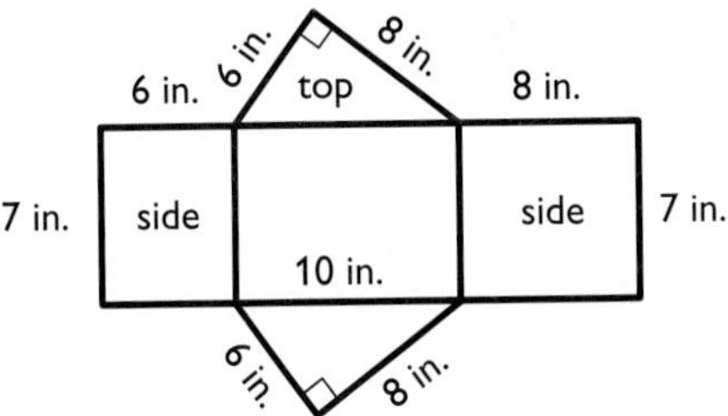

To find the surface area of this solid, use the area formulas for a rectangle ($A = lw$) and a triangle ($A = \frac{1}{2}bh$) to find the areas of the five faces and then find the sum of the areas.

Below are two pyramids and their nets. For these polyhedrons, you would again use the area formulas for a rectangle ($A = lw$) and a triangle ($A = \frac{1}{2}bh$) to find the areas of the faces and then find the sum of the areas.

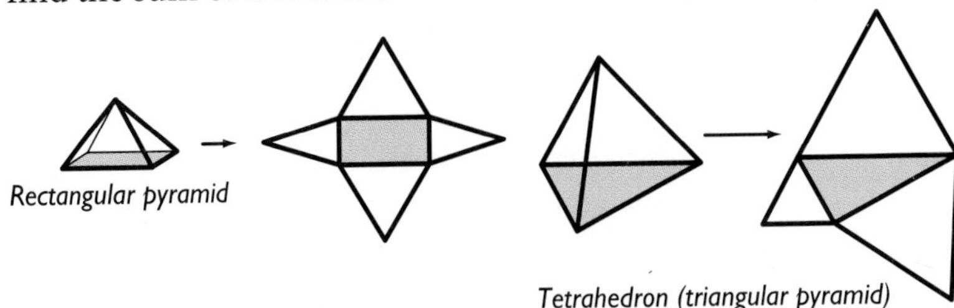

Rectangular pyramid

Tetrahedron (triangular pyramid)

The surface area of a **cylinder** is the sum of the areas of two **circles** and a rectangle.

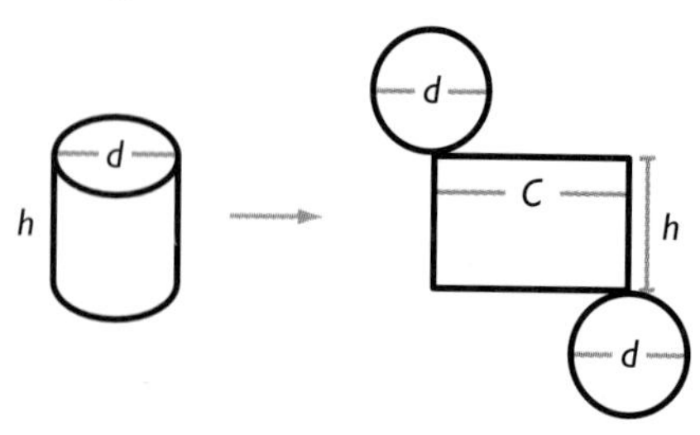

The two bases of a cylinder are equal in area. The height of the rectangle is the height of the cylinder. Its length is the **circumference** of the cylinder.

To find the surface area of a cylinder, you would:

- Use the formula for the area of a circle to find the area of each base.

  $A = \pi r^2$

- Find the area of the rectangle using the formula $h \times (2\pi r)$.

For a review of *circles*, see page 388.

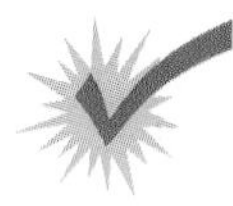

Check It Out

3. Unfold the triangular prism and find the surface area.

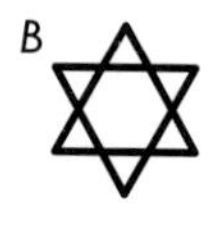

4. Which unfolded figure represents the pyramid?

5. Find the surface area of the cylinder. Use $\pi = 3.14$.

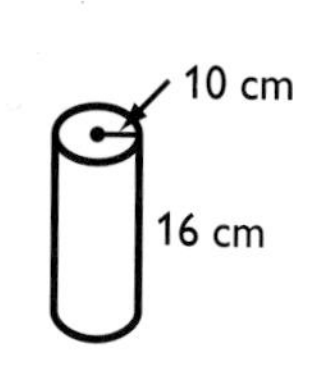

7·6 EXERCISES

Find the surface area of each shape. Round to the nearest tenth.

1.

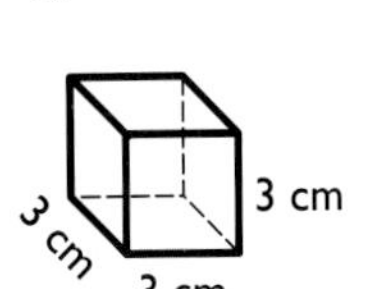

2.

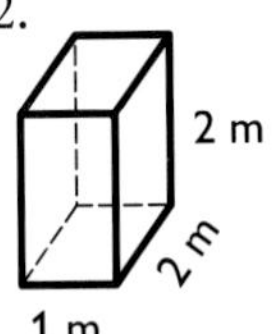

3.

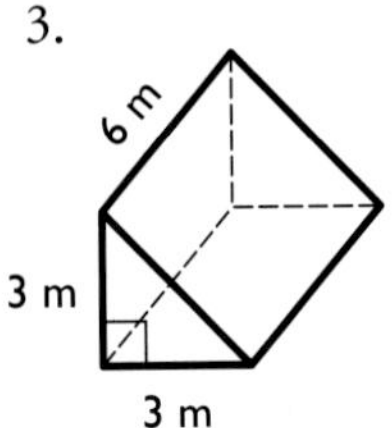

4.

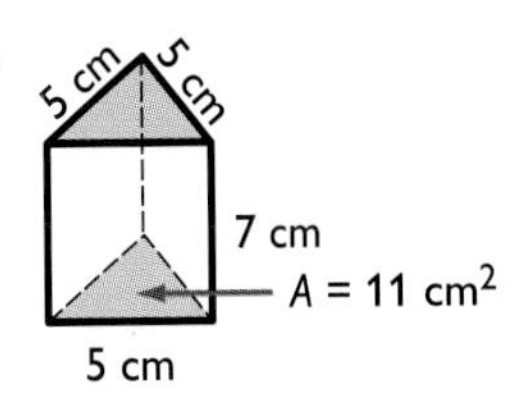

5.

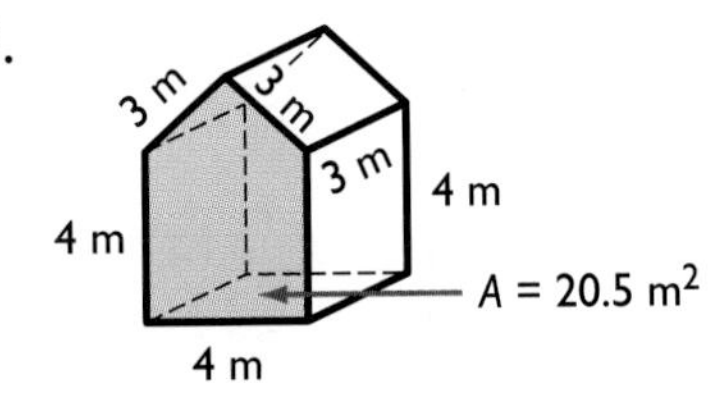

6. A rectangular prism is 8 in. by 6 in. by 2 in. Find the surface area.
7. The surface area of a cube is 294 ft^2. What is the length of each edge?

   A. 5 ft B. 6 ft C. 7 ft D. 8 ft

Find the surface area of each cylinder. Round to nearest tenth.

8.

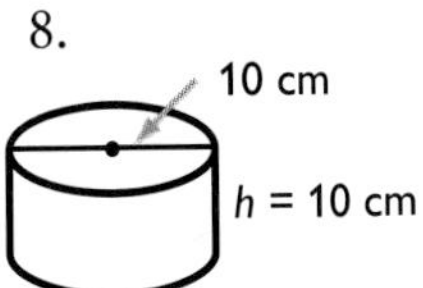

9.

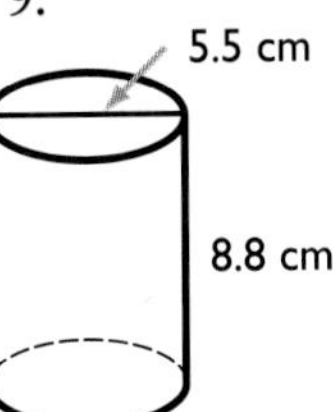

10. Rita and Derrick are building a 3 ft by 3 ft by 6 ft platform to use for skateboarding. They plan to waterproof all six sides of the platform, using sealant that covers about 50 square feet per quart. How many quarts of sealant will they need?

7·7 Volume

What Is Volume?

Volume is the space inside a figure. One way to measure volume is to count the number of cubic units that would fill the space inside a figure.

The volume of this small cube is 1 **cubic inch.**

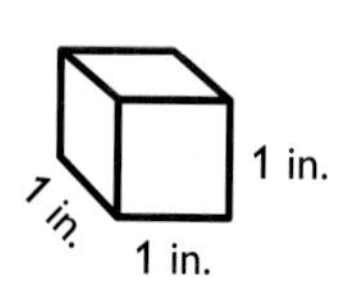

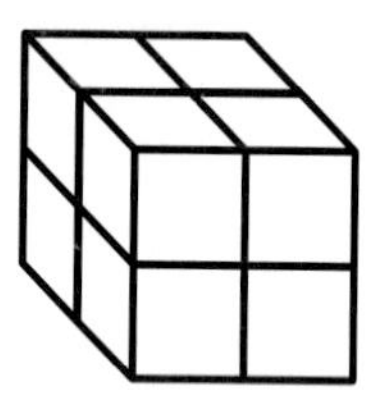

The number of smaller cubes that it takes to fill the space inside the larger cube is 8, and so the volume of the larger cube is 8 cubic inches.

You measure the volume of shapes in *cubic* units. For example, 1 cubic inch is written as 1 $in.^3$, and 1 **cubic meter** is written as 1 m^3.

For a review of *cubes*, see page 356.

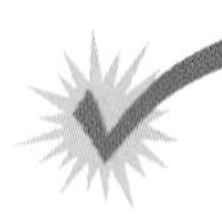

Check It Out

What is the volume of each shape?

1. 1 cube = 1 cm^3

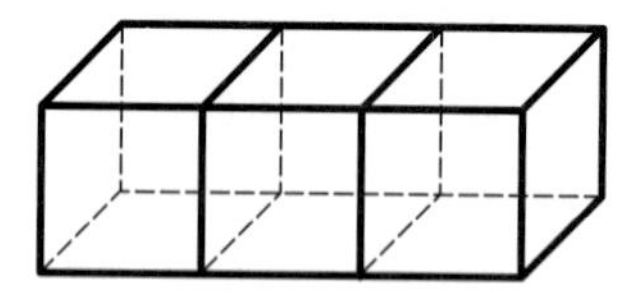

2. 1 cube = 1 ft^3

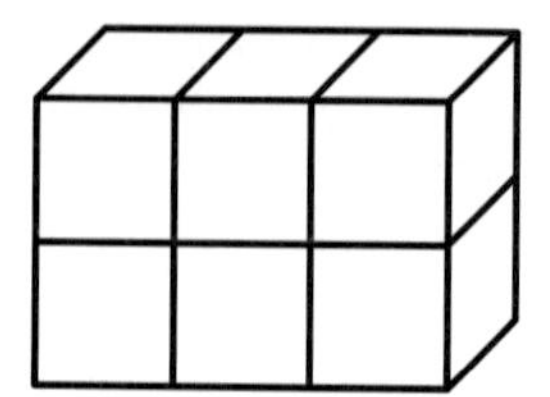

Volume of a Prism

The volume of a prism can be found by multiplying the *area* (pp. 372–377) of the *base, B,* and the *height, h.*

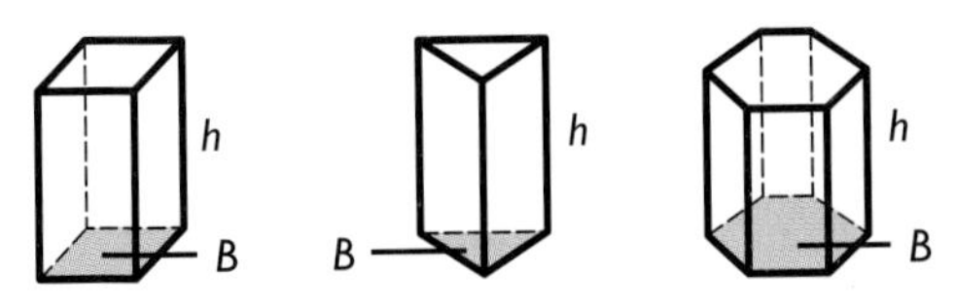

Volume $= Bh$
See *formulas,* pages 62–63.

FINDING THE VOLUME OF A PRISM

Find the volume of the rectangular prism. The base is 12 in. long and 10 in. wide; the height is 15 in.

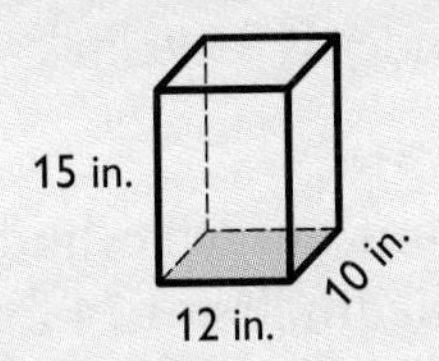

base $A = 12 \text{ in.} \times 10 \text{ in.}$
$= 120 \text{ in.}^2$
- Find the area of the base.

$V = 120 \text{ in.}^2 \times 15 \text{ in.}$
$= 1{,}800 \text{ in.}^3$
- Multiply the base and the height.

The volume of the prism is $1{,}800 \text{ in.}^3$

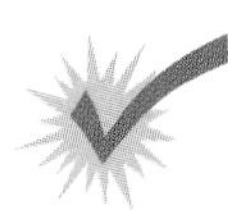

Check It Out!

Find the volume of each shape.

3. 4.

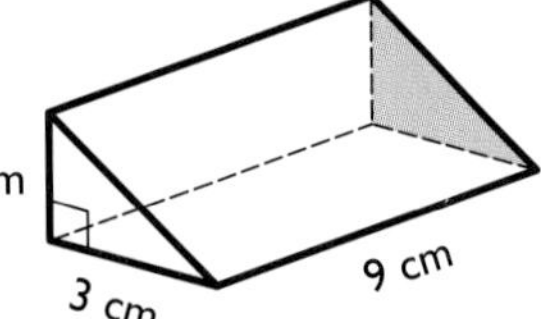

Volume of a Cylinder

You can find the volume of a cylinder the same way you found the volume of a prism, using the formula $V = Bh$. *Remember:* The base of a cylinder is a circle (p. 388).

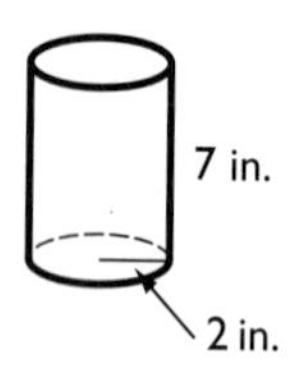

The base has a radius of 2 in. Estimating **pi** (π) at 3.14, you will find that the area of the base is about 12.56 sq in. Since you also know the height, you can use the formula $V = Bh$.

$$V = 12.56 \text{ in.}^2 \times 7 \text{ in.}$$
$$= 87.92 \text{ in.}^3$$

The volume of the cylinder is 87.92 in^3.

Check It Out

Find the volume of each cylinder. Round to the nearest hundredth. Use 3.14 for π.

5.

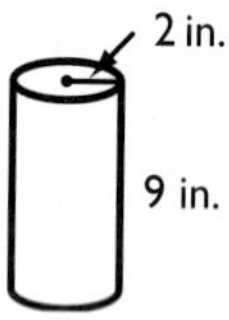

6.

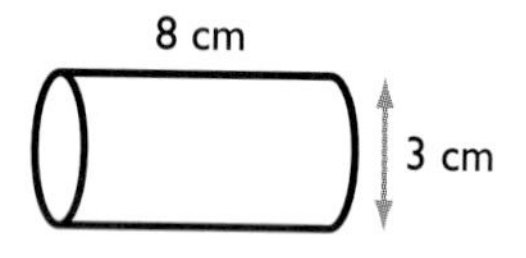

Volume of a Pyramid and a Cone

The formula for the volume of a pyramid or a cone is $V = \frac{1}{3}Bh$.

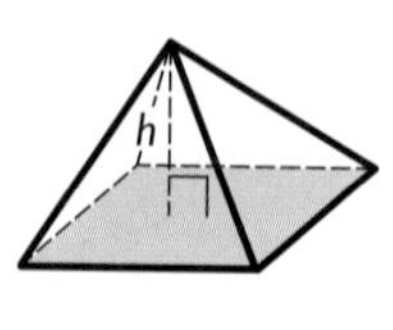

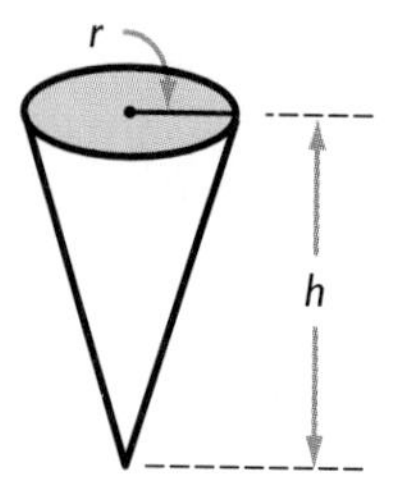

FINDING THE VOLUME OF A PYRAMID

Find the volume of the pyramid. The base is 175 cm long and 90 cm wide; the height is 200 cm.

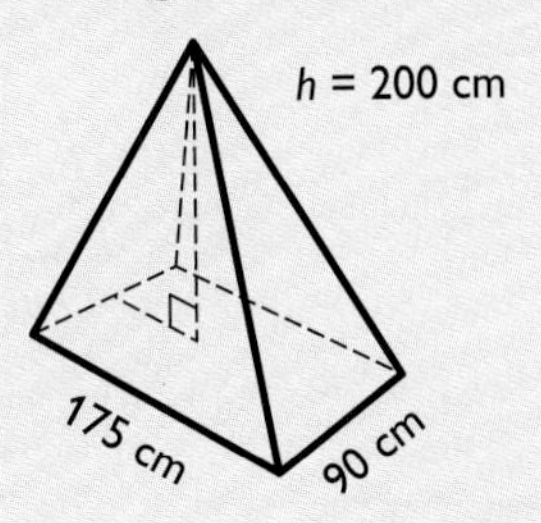

base $A = (175 \times 90)$ • Find the area of the base.

$= 15{,}750 \text{ cm}^2$

$V = \frac{1}{3}(15{,}750 \times 200)$ • Multiply the base by the height and then by $\frac{1}{3}$.

$= 1{,}050{,}000$

The volume is $1{,}050{,}000 \text{ cm}^3$.

To find the volume of a cone, you follow the same procedure as above. You may use your calculator to help find the area of the base of the cone. For example, a cone has a base with a radius of 3 cm and a height of 10 cm. What is the volume of the cone to the nearest tenth?

Square the radius and multiply by π to find the area of the base. Then multiply by the height and divide by 3 to find the volume. The volume of the cone is 94.2 cm^3.

Press [π] [×] 9 [=] [28.27433] [×] 10 [÷] 3 [=] [94.24778]

For other volume *formulas,* see page 62.

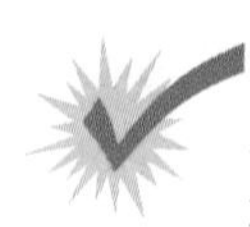

Check It Out

Find the volume of the shapes below, rounded to the nearest tenth.

7. 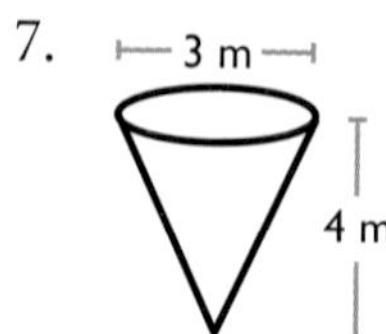

8. 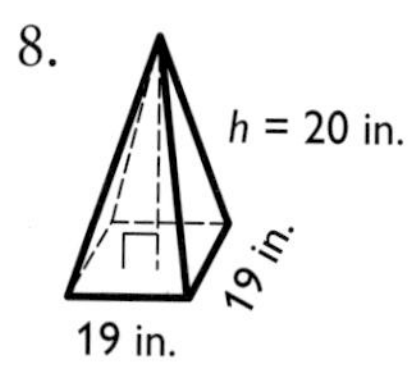

Good Night, T. Rex

Why did the dinosaurs disappear? New evidence from the ocean floor points to a giant asteroid that collided with Earth some 65 million years ago.

The asteroid, 6 to 12 mi in diameter, hit the earth somewhere in the Gulf of Mexico. It was traveling at a speed of thousands of miles per hour.

The collision sent billions of tons of debris into the atmosphere. The debris rained down on the planet, obscuring the sun. Global temperatures plummeted. The fossil record shows that most of the species that were alive before the collision disappeared.

Assume the crater left by the asteroid had the shape of a hemisphere with a diameter of 165 mi. How many cubic miles of debris would have been flung from the crater into the air? For formula for volume of sphere, see p. 62. See Hot Solutions for answer.

7·7 EXERCISES

Use the rectangular prism to answer items 1–4.

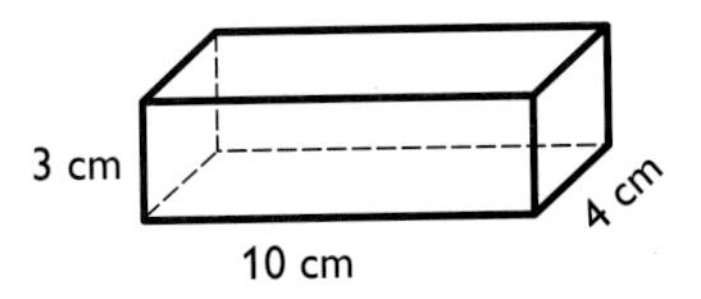

1. How many cubic centimeter cubes would it take to make one layer in the bottom of the prism?
2. How many layers of cubes would you need to fill the prism?
3. How many cubes do you need to fill the prism?
4. Each cube has a volume of 1 cm^3. What is the volume of the prism?
5. Find the volume of a rectangular prism with base 10 cm, width 10 cm, and height 8 cm.
6. The base of a cylinder has an area of 5 cm^2. Its height is 7 cm. What is its volume?
7. Find the volume of a cylinder 8.2 m high when its base has a radius of 2.1 m. Round your answer to the nearest tenth.
8. Find the volume of a pyramid with a height of 4 in. and a rectangular base that measures 6 in. by 3.5 in.
9. Look at the cone and triangular pyramid below. Which has the greater volume? by how many cubic inches?

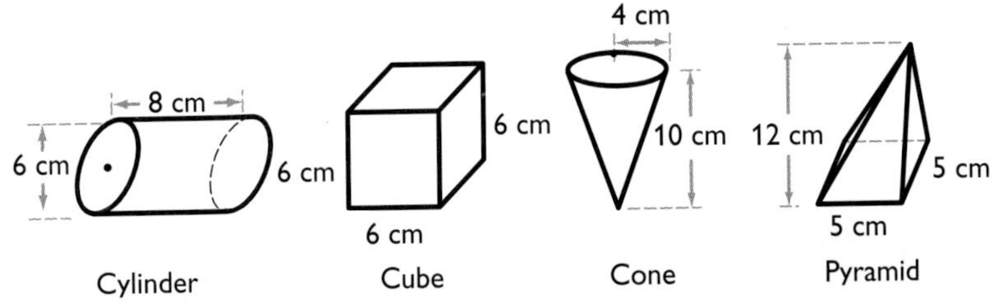

10. List the shapes above from the least volume to the greatest volume.

7·8 Circles

Parts of a Circle

Of the many shapes you may encounter in geometry, circles are among the most unique. They differ from other geometric shapes in several ways. For instance, while all circles are the same shape, polygons vary in shape. Circles do not have any sides, while polygons are named and classified by the number of sides they have. The *only* thing that makes one circle different from another is size.

A circle is a set of points equidistant from a given point. That point is the **center of the circle**. A circle is named by its center point.

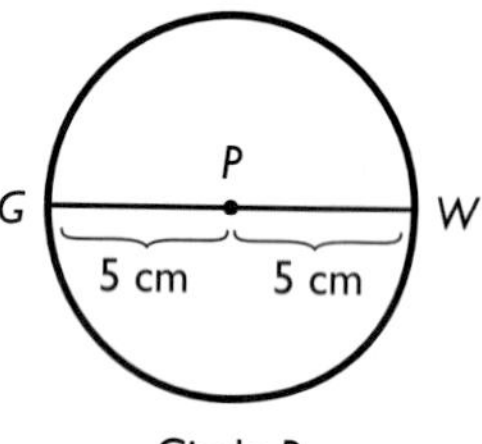

Circle *P*

A **radius** is a **segment** that has one endpoint at the center and the other endpoint on the circle. In circle *P*, $\overline{PW}$ is a *radius,* and so is $\overline{PG}$.

A **diameter** is a line segment that passes through the center of the circle and has both endpoints on the circle. $\overline{GW}$ is a diameter of circle *P*. Notice that the length of the diameter $\overline{GW}$ is equal to the sum of $\overline{PW}$ and $\overline{PG}$. So the diameter is twice the length of the radius. If d equals the diameter and r equals the radius, d is twice the radius, r. So the diameter of circle P is 2(5) or 10 cm.

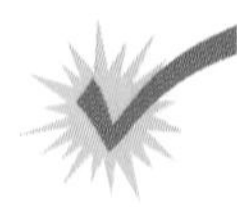

Check It Out

1. Find the radius of a circle with diameter 18 in.
2. Find the radius of a circle with diameter 3 m.
3. Find the radius of a circle in which $d = x$.
4. Find the diameter of a circle with radius 6 cm.
5. Find the diameter of a circle with radius 16 m.
6. Find the diameter of a circle where $r = y$.

Circumference

The circumference of a circle is the distance around the circle. The **ratio** (p. 308) of every circle's circumference to its diameter is always the same. That ratio is a number that is close to 3.14. In other words, in every circle, the circumference is about 3.14 times the diameter. The symbol π, which is read as *pi*, is used to represent the ratio $\frac{C}{d}$.

$\frac{C}{d} = 3.141592 \ldots$

Circumference = pi × diameter, or $C = \pi d$

Look at the illustration below. The circumference of the circle is about the same length as three diameters.This is true for any circle.

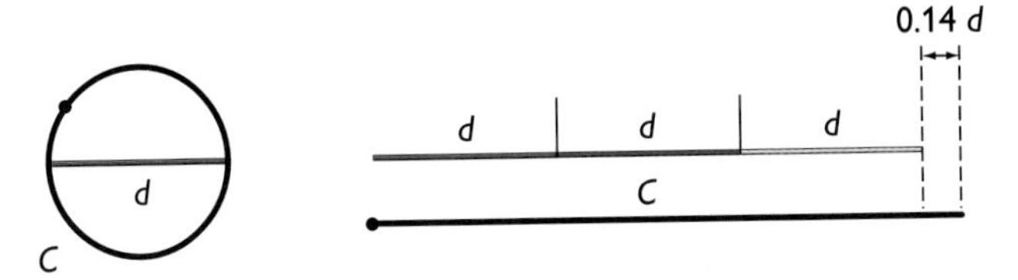

Since $d = 2r$, Circumference = two × pi × radius, or $C = 2\pi r$.

If you are using a calculator that has a π key, hit it, and you will get an approximation for π to several more decimal places: $\pi = 3.141592 \ldots$. For practical purposes, however, when you are finding the circumference of a circle, round π to 3.14, or simply leave the answer in terms of π.

FINDING THE CIRCUMFERENCE OF A CIRCLE

Find the circumference of a circle with radius 8 m.

- Use the formula $C = \pi d$. Remember to multiply the radius by 2 to get the diameter. Round the answer to the nearest tenth.

$$d = 8 \times 2 = 16$$

$$C = 16\pi$$

The exact circumference is 16π m.

$$C = 16 \times 3.14$$

$$= 50.24$$

So, to the nearest tenth, the circumference is 50.2 m.

You can find the diameter if you know the circumference. Divide both sides by π.

$$C = \pi d$$

$$\frac{C}{\pi} = \frac{\pi d}{\pi}$$

$$\frac{C}{\pi} = d$$

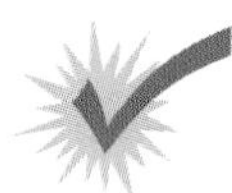

Check It Out

7. Find the circumference of a circle with a diameter of 5 in. Give the answer in terms of π.
8. Find the circumference of a circle with a radius of 3.2 cm. Round to the nearest tenth.
9. Find the diameter of a circle with circumference 25 m. Round to the nearest hundredth of a meter.
10. Using the π key on your calculator or $\pi = 3.141592$, find the radius of a circle with a circumference of 35 in. Round your answer to the nearest half inch.

Central Angles

A central angle is an angle whose vertex is at the center of a circle. The sum of the central angles in any circle is 360°. For a review of *angles*, see page 343.

The part of a circle where a central angle **intercepts** the circle is called an **arc.** The measure of the arc, in **degrees,** is equal to the measure of the central angle.

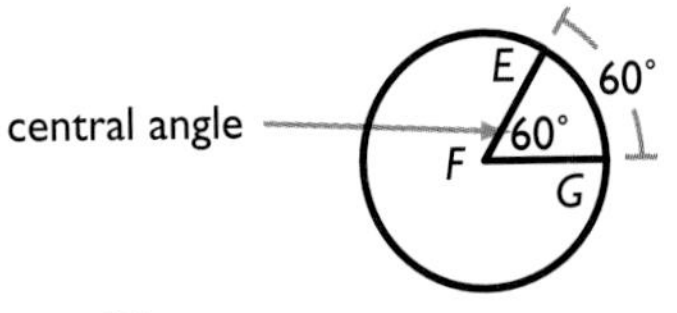

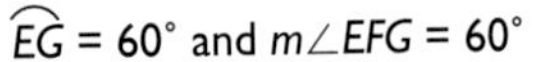
$\widehat{EG} = 60°$ and $m\angle EFG = 60°$

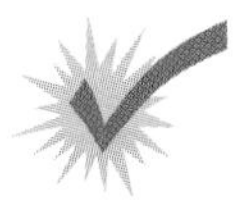

Check It Out

11. Name a central angle in circle *B*.
12. What is the measure of $\widehat{AC}$?

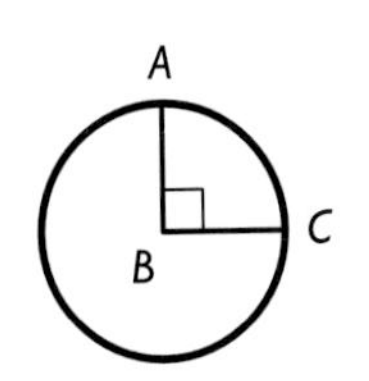

13. What is the measure of $\angle LMN$?

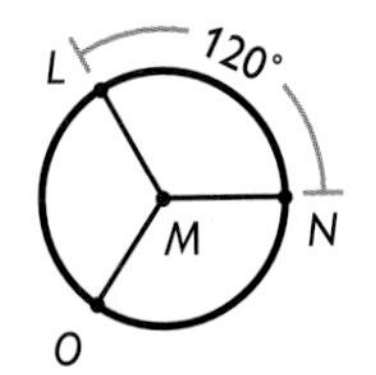

Around the World

Your blood vessels are a network of arteries and veins that carry oxygen to every part of the body and return blood with carbon dioxide to the lungs. Altogether, there are approximately 60,000 miles of blood vessels in the human body.

Just how far is 60,000 miles? The circumference of the earth is about 25,000 miles. If the blood vessels in one human body were placed end to end, approximately how many times would they wrap around the equator? See Hot Solutions for answer.

Area of a Circle

To find the area of a circle, you use the formula Area = pi × radius^2, or $A = \pi r^2$. As with the area of polygons, the area of a circle is expressed in square units.

For a review of *area* and *square units*, see page 372.

FINDING THE AREA OF A CIRCLE

Find the area of the circle Q to the nearest whole number.

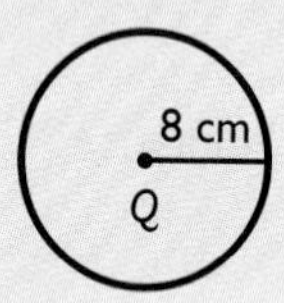

$A = \pi \times 8^2$ • Use the formula $A = \pi r^2$.

$= 64\pi$ • Square the radius.

≈ 200.96 • Multiply by 3.14, or use the calculator key for π for a more exact answer.

$\approx 201 \text{ cm}^2$

The area of circle Q is about 201 cm^2.

If you are given the diameter instead of the radius, remember to divide the diameter by two.

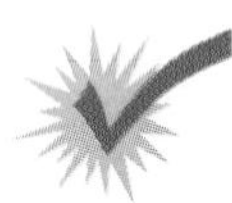

Check It Out

14. The diameter of a circle is 9 in. Find the area. Give your answer in terms of π; then multiply and round to the nearest tenth.
15. Use your calculator to find the area of a circle with a diameter of 15 cm. Use the calculator key for π, or $\pi = 3.14$, and round to the nearest square centimeter.

7·8 EXERCISES

Find the diameter of each circle with the given radius.

1. $r = 11$ ft
2. $r = 7.2$ cm
3. $r = x$

Find the radius with the given diameter.

4. $d = 7$ in.
5. $d = 2.6$ m
6. $d = y$

Given the r or d, find the circumference to the nearest tenth.

7. $d = 1$ m
8. $d = 7.9$ cm
9. $r = 18$ in.

The circumference of a circle is 47 cm. Find the following to the nearest tenth.

10. the diameter
11. the radius

Find the measure of each arc.

12. Arc AB
13. Arc CB
14. Arc AC

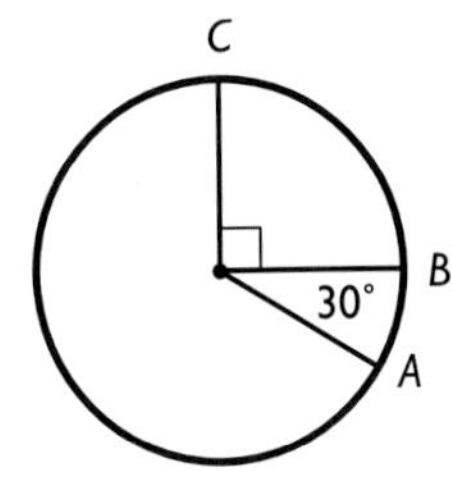

Find the area of each circle given r or d. Round to the nearest whole number.

15. $r = 2$ m
16. $r = 35$ in.
17. $d = 50$ cm
18. $d = 10$ ft.

19. A dog is tied to a stake. The rope is 20 m long, so the dog can roam up to 20 m from the stake. Find the area within which the dog can roam. (If you use a calculator, round to the nearest whole number.)

20. Tony's Pizza Palace sells a large pizza with a diameter of 14 in. Pizza Emporium sells a large pizza with a diameter of 15 in. for the same price. How much more pizza are you getting for your money at Pizza Emporium?

7·9 Pythagorean Theorem

Right Triangles

The illustration below left shows a right triangle on a geoboard. The triangle has an area of $\frac{1}{2}$ square unit and each leg is one unit long.

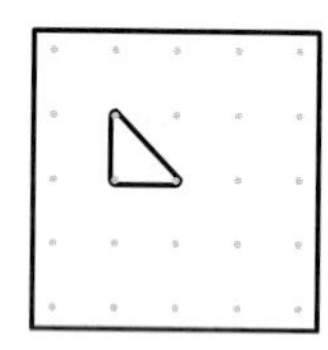

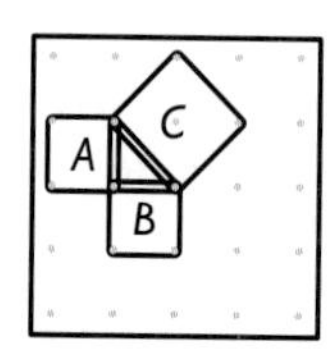

Now look at the squares on each of the three sides of the triangle. Call the squares *A*, *B*, and *C*.

Area $A = 1 \times 1 = 1$ square unit

Area $B = 1 \times 1 = 1$ square unit

When you look at the pegs of the geoboard, the area of *C* is not easy to determine, but it is clear that it is equal to four of the original triangles.

Area $C = 4 \times \frac{1}{2}$

$C = 2$

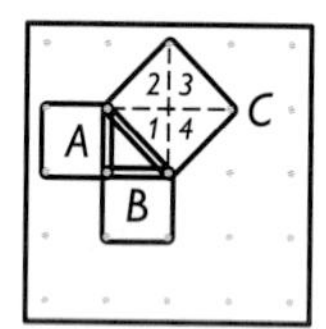

The area of *C* is 2 square units.

Note the relationship among the three areas:

Area *A* + Area *B* = Area *C*

This relationship holds true for all right triangles.

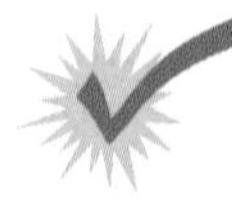

Check It Out

1. What is the area of each of the squares?
2. Does the sum of the areas of the two smaller squares equal the area of the third square?

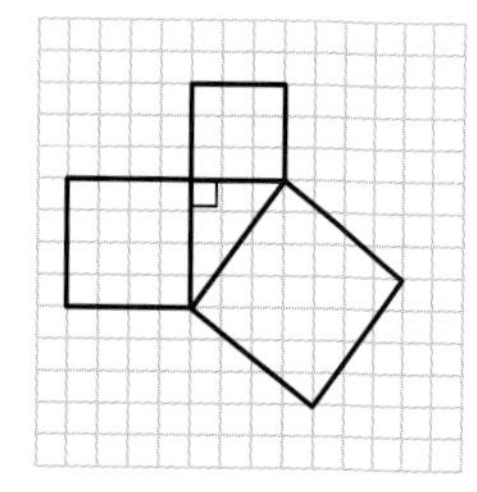

The Pythagorean Theorem

When you look at the areas of the squares on each of the sides of the triangles shown on the geoboards, you see the relationship of the area of the square of the **hypotenuse,** the side opposite the **right angle,** to the areas of the squares of the legs. That relationship, or pattern, is based on the lengths of all three legs. A Greek mathematician named Pythagoras noticed the relationship about 2,500 years ago and drew a conclusion. That conclusion, known as the Pythagorean Theorem, can be stated as follows: In a right triangle, the square of the length of the hypotenuse is equal to the sum of the squares of the lengths of the legs.

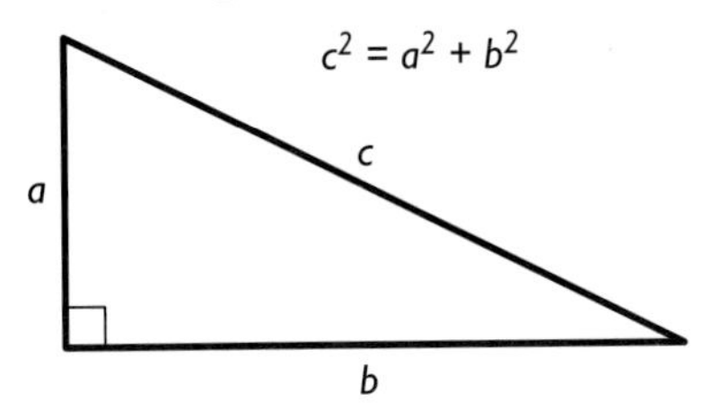

You can use the Pythagorean Theorem to find the third side of a right triangle if you know two sides.

USING THE PYTHAGOREAN THEOREM TO FIND THE HYPOTENUSE

Use the Pythagorean Theorem to find the length of the hypotenuse, c, of $\triangle EFG$.

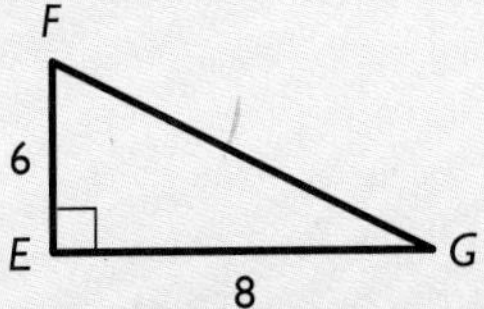

$c^2 = a^2 + b^2$

$c^2 = 6^2 + 8^2$
- Substitute the two known lengths for a and b.

$c^2 = 36 + 64$
- Square the two known lengths.

$c^2 = 100$
- Find the sum of the squares of the two legs.

$c = 10$
- Take the square root of the sum.

USING THE PYTHAGOREAN THEOREM TO FIND A SIDE LENGTH

Use the Pythagorean Theorem, $c^2 = a^2 + b^2$, to find the length of the leg, b, of a right triangle with a hypotenuse of 14 in. and one leg measuring 5 in.

14^2	$= 5^2 + b^2$	• Use $c^2 = a^2 + b^2$.
196	$= 25 + b^2$	• Square the known lengths.
$196 - 25$	$= (25 - 25) + b^2$	• Subtract to isolate the unknown.
171	$= b^2$	
$13.076696 \ldots$	$= b$	• Use your calculator to find the square root. Round to nearest tenth.

13.1 in. is the length of the unknown side.

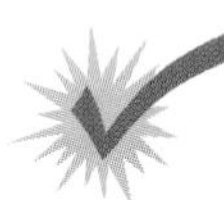

Check It Out

3. Find the length to the nearest whole number of the hypotenuse of a right triangle with legs measuring 9 cm and 11 cm.
4. Find the length of $\overline{SR}$ to the nearest whole number.

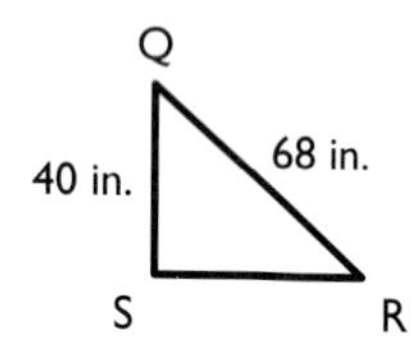

Pythagorean Triples

The numbers 3, 4, and 5 form a **Pythagorean triple** because $3^2 + 4^2 = 5^2$. Pythagorean triples are formed by whole numbers, so that $a^2 + b^2 = c^2$. There are many Pythagorean triples. Here are three:

5, 12, 13   8, 15, 17   7, 24, 25

If you multiply each number of a Pythagorean triple by the same number, you form another Pythagorean triple. 6, 8, 10 is a triple because it is 2(3), 2(4), 2(5).

7•9 EXERCISES

Each side of the dot-paper triangle has a square on it. The squares are labeled regions I, II, and III.

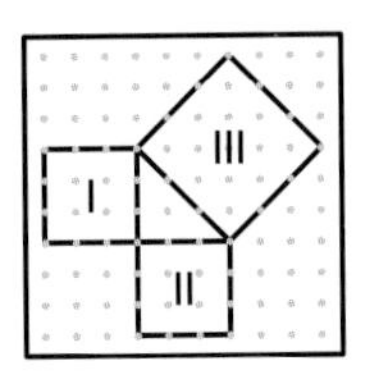

1. Find the areas of regions I, II, and III.
2. What relationship exists among the areas of regions I, II, and III?

Find the missing length in each right triangle. Round to the nearest tenth.

3.
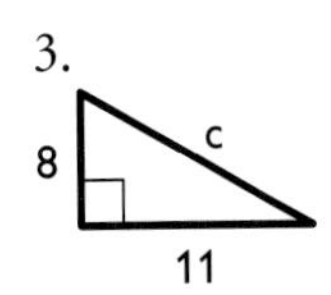

4.
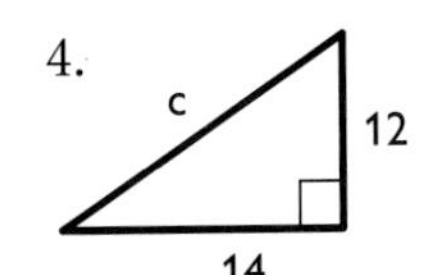

5.
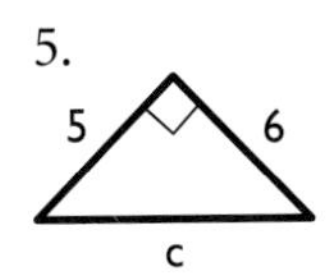

Which numbers are Pythagorean triples? Write yes or no.

6. 3, 4, 5
7. 4, 5, 6
8. 24, 45, 51
9. Find the length, to the nearest tenth, of the unknown leg of a right triangle with a hypotenuse of 16 in. and one leg measuring 9 in.
10. To the nearest tenth, find the length of the hypotenuse of a right triangle with legs measuring 39 cm and 44 cm.

7·10 Tangent Ratio

Sides and Angles in a Right Triangle

In every right triangle there is one right angle and two **acute angles.** The hypotenuse, which is the longest side, is opposite the right angle. The other two sides are called the *legs of a triangle.*

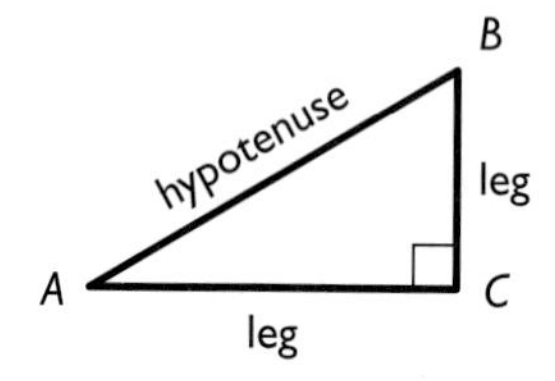

$m\angle A$ and $m\angle B < 90°$

Sometimes the legs are called the *opposite* and *adjacent* sides to describe where they are in relation to one of the acute angles of the right triangle.

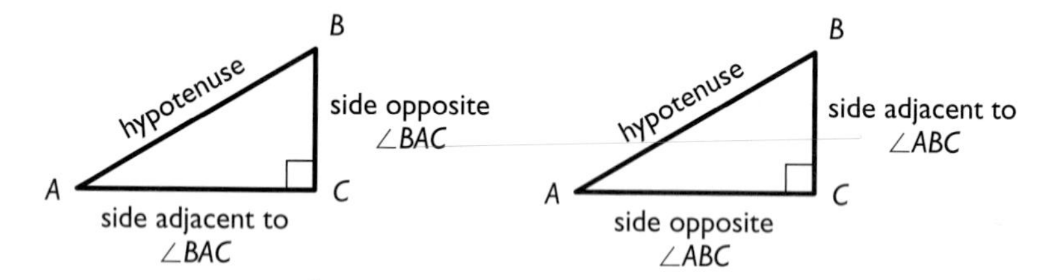

Tangent of an Angle

For any acute angle of a right triangle, the ratio of the length of the opposite side and the length of the adjacent side is called the **tangent** of the angle.

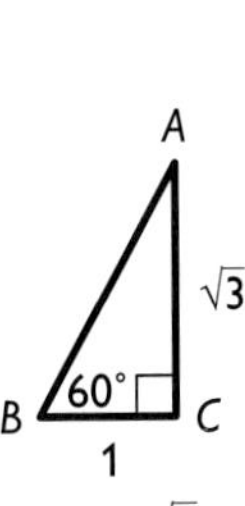

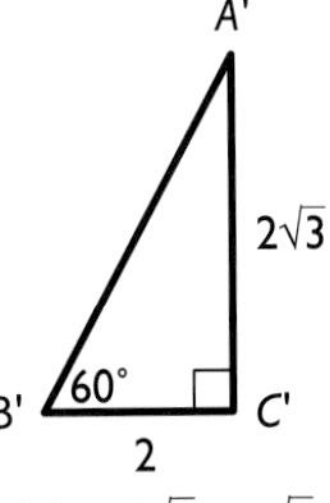

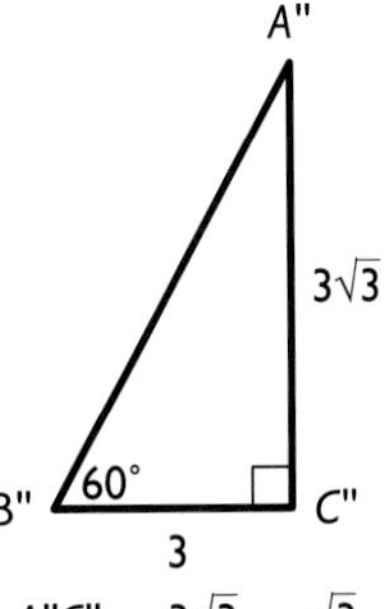

$\frac{AC}{BC} = \frac{\sqrt{3}}{1}$ $\frac{A'C'}{B'C'} = \frac{2\sqrt{3}}{2} = \frac{\sqrt{3}}{1}$ $\frac{A''C''}{B''C''} = \frac{3\sqrt{3}}{3} = \frac{\sqrt{3}}{1}$

This ratio is called the tangent of an acute angle of a right triangle. Using *tan A* as an abbreviation, the ratio is:

$$\tan A = \frac{\text{length of the side opposite } \angle A}{\text{length of the side adjacent } \angle A}$$

FINDING THE TANGENT OF AN ANGLE

Find tan *P*.

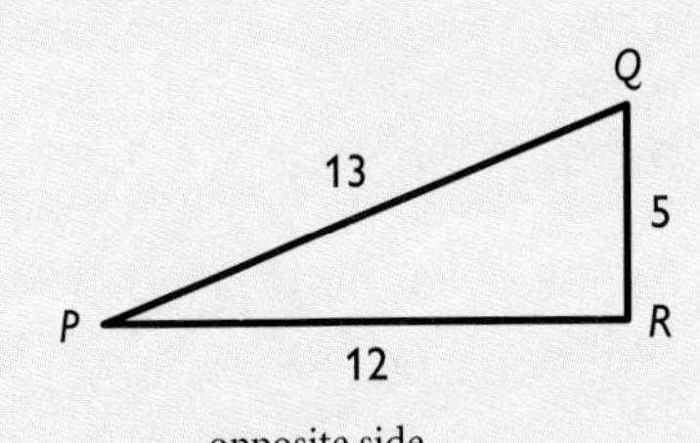

$$\tan P = \frac{\text{opposite side}}{\text{adjacent side}}$$
$$= \frac{5}{12}$$
$$= 0.42$$

- Identify the sides opposite and adjacent to the angle.
- Use the lengths of the opposite and adjacent sides to write the tangent ratio as a fraction.
- Write the tangent of the angle as a decimal.

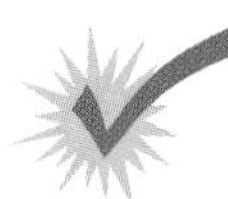

Check It Out

1. Find tan Q for $\triangle PQR$ above.
2. Find tan M.

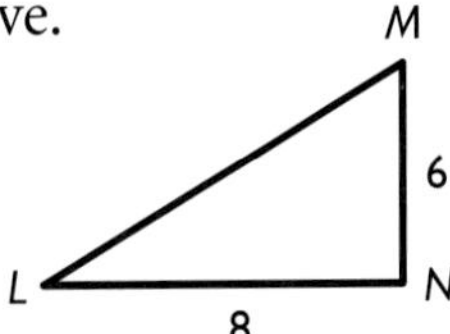

Tangent Table

Since the **tangent ratio** always remains the same for an angle measure, you will find a table of tangents useful for solving problems with right triangles.

To find the tangent of an angle, find the measure of the angle in the Angle column and then read the tangent ratio in the Tangent column.

Angle	Tangent	Angle	Tangent	Angle	Tangent
1°	0.0175	31°	0.6009	61°	1.8040
2°	0.0349	32°	0.6249	62°	1.8807
3°	0.0524	33°	0.6494	63°	1.9626
4°	0.0699	34°	0.6754	64°	2.0503
5°	0.0875	35°	0.7002	65°	2.1445
6°	0.1051	36°	0.7265	66°	2.2460
7°	0.1228	37°	0.7536	67°	2.3559
8°	0.1405	38°	0.7813	68°	2.4751
9°	0.1584	39°	0.8098	69°	2.6051
10°	0.1763	40°	0.8391	70°	2.7475
11°	0.1944	41°	0.8693	71°	2.9042
12°	0.2126	42°	0.9004	72°	3.0777
13°	0.2309	43°	0.9325	73°	3.2709
14°	0.2493	44°	0.9657	74°	3.4874
15°	0.2679	45°	1.0000	75°	3.7321
16°	0.2867	46°	1.0355	76°	4.0108
17°	0.3057	47°	1.0724	77°	4.3315
18°	0.3249	48°	1.1106	78°	4.7046
19°	0.3443	49°	1.1504	79°	5.1446
20°	0.3640	50°	1.1918	80°	5.6713
21°	0.3839	51°	1.2349	81°	6.3138
22°	0.4040	52°	1.2799	82°	7.1154
23°	0.4245	53°	1.3270	83°	8.1443
24°	0.4452	54°	1.3764	84°	9.5144
25°	0.4663	55°	1.4281	85°	11.4301
26°	0.4877	56°	1.4826	86°	14.3007
27°	0.5095	57°	1.5399	87°	19.0811
28°	0.5317	58°	1.6003	88°	28.6363
29°	0.5543	59°	1.6643	89°	57.2900
30°	0.5774	60°	1.7321		

7·10 EXERCISES

Find the value of each tangent. Give answers to the nearest hundredth.

1. tan A
2. tan B
3. tan D
4. tan E

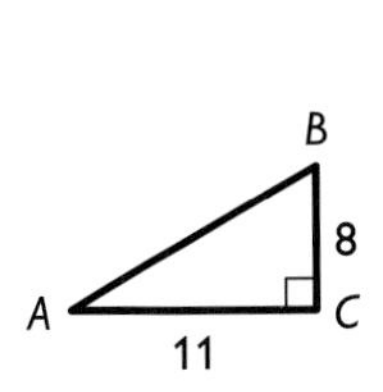

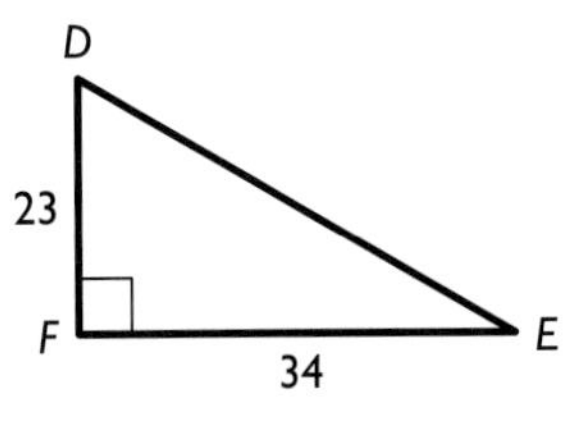

5. tan 45°
6. tan 30°
7. tan 74°
8. tan 17°
9. tan 53°

10. Find the measure of $\angle J$ to the nearest degree.

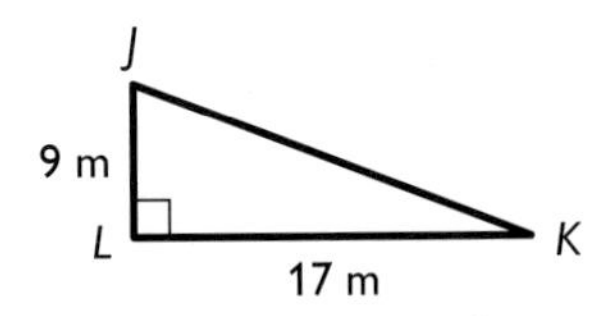

11. A flagpole stands 6 m high. A wire is fastened to the ground 15 m away and is attached to the top of the pole. What angle does the wire form with the ground?

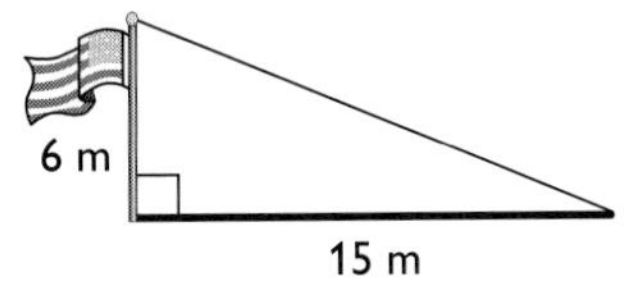

Use $\triangle MNO$ to answer items 12–15.

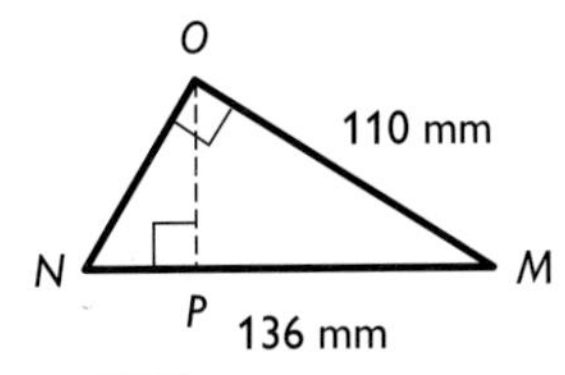

12. What is the length of $\overline{NO}$?
13. What is $\angle M$?
14. What is $\angle N$?
15. What is the length of $\overline{OP}$?

What have you learned?

You can use the problems and the list of words that follow to see what you have learned in this chapter. You can find out more about a particular problem or word by referring to the boldfaced topic number (for example, **7•2**).

Problem Set

Refer to this figure to answer items 1–3. **7•1**

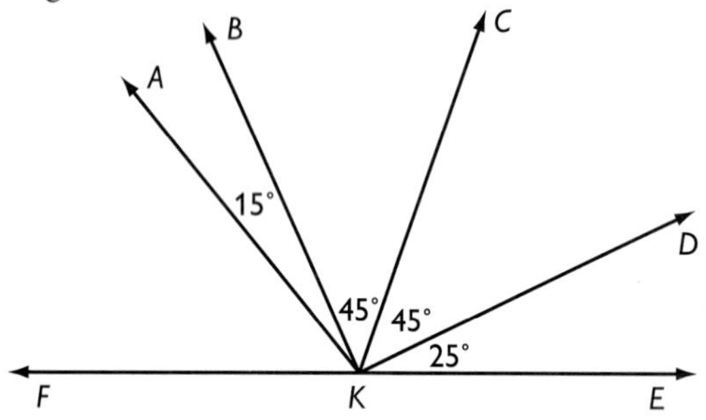

1. Name the angle adjacent to $\angle BKE$.
2. Name the right angle in this figure.
3. $\angle FKE$ is a straight angle. What is the $m\angle FKA$?
4. Find the measure of each angle of a regular hexagon. **7•2**
5. If a figure is a square, must it also be a rhombus? **7•2**
6. A solid figure has a 12-sided base and triangular sides that meet in a point. Is it a prism or a pyramid? **7•3**
7. A right triangle has legs of 3 cm and 4 cm. What is the perimeter of the triangle? **7•4**
8. What is the perimeter of a regular octagon whose sides are 16 cm in length? **7•4**
9. Find the area of a right triangle with legs 20 m and 5 m. **7•5**
10. A rectangle measures 18 ft by 6 ft. What is its area? **7•5**
11. A trapezoid has bases of 12 ft and 20 ft. The distance between these bases is 6 ft. What is the area of the trapezoid? **7•5**
12. Each face of a hexagonal prism is 10 cm by 12 cm. The area of each base is 45 cm^2. Find the surface area of the prism. **7•6**
13. Find the surface area of a cylinder if its height is 10 ft and circumference is 4.5 ft. Round to the nearest square ft. **7•6**

14. Find the volume of a cylinder that has an 8 in. diameter and is 6 in. high. Round to the nearest cubic inch. **7•7**
15. A cone, a rectangular prism, and a cylinder all have the same base area and height. Which figure has less volume? **7•7**
16. What is the circumference and area of a circle with radius 30 m? Round to the nearest m or m^2. **7•7**
17. A triangle has sides of 15 cm, 16 cm, and 23 cm. Use the Pythagorean Theorem to determine whether this triangle is a right triangle. **7•9**
18. Find the length of the unknown leg of a right triangle that has hypotenuse 18 m and one leg that is 10 m. **7•9**
19. In right triangle RST, $\overline{RS}$ is the hypotenuse and $\angle T$ is the right angle. Which side is opposite $\angle S$, which side is adjacent to $\angle S$, and what is the tangent ratio for $\angle S$? **7•10**
20. In the right triangle of problem 19, if TS is 15 m and RT is 11.3 m, what is the measure of $\angle S$? **7•10**

hot words

WRITE THE DEFINITIONS FOR THE FOLLOWING WORDS:

acute angle **7•10**
angle **7•1**
arc **7•8**
base **7•5**
circle **7•6**
circumference **7•6**
congruent **7•1**
cube **7•2**
cylinder **7•6**
degree **7•1**
diagonal **7•2**
diameter **7•8**
face **7•2**
hexagon **7•2**
hypotenuse **7•9**
isosceles triangle **7•4**
legs of a triangle **7•5**
line **7•1**
opposite angle **7•2**
parallel **7•2**
parallelogram **7•2**
pentagon **7•2**
perimeter **7•4**
perpendicular **7•5**
pi **7•7**
point **7•1**
polygon **7•1**
polyhedron **7•2**
prism **7•2**
pyramid **7•2**
Pythagorean Theorem **7•4**
Pythagorean triple **7•9**
quadrilateral **7•2**
radius **7•8**
ray **7•1**
rectangular prism **7•2**
reflection **7•3**
regular shape **7•2**
rhombus **7•2**
right angle **7•1**
right triangle **7•4**
rotation **7•3**
segment **7•8**
surface area **7•6**
symmetry **7•3**
tangent **7•10**
tetrahedron **7•2**
transformation **7•3**
translation **7•3**
trapezoid **7•2**
triangular prism **7•6**
vertex **7•1**
volume **7•7**

hot topics 8

Measurement

8•1	Systems of Measurement	408
8•2	Length and Distance	412
8•3	Area, Volume, and Capacity	416
8•4	Mass and Weight	420
8•5	Time	422
8•6	Size and Scale	424

What do you already know?

You can use the problems and the list of words that follow to see what you already know about this chapter. The answers to the problems are in Hot Solutions at the back of the book, and the definitions are in Hot Words at the front of the book. You can find out more about a particular problem or word by referring to the boldfaced topic number (for example, **8•2**).

Problem Set

Give the meaning for each metric system prefix. **8•1**

1. centi-
2. kilo-
3. milli-

Complete each of the following conversions. Round your answers to the nearest hundredth. **8•2**

4. 800 mm = ? m
5. 5,500 m = ? km
6. 3 mi = ? ft
7. 468 in. = ? yd

Items 8–13 refer to the rectangle. Round to the nearest whole unit.

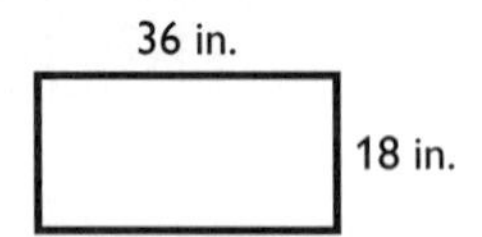

8. What is the perimeter in inches? **8•2**
9. What is the perimeter in yards? **8•2**
10. What is the perimeter in centimeters? **8•2**
11. What is the approximate perimeter in meters? **8•2**
12. What is the area in square inches? **8•3**
13. What is the area in square centimeters? **8•3**

Convert the following area and volume measurements as indicated. **8•3**

14. $5\ m^2 = ?\ cm^2$
15. $10\ yd^2 = ?\ ft^2$
16. $3\ ft^3 = ?\ in.^3$
17. $4\ cm^3 = ?\ mm^3$

18. You pour 6 pints of water into a gallon jar. What fraction of the jar is filled? **8•3**
19. A perfume bottle holds $\frac{1}{2}$ fl oz. How many bottles would you need to fill 1 cup? **8•3**
20. A can of juice holds 385 mL. About how many cans will it take to fill a 5 liter container? **8•3**
21. You are allowed to take a 20 kg suitcase on a small airplane in Africa. About how many pounds will your suitcase weigh? **8•4**
22. Your cookie recipe calls for 4 oz of butter. To make 12 batches of cookies for the bake sale, how many pounds of butter do you buy? **8•4**
23. How many seconds are in 2 days? **8•5**

A picture 5 in. high and 8 in. wide was enlarged to make a poster. The width of the poster is 1.5 ft.

24. What is the ratio of the width of the poster to the width of the original photo? **8•6**
25. What is the scale factor? **8•6**

CHAPTER 8

hot words

accuracy **8•1**
area **8•1**
customary system **8•1**
distance **8•2**
factors **8•1**
fractions **8•1**
length **8•2**
metric system **8•1**
perimeter **8•1**
power **8•1**
ratio **8•6**
rectangle **8•1**
rounding **8•1**
scale factor **8•6**
side **8•1**
similar figures **8•6**
square **8•1**
time **8•5**
volume **8•3**

8•1 Systems of Measurement

If you have ever watched the Olympic Games, you may have noticed that the distances are measured in meters or kilometers, and weights are measured in kilograms. That is because the most common system of measurement in the world is the **metric system.** In the United States, we use the **customary system** of measurement. It will be useful for you to be able to make conversions from one unit of measurement to another within each system, as well as convert units between the two systems.

The Metric and Customary Systems

The metric system of measuring is based on **powers** of ten, such as 10, 100, and 1,000. Converting within the metric system is simple because it is easy to multiply and divide by powers of ten.

Prefixes in the metric system have consistent meanings.

Prefix	Meaning	Example
milli-	one thousandth	1 *milli*liter is 0.001 liter.
centi-	one hundredth	1 *centi*meter is 0.01 meter.
kilo-	one thousand	1 *kilo*gram is 1,000 grams.

BASIC MEASURES

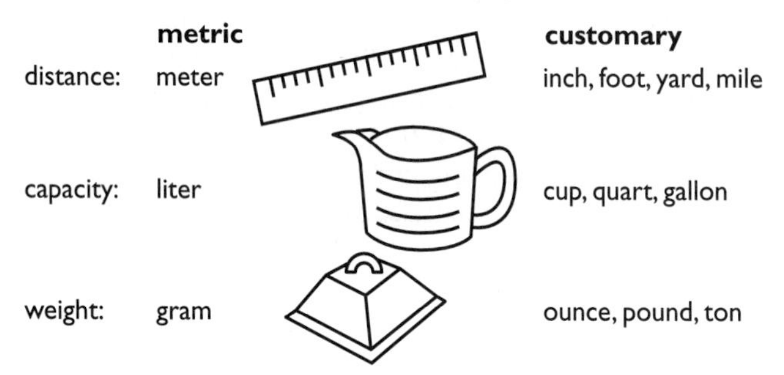

The customary system of measurement is not based on powers of ten. It is based on numbers like 12 and 16, which have many **factors.** This makes it easy to find, say, $\frac{2}{3}$ ft or $\frac{3}{4}$ lb. While the metric system uses decimals, you will frequently encounter **fractions** in the customary system.

Unfortunately, there are no convenient prefixes as in the metric system, so you will have to memorize the basics: 16 oz = 1 lb; 36 in. = 1 yd; 4 qt = 1 gal; and so on.

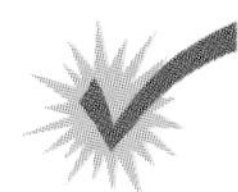

Check It Out

1. Which system is based on multiples of 10?
2. Which system uses fractions?

From Boos to Cheers

It took 200 skyjacks two years and 2.5 million rivets to put together the Eiffel Tower. When it was completed in 1899, the art critics of Paris considered it a blight on the landscape. Today, it is one of the most familiar and beloved monuments in the world.

The tower's height, not counting its TV antennas, is 300 meters—that's about 330 yards or 3 football fields. On a clear day, the view can extend for 67 km. Visitors can take elevators to the platforms or climb up the stairs—all 1,652 of them!

Accuracy

Accuracy has to do with both reasonableness and **rounding.** The length of each **side** of the **square** below is measured accurately to the nearest tenth of a meter. But the actual length could be anywhere from 12.15 m to 12.24 m. (These are the numbers that all round to 12.2.)

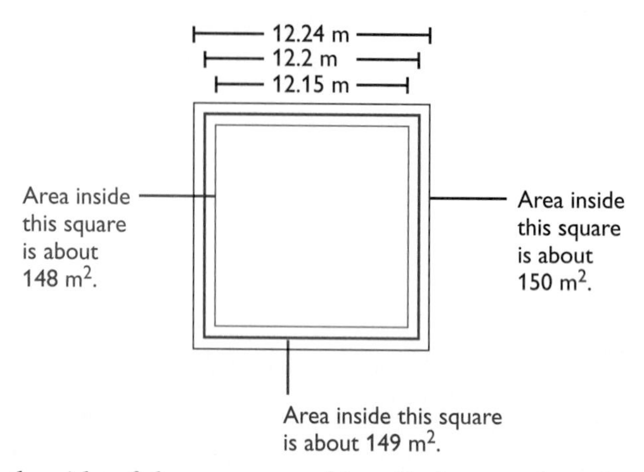

Since the side of the square could really be anywhere between 12.15 m and 12.24 m, the actual **area** may range anywhere between 148 m^2 and 150 m^2. So is it reasonable to square the side $(12.2)^2$ to get an area of 148.84 m^2? No, it isn't. Here is why. The actual length is between 12.15 m and 12.24 m. The area is between 148 m^2 and 150 m^2. Therefore 149 m^2 is reasonable, but the last two digits in 148.84 are meaningless.

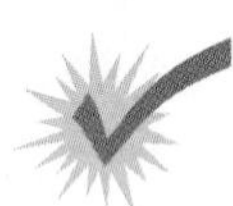

Check It Out

3. In the 12.2 m × 12.2 m square, find the **perimeter** and discuss the accuracy of your answer. (Assume 12.2 is rounded to the nearest tenth.)
4. If your friend's car can drive approximately 300.5 mi on 17 gal of gas, would it be accurate to say the car gets 17.67 mi/gal? Why or why not?

8•1 EXERCISES

Give the meaning of each prefix.
1. centi-
2. kilo-
3. milli-

Write the system of measurement used for the following.
4. Inches and pounds
5. Quarts and gallons
6. Liters and grams
7. Which system of measurement is based on powers of ten?

In each **rectangle,** the dimensions are given to the nearest tenth. Find the perimeter, and discuss the level of accuracy.

8. 8.1 m; 12.8 m

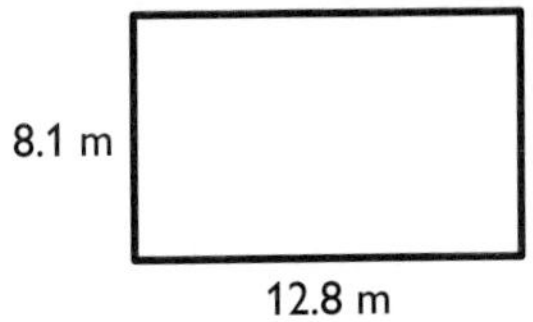

9. 30.1 ft; 4.4 ft

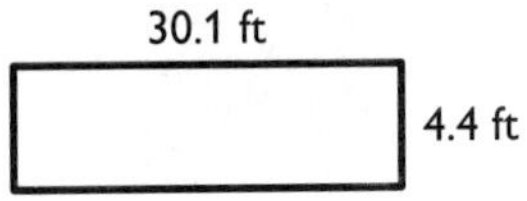

10. 30.8 cm; 7.4 cm

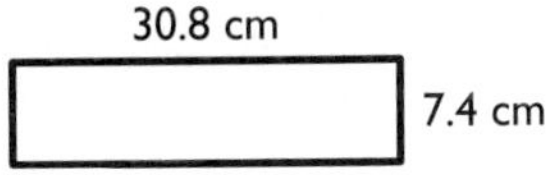

8•2 Length and Distance

About What Length?

When you get a feel for "about how long" or "around how far," it's easier to make estimations about length and distance. Here are some everyday items that will help you keep in mind what metric and customary units mean.

METRIC UNITS	CUSTOMARY UNITS
centimeter 1 cm about the width of a small paper clip	inch 1 in. about the length of a small paper clip
millimeter 1 mm a little less than the width of a dime	foot 1 ft a little taller than a binder
meter 1 m about the height of a standard doorknob	yard 1 yd about the height of a stool

Check It Out

1. Use a metric rule or meter stick to measure common items. Name an item that is about a millimeter; about a centimeter; about a meter.
2. Use a customary rule or yardstick to measure common items. Name an item that is about an inch; about a foot; about a yard.

Metric and Customary Units

When you are calculating **length** and **distance,** you may encounter two different *systems of measurement* (p. 408). One is the metric system, and the other is the customary system. The commonly used metric measures for length and distance are millimeter (mm), centimeter (cm), meter (m), and kilometer (km). The customary system uses inch (in.), foot (ft), yard (yd), and mile (mi).

Metric Equivalents

1 km	=	1,000 m	=	100,000 cm	=	1,000,000 mm
0.001 km	=	1 m	=	100 cm	=	1,000 mm
		0.01 m	=	1 cm	=	10 mm
		0.001 m	=	0.1 cm	=	1 mm

Customary Equivalents

1 mi	=	1,760 yd	=	5,280 ft	=	63,360 in.
$\frac{1}{1,760}$ mi	=	1 yd	=	3 ft	=	36 in.
		$\frac{1}{3}$ yd	=	1 ft	=	12 in.
		$\frac{1}{36}$ yd	=	$\frac{1}{12}$ ft	=	1 in.

CHANGING UNITS WITHIN A SYSTEM

How many inches are in $\frac{1}{4}$ mile?

units you have
1 mi = 63,360 in.
conversion factor for new units

- Find the units you have where they equal 1 on the equivalents chart.
- Find the conversion factor.

$\frac{1}{4} \times 63,360 = 15,840$

- Multiply to get new units.

There are 15,840 inches in $\frac{1}{4}$ mile.

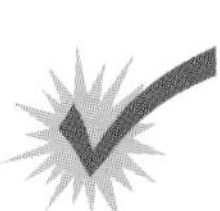

Check It Out

3. 8 m = cm
4. 3,500 m = km
5. 48 in. = ft
6. 2 mi = ft

Conversions Between Systems

Once in a while, you may want to convert between the metric system and the customary system. You can use this conversion table to help.

CONVERSION TABLE

1 inch	=	25.4 millimeters	1 millimeter	=	0.0394 inch
1 inch	=	2.54 centimeters	1 centimeter	=	0.3937 inch
1 foot	=	0.3048 meter	1 meter	=	3.2808 feet
1 yard	=	0.914 meter	1 meter	=	1.0936 yards
1 mile	=	1.609 kilometers	1 kilometer	=	0.621 mile

To make a conversion, find the listing where the unit you have is one. Multiply the number of units you have by the conversion factor for the new units.

Your friend in Costa Rica says he can jump 127 cm. Should you be impressed? 1 cm = 0.3937 in. So $127 \times 0.3937 =$ about 50 in. How far can you jump?

Most of the time you just need to estimate the conversion from one system to the other to get an idea of the size of your item. Round numbers in the conversion table to simplify your thinking. Think that 1 meter is just a little more than 1 yard, 1 inch is between 2 and 3 centimeters, 1 mile is about $1\frac{1}{2}$ kilometers. So now when your friend in Argentina says she caught a fish 60 cm long, you know that the fish is between 20 in. and 30 in. long.

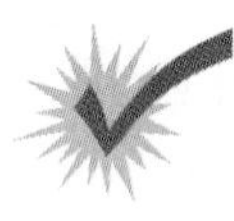

Check It Out

Make exact conversions. Use a calculator and round to the nearest tenth.

7. change 28 in. to centimeters.
8. change 82 m to yards.
9. 9 km is about A. 9 mi, B. 6 mi, C. 15 mi
10. 66 in. is about A. 140 cm, B. 210 cm, C. 167.6 cm
11. 100 m is about A. 100 yd, B. 110 yd, C. 360 ft

8·2 EXERCISES

Complete the conversions.

1. 10 cm = ____ mm
2. 200 mm = ____ m
3. 3,000 mm = ____ cm
4. 2.4 km = ____ m
5. 11 yd = ____ in.
6. 7 mi = ____ ft
7. 400 in. = ____ ft
8. 3,024 in. = ____ yd
9. 0.5 yd = ____ ft
10. 520 yd = ____ mi

Make exact conversions. Use a calculator, and round to the nearest tenth.

11. Change 6 in. to cm.
12. Change 215 cm to in.
13. Change 2 ft to cm.
14. Change 4 ft to m.
15. Change 200 mm to in.
16. Change 3 km to mi.

Choose the conversion you estimate to be about right.

17. 5 mm is about
    A. 5 in. B. 2 in. C. 5 yd D. $\frac{1}{5}$ in.
18. 1 ft is about
    A. 30 cm B. 1 m C. 50 cm D. 35 mm
19. 25 in. is about
    A. 25 cm B. 1 m C. 0.5 m D. 63.5 cm
20. 300 m is about
    A. $\frac{1}{2}$ mi B. 300 yd C. 600 ft D. 100 yd
21. 100 km is about
    A. 200 mi B. 1,000 yd C. 60 mi D. 600 mi
22. 36 in. is about
    A. 1 cm B. 1 mm C. 1 km D. 1 m
23. 6 ft is about
    A. 6 m B. 200 cm C. 600 cm D. 60 cm
24. 1 cm is about
    A. $\frac{1}{2}$ in. B. 1 in. C. 2 in. D. 1 ft
25. 2 mi is about
    A. 300 m B. 2,000 m C. 2 km D. 3 km

8·3 Area, Volume, and Capacity

Area

Area is the measure of a surface. The walls in your room are surfaces. The large surface of the United States takes up an area of 3,787,319 square miles. The area of contact between a tire and a wet road makes the difference between skidding and staying in control. Area is given in square units.

Area can be measured in metric units or customary units. Sometimes you might want to convert within a measurement system. You can figure out the conversions by going back to the basic *dimensions* (p. 413). Below is a chart that provides the most common conversions.

Metric	Customary
$100 \text{ mm}^2 = 1 \text{ cm}^2$	$144 \text{ in.}^2 = 1 \text{ ft}^2$
$10{,}000 \text{ cm}^2 = 1 \text{ m}^2$	$9 \text{ ft}^2 = 1 \text{ yd}^2$
	$4{,}840 \text{ yd}^2 = 1 \text{ acre}$
	$640 \text{ acre} = 1 \text{ mi}^2$

To convert to a new unit, multiply the units you have by the conversion factor for the new units. If the United States covers an area of about $3{,}800{,}000 \text{ mi}^2$, how many acres is it?

$1 \text{ mi}^2 = 640$ acres,

so $3{,}800{,}000 \text{ mi}^2 = 3{,}800{,}000 \times 640 = 2{,}432{,}000{,}000$ acres

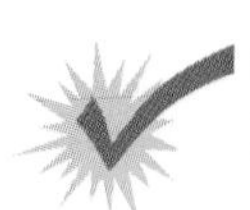

Check It Out

1. How many square millimeters are equal to 16 cm^2?
2. How many square inches are equal to 2 ft^2?

Volume

Volume is expressed in cubic units. Here are the basic relationships among units of volume.

Metric	Customary
$1{,}000 \text{ mm}^3 = 1 \text{ cm}^3$	$1{,}728 \text{ in.}^3 = 1 \text{ ft}^3$
$1{,}000{,}000 \text{ cm}^3 = 1 \text{ m}^3$	$27 \text{ ft}^3 = 1 \text{ yd}^3$

CONVERTING VOLUME WITHIN A SYSTEM OF MEASUREMENT

Express the volume of the carton in cubic meters.

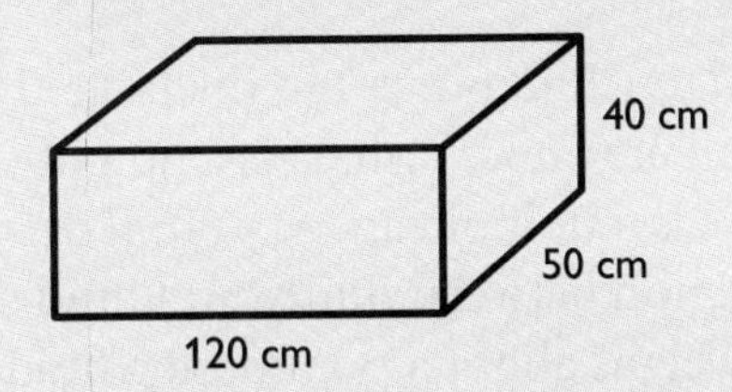

$V = lwh$
$= 120 \times 50 \times 40$
$= 240{,}000 \text{ cm}^3$

- Use a formula to find the *volume* (p. 62), using the units of the dimensions.

$1{,}000{,}000 \text{ cm}^3 = 1 \text{ m}^3$

- Find the conversion factor.

$240{,}000 \div 1{,}000{,}000 = 0.24 \text{ m}^3$

- Multiply to convert to smaller units. Divide to convert to larger units.

So the volume of the carton is 0.24 m^3.

- Include the unit of measurement in your answer.

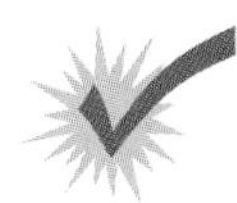

Check It Out

3. Find the volume of a box that measures $9 \text{ ft} \times 6 \text{ ft} \times 6 \text{ ft}$. Convert to cubic yards.
4. Find the volume of a cube that measures 8 cm on a side. Convert to cubic millimeters.

Capacity

Capacity is closely related to volume, but there is a difference. A block of wood has volume but no capacity to hold liquid. The capacity of a container is a measure of the volume of liquid it will hold.

Metric	Customary
1 liter (L) = 1,000 milliliters (mL)	8 fl oz = 1 cup (c)
1 L = 1.057 qt	2 c = 1 pint (pt)
	2 pt = 1 quart (qt)
	4 qt = 1 gallon (gal)

Note the use of *fl oz* (fluid ounce) in the table. This is to distinguish it from *oz* (ounce) which is a unit of weight (16 oz = 1 lb). Fluid ounce is a unit of capacity (16 fl oz = 1 pint). There is a connection between ounce and fluid ounce, however. A pint of water weighs about a pound, so a fluid ounce of water weighs an ounce. For water, as well as for most other liquids used in cooking, *fluid ounce* and *ounce* are equivalent, and the "fl" is sometimes omitted (for example, "8 oz = 1 cup"). To be correct, though, use *ounce* for weight only and *fluid ounce* for capacity. For liquids that weigh considerably more or less than water, the difference is significant.

In the metric system, the basic units of capacity are related.

1 liter (L) = 1,000 milliliters (mL)
Think of a liter as being about a quart.
1 L = 1.057 qt

Gasoline is priced at \$0.39/L. How much is that per gallon? There are 4 quarts in a gallon, so there are 4 × 1.057 or 4.228 liters in a gallon. So a gallon costs \$0.39 × 4.228 or \$1.649 per gallon.

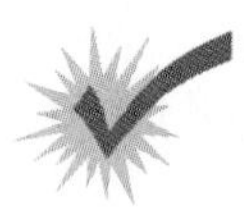

Check It Out

5. If liters of cola are on sale for \$.69 each and you can buy a can of juice that makes 1 gallon for \$2.49, which is the better buy?

8·3 EXERCISES

Tell whether the unit is a measure of distance, area, or volume.

1. cm
2. in^3
3. acre
4. mm^2

Give the volume of the cartons in each measurement unit below.

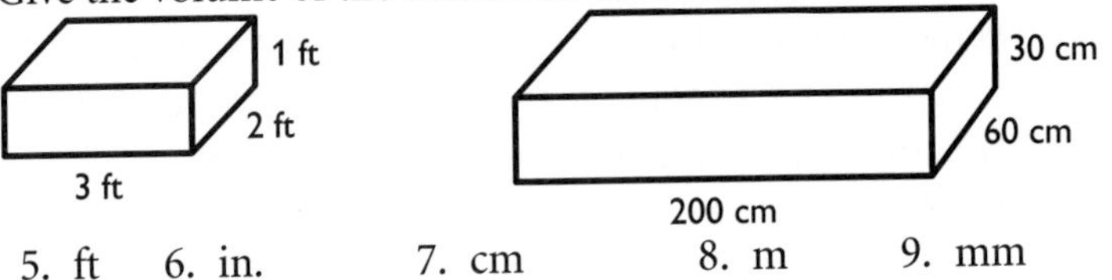

5. ft
6. in.
7. cm
8. m
9. mm

Convert to new units.

10. 1 gal = ? c
11. 2 qt = ? fl. oz
12. 160 fl. oz = ? qt
13. 4 gal = ? qt
14. 3 pt = ? gal
15. 4 fl. oz = ? pt
16. 8 L = ? mL
17. 24,500 mL = ? L
18. 10 mL = ? L
19. Krutika has a fish tank that holds 15 L of water. One liter of water has evaporated. She has a 200-mL measuring cup. How many times will she have to fill the cup in order to refill the tank?
20. Estimate, to the nearest dime, the price per liter of gasoline selling for $1.20/gal.

In the Soup!

One morning on a California freeway, a big-rig truck tipped over on its side. The truck was carrying 43,000 cans of cream of mushroom soup.

At 24 cans per carton, how many cartons of soup was the truck carrying? If each carton had a width of 11 in., a length 16 in., and a height of 5 in., what was the approximate carrying capacity of the truck in cubic feet? See Hot Solutions for answer.

8·3 EXERCISES

8•4 Mass and Weight

Technically, mass and weight are different. Mass is the amount of substance you have. Weight is the pull of gravity on the amount of substance. On Earth, mass and weight are equal at sea level and about equal at other elevations. But on the moon, mass and weight can be quite different. Your mass would be the same on the moon as it is here on Earth. But, if you weigh 100 pounds on Earth, you would weigh about $16\frac{2}{3}$ pounds on the moon. That is because the gravitational pull of the moon is only $\frac{1}{6}$ that of the Earth.

The customary system measures weight. The metric system measures mass.

Metric	Customary
1 kg = 1,000 g = 1,000,000 mg	1 T = 2,000 lb = 32,000 oz
0.001 kg = 1 g = 1,000 mg	0.0005 T = 1 lb = 16 oz
0.000001 kg = 0.001 g = 1 mg	0.0625 lb = 1 oz

1 pound ≈ 0.4536 of the weight of a kilogram at sea level
a 1 kilogram mass weighs ≈ 2.205 pounds

To convert from one unit of mass or weight to another, first find the 1 for the units you have in the equivalents chart. Then multiply the number of units you have by the conversion factor for the new units.

If you have 64 oz of peanut butter, how many pounds do you have? 1 oz = 0.0625 lb, so 64 oz = 64 × 0.0625 lb = 4 lb. You have 4 pounds of peanut butter.

Check It Out

Complete the following conversions.

1. 5 lb = ? oz
2. 7,500 lb = ? T
3. 8 kg = ? mg
4. 375 mg = ? g

8·4 EXERCISES

Convert to the indicated units.

1. 1.2 kg = ? mg
2. 250 mg = ? g
3. 126,500 lb = ? T
4. 24 oz = ? lb
5. 8,000 mg = ? kg
6. 2.3 T = ? lb
7. 8 oz = ? lb
8. 250 g = ? oz
9. 100 kg = ? lb
10. 25 lb = ? kg
11. 200 oz = ? lb
12. 880 oz = ? kg
13. 880 g = ? lb
14. 8 g = ? oz
15. 16 oz = ? kg
16. 1.5 T = ? kg
17. Your cookie recipe calls for 12 oz of butter. For your party, you will make 4 batches of cookies. How many pounds of butter do you need to buy?
18. Two brands of laundry soap are on sale. A 2-lb box of Brand A is selling for $12.50. A 20-oz box of Brand X is on sale for $7.35. Which is the best deal?
19. French chocolates sell for $18.50 per kilogram. You find a 10-oz box of chocolates for $7.75. Which is the best deal?
20. If an elephant weighs about 3,500 kg on Earth, how many pounds would it weigh on the moon? Could you lift it? Round your answer to the nearest pound.

Poor SID

SID is a crash-test dummy. After a crash, SID goes to the laboratory for a readjustment of sensors and perhaps a replacement head or other body parts. Because of the forces at work when a car crashes, body parts weigh as much as 20 times their normal weight.

The weight of a body changes during a crash. Does the mass of the body also change? **See Hot Solutions for the answer.**

8•5 Time

Time measures the interval between two or more events. You can measure time with a very short unit—a second—a very long unit—a millennium—and many units in between.

1,000,000 seconds before 12:00 A.M. January 1, 2000, will be 10:13:20 A.M. on December 20, 1999. 1,000,000 hours before 12:00 A.M.January 1, 2000, is 8 A.M. December 8, 1885.

60 seconds (sec) = 1 minute (min)
60 min = 1 hour (hr)
24 hr = 1 day (da)
7 da = 1 week (wk)
365 da = 1 year (yr)
10 yr = 1 decade
100 yr = 1 century
1,000 yr = 1 millennium

Working with Time Units

Like other kinds of measurement, you can convert one unit of time to another, by using the information in the table above.

Hulleah is 13 years old. Her age in months is 13 × 12, or 156 months.

Leap Years

Every four years, February has an extra day. These 366 day years are called *leap years.* Leap years are divisible by 4, but not by 100. However, years that are divisible by 400 are leap years. The year 1996 is a leap year, but the year 1900 is not. The year 2000 is also a leap year.

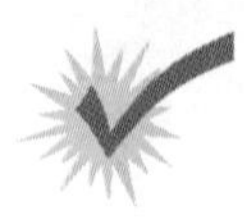

Check It Out

1. How many months old will you be on your 16th birthday?
2. What day will it be 5,000 days from January 1, 2000?

8·5 EXERCISES

Convert each time to the units indicated.

1. 4 min = ? sec
2. 72 hr = ? da
3. 2 yr = ? da
4. 400 sec = ? min
5. 500 yr = ? centuries
6. 20 centuries = ? millennia
7. When you are 14 years old, approximately how many minutes (not including leap years) have you been alive?
8. How many seconds are in a century?
9. How many years is 94,608,000 seconds?
10. How many hours in a month of 30 days?

The World's Largest Reptile

Would it surprise you to learn that the world's largest reptile is a turtle? The leatherback turtle can weigh as much as 2,000 pounds. By comparison, an adult male crocodile weighs about 1,000 pounds.

The leatherback has existed in its current form for over 20 million years, but this prehistoric giant is now endangered. If after 20 million years of existence the leatherback was to become extinct, how many times longer than Homo sapiens will it have existed? Assume Homo sapiens has been around 4,000 millennia. **See Hot Solutions for answer.**

8·6 Size and Scale

Similar Figures

Similar figures are figures that have exactly the same shape. When two figures are similar, one may be larger than the other.

DECIDING IF TWO FIGURES ARE SIMILAR

Are these two rectangles similar?

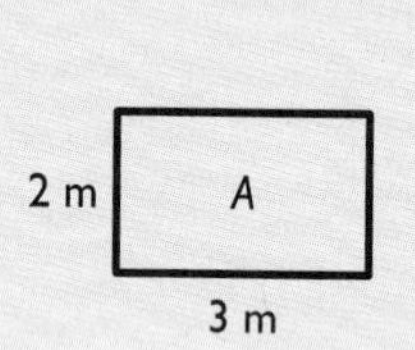

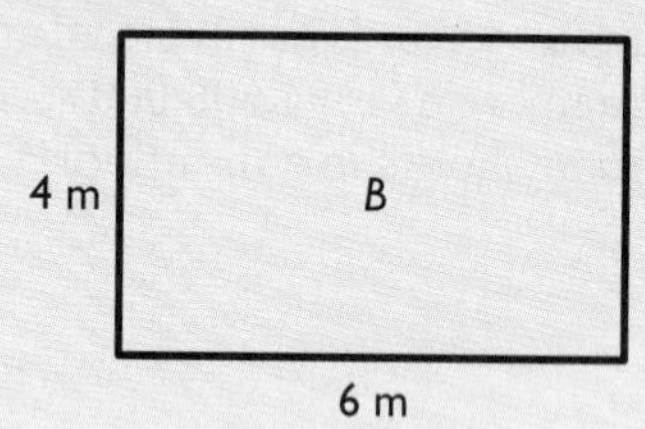

$\frac{3}{6} \stackrel{?}{=} \frac{2}{4}$

$12 = 12$

So the rectangles are similar.

- Set up the **ratios:** $\frac{\text{length } A}{\text{length } B} \stackrel{?}{=} \frac{\text{width } A}{\text{width } B}$
- Cross multiply to see if ratios are equal.
- If all sides have equal ratios, the figures are similar.

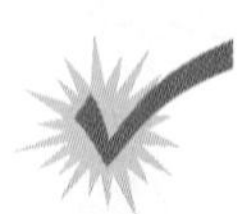

Check It Out

1. Which figures are similar?

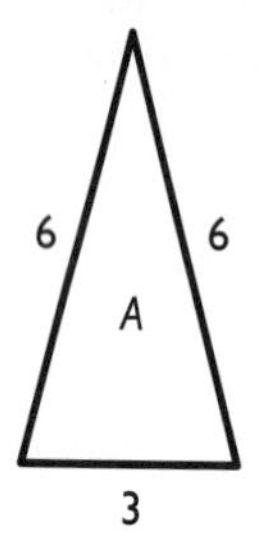

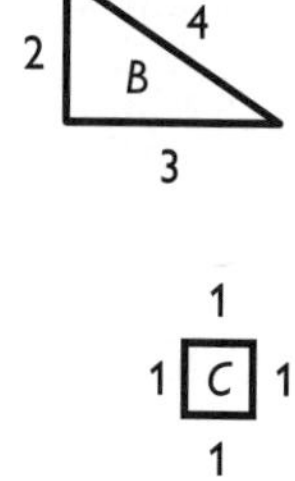

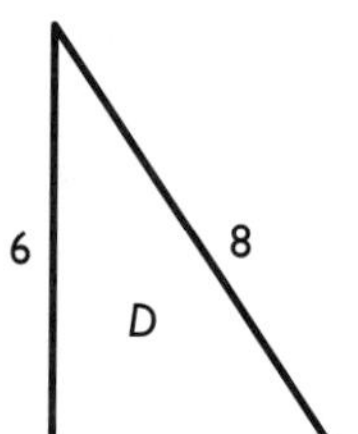

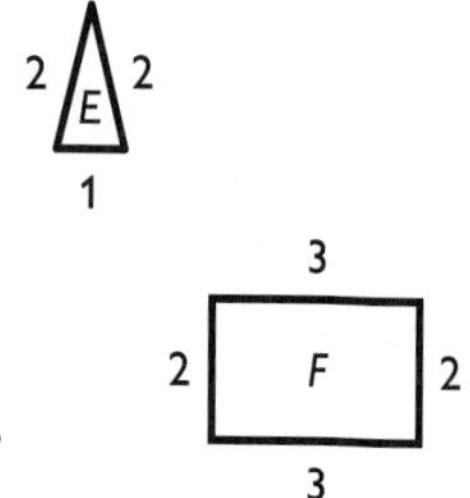

Scale Factors

A **scale factor** indicates the ratio of sizes of two similar figures.

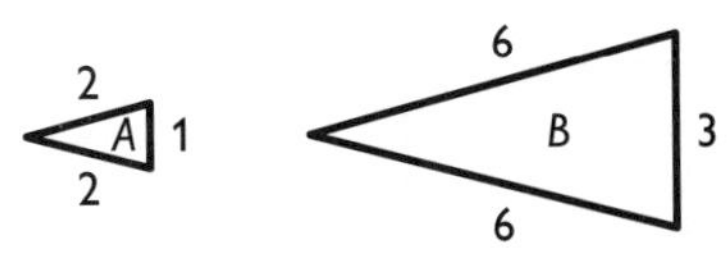

Triangle A is similar to triangle B. $\triangle B$ is 3 times larger than $\triangle A$. The scale factor is 3.

FINDING SCALE FACTOR

What is the scale factor for the similar pentagons?

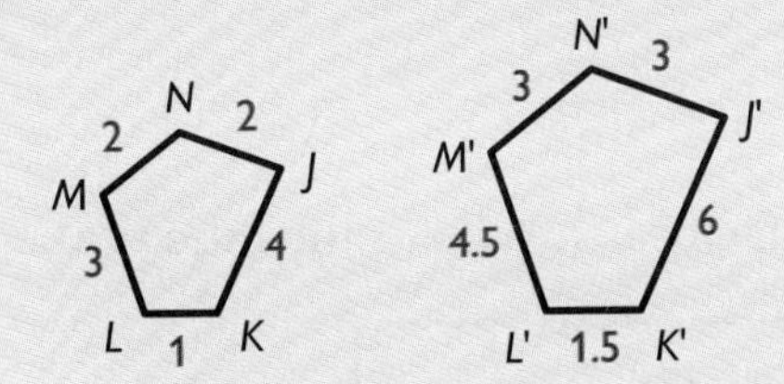

- Decide which figure is the "original figure."
- Make a ratio of corresponding sides: $\frac{\text{new figure}}{\text{original figure}}$ — $\frac{K'J'}{KJ} = \frac{6}{4}$
- Reduce if possible. — $= \frac{3}{2}$

The scale factor of the two pentagons is $\frac{3}{2}$.

When a figure is enlarged, the scale factor is greater than 1. When two similar figures are identical in size, the scale factor is equal to 1. When a figure is reduced, the scale factor is less than 1.

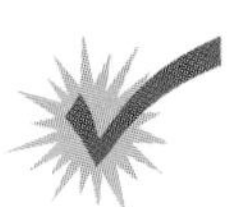

Check It Out

What is the scale factor?

2. 4, 4, 5.5 → 8, 8, 11

3.

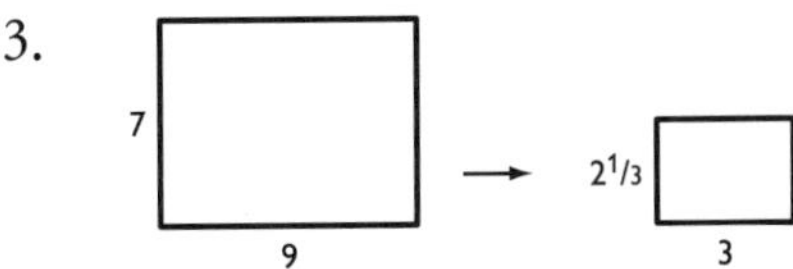

Scale Factors and Area

Scale factor refers to a ratio of lengths only, not of the areas.

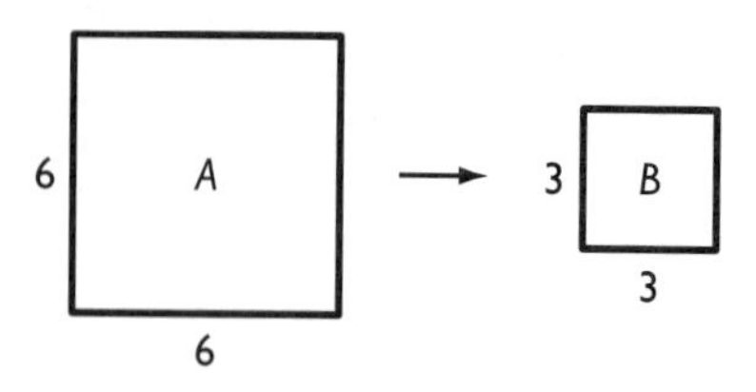

For the squares, the scale factor is $\frac{1}{2}$ because the ratio of sides is $\frac{3}{6} = \frac{1}{2}$. Notice that, while the scale factor is $\frac{1}{2}$, the ratio of the areas is $\frac{1}{4}$.

$$\frac{\text{Area of } B}{\text{Area of } A} = \frac{3^2}{6^2} = \frac{9}{36} = \frac{1}{4}$$

The scale factor is $\frac{1}{3}$. What is the ratio of the areas?

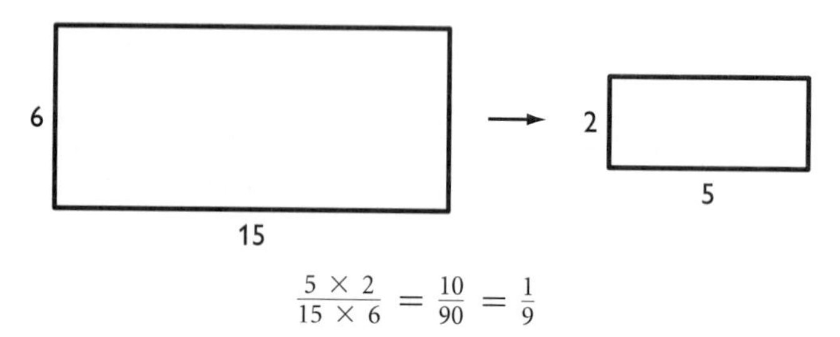

$$\frac{5 \times 2}{15 \times 6} = \frac{10}{90} = \frac{1}{9}$$

The ratio of the areas is $\frac{1}{9}$.

In general, the ratio of the areas of two similar figures is the *square* of the scale factor.

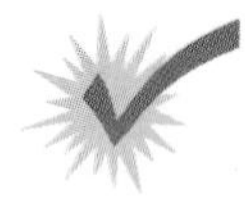

Check It Out

4. The scale factor for two similar figures is $\frac{3}{2}$. What is the ratio of the areas?
5. The scale on a blueprint for a garage is 1 ft = 4 ft. An area of 1 ft^2 on the blueprint represents how much area on the garage floor?

8·6 EXERCISES

Give the scale factor.

1. 12.5 × 5 → 5 × 2

2. 6 → 8

3. A 3 in. × 5 in. photograph is enlarged by a scale factor of 3. What are the dimensions of the enlarged photo?
4. A document 11 in. long and $8\frac{1}{2}$ in. wide is reduced. The reduced document is $5\frac{1}{2}$ in. long. How wide is it?
5. The triangles are similar. If the scale factor is 4, what are the dimensions of the larger triangle?

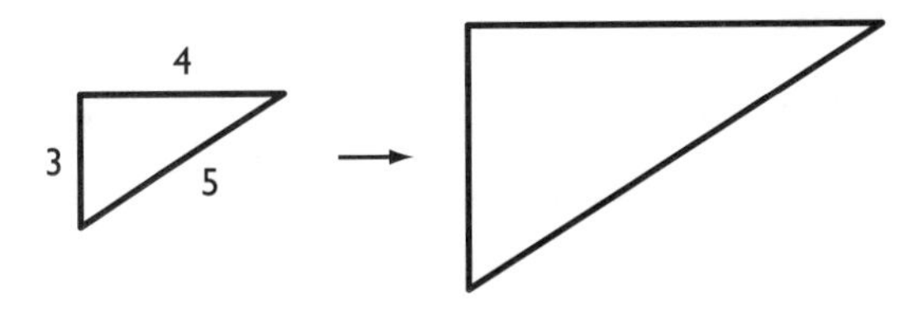

6. A map shows a scale of 1 cm = 20 km. If two towns are about 50 km apart, how far apart will they be on the map?
7. If a map's scale is 1 in. = 5 mi, and the map is a rectangle 12 in. × 15 in., what is the area shown on the map?
8. A photo is enlarged by a scale factor of 1.5. The area of the larger photo is how many times the area of the smaller photo?
9. On a map marked Scale: $\frac{1}{2}$ in. = 10 mi, a road appears to be $2\frac{3}{4}$ in. long. About how many miles long is the road?
10. The triangles are similar, with a scale factor of 2. Find the value of x and y.

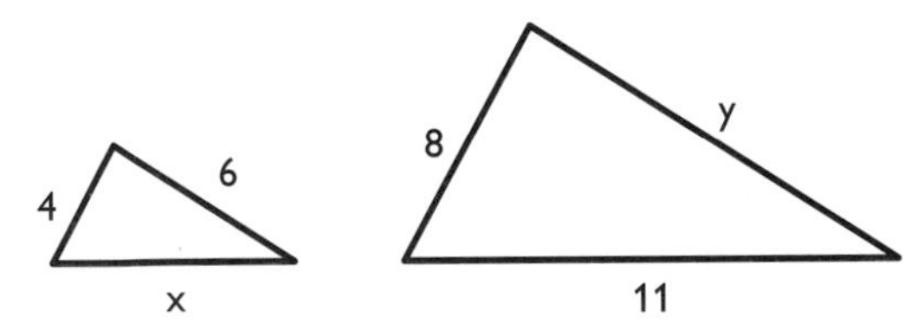

What have you learned?

You can use the problems and the list of words that follow to see what you have learned in this chapter. You can find out more about a particular problem or word by referring to the boldfaced topic number (for example **8•1**).

Problem Set

Give the meaning for each metric system prefix. **8•1**

1. centi-
2. kilo-
3. milli-

Complete each of the following conversions. Round your answers to the nearest hundredth. **8•2**

4. 600 mm = ? m
5. 367 m = ? km
6. 2.5 mi = ? ft
7. 288 in. = ? yd

Items 8–13 refer to the rectangle. Round to the nearest whole unit.

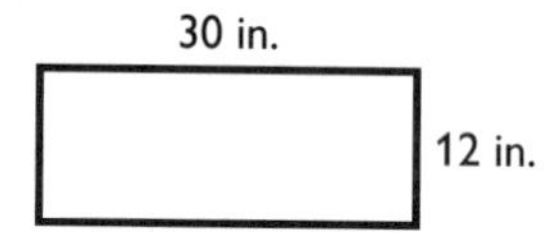

8. What is the perimeter in inches? **8•2**
9. What is the perimeter in yards? **8•2**
10. What is the perimeter in centimeters? **8•2**
11. What is the approximate perimeter in meters? **8•2**
12. What is the area in square inches? **8•3**
13. What is the area in square centimeters? **8•3**

Convert the following area and volume measurements as indicated. **8•3**

14. $42.5\ m^2$ = ? cm^2
15. $7\ yd^2$ = ? ft^2
16. $10\ ft^3$ = ? $in.^3$
17. $6.5\ cm^3$ = ? mm^3

18. You pour 3 pints of water into a gallon jar. What fraction of the jar is filled? **8•3**
19. A perfume bottle holds $\frac{1}{2}$ fl oz. How many bottles would you need to fill $\frac{1}{2}$ cup? **8•3**
20. A can of juice holds 1,250 mL. How many cans will it take to fill a 15 liter container? **8•3**
21. About how many kilograms in 17 lb? **8•4**
22. How many ounces in 8 lb? **8•4**
23. If you count one number per second, how many days would it take you to count to 1,000,000? **8•5**

A photograph 5 in. high and 3 in. wide was enlarged to make a poster. The width of the poster is 1 ft.

24. What is the ratio of the width of the poster to the width of the original photo? **8•6**
25. What is the scale factor? **8•6**

WRITE DEFINITIONS FOR THE FOLLOWING WORDS.

hot words

accuracy **8•1**
area **8•1**
customary system **8•1**
distance **8•2**
factors **8•1**
fractions **8•1**
length **8•2**
metric system **8•1**
perimeter **8•1**
power **8•1**
ratio **8•6**
rectangle **8•1**
rounding **8•1**
scale factor **8•6**
side **8•1**
similar figures **8•6**
square **8•1**
time **8•5**
volume **8•3**

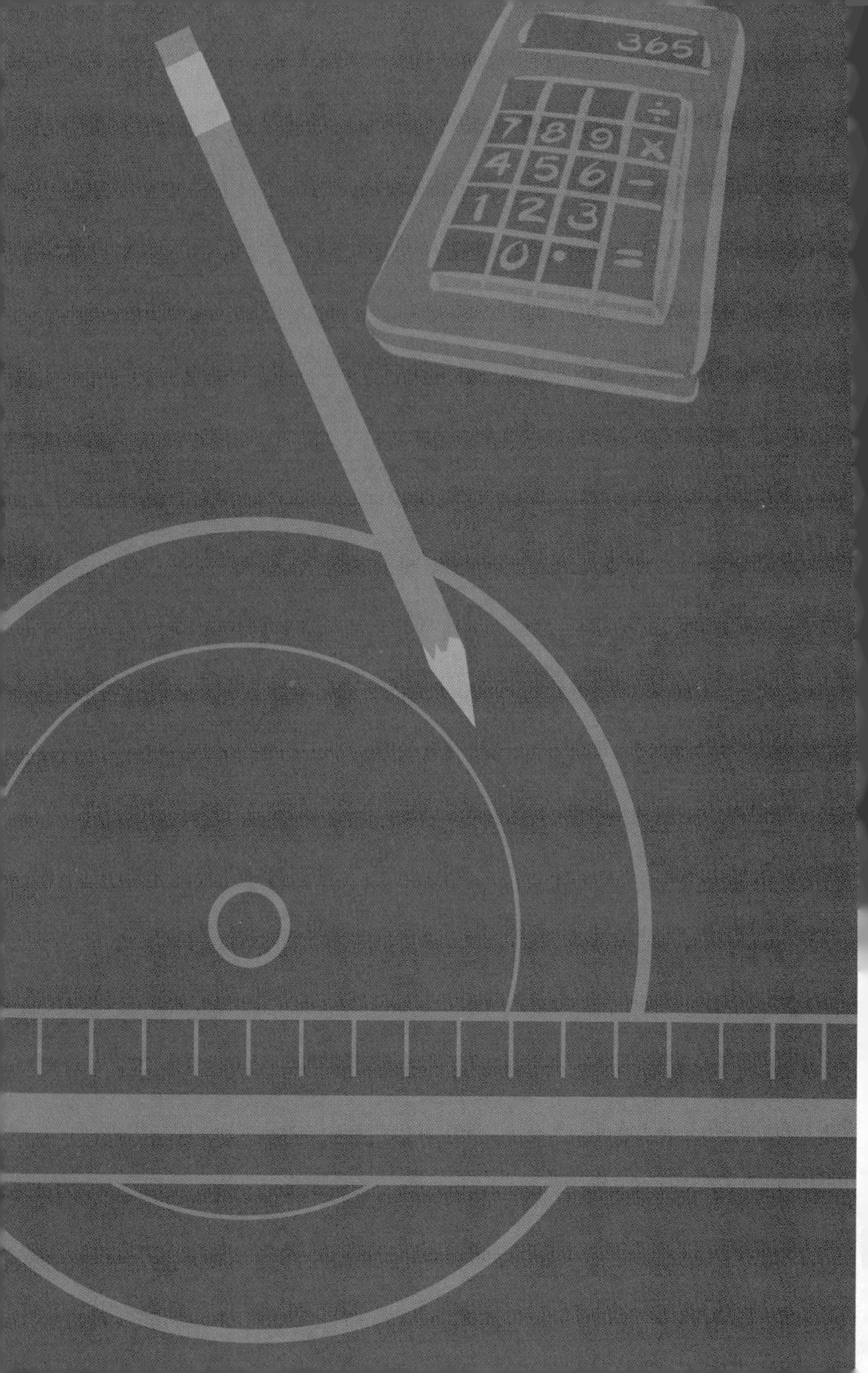

Tools

9•1	Four-Function Calculator	434
9•2	Scientific Calculator	440
9•3	Geometry Tools	444
9•4	Spreadsheets	452

What do you already know?

You can use the problems and the list of words that follow to see what you already know about this chapter. The answers to the problems are in Hot Solutions at the back of the book, and the definitions of the words are in Hot Words at the front of the book. You can find out more about a particular problem or word by referring to the boldfaced topic number (for example, **9•2**).

Problem Set

Use your calculator for items 1–6. **9•1**

1. $82 + 67 \times 14$
2. 225% of 3,500

Round answers to the nearest tenth.

3. $42 \times \sqrt{33} + 27.25$
4. $2 \times \pi - \sqrt{.036}$

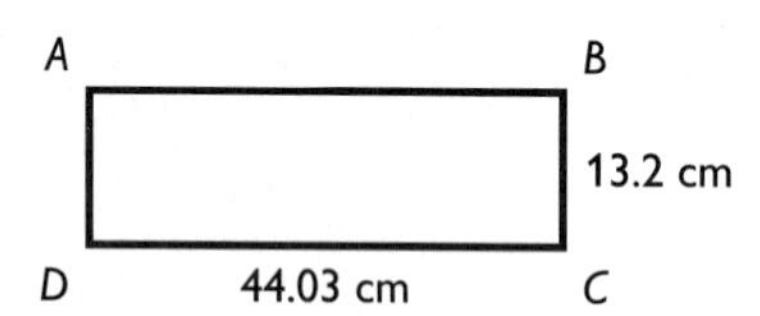

5. Find the perimeter of rectangle *ABCD*.
6. Find the area of rectangle *ABCD*.

Use a scientific calculator for items 7–12. Round decimal answers to the nearest hundredth. **9•2**

7. 8.9^5
8. Find the reciprocal of 3.4.
9. Find the square of 4.5.
10. Find the square root of 4.5.
11. $(8 \times 10^4) \times (4 \times 10^8)$
12. $0.7 \times (4.6 + 37)$

13. What is the measure of $\angle VRT$? **9•3**
14. What is the measure of $\angle VRS$? **9•3**
15. What is the measure of $\angle SRT$? **9•3**
16. Does $\overrightarrow{RT}$ divide $\angle VRS$ into two equal angles? **9•3**

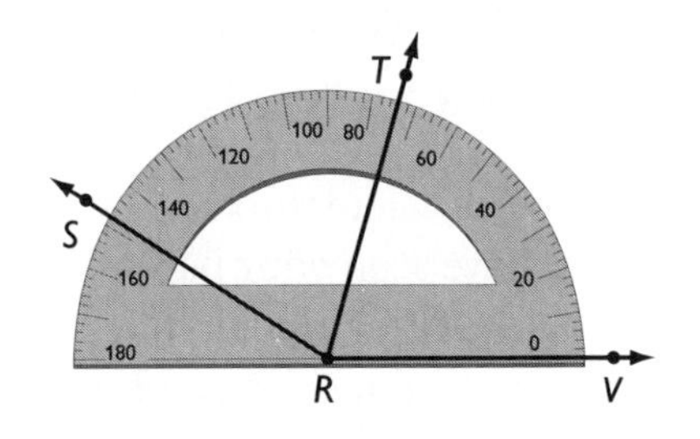

17. What are the basic construction tools in geometry? **9•3**

For items 18–20, refer to the spreadsheet below. **9•4**

	A	B	C	D
1	34	68	100	66
2	14	28	200	
3	20	40	300	
4				

18. Name the cell holding 14.
19. A formula for cell C3 is 3 × C1. Name another formula for cell C3.
20. Cell D1 contains the number 66 and no formula. After using the command fill down, what number will be in cell D10?

CHAPTER 9

hot words

angle **9•2**
arc **9•3**
cell **9•4**
circle **9•1**
column **9•4**
cube **9•2**
cube root **9•2**
decimal **9.1**
degree **9•2**
distance **9•3**
factorial **9•2**
formula **9•4**
horizontal **9•4**
negative number **9•1**
parentheses **9•2**
percent **9•1**
perimeter **9•4**
pi **9•1**
point **9•3**
power **9•2**
radius **9•1**
ray **9•3**
reciprocal **9•2**
roots **9•2**
row **9•4**
spreadsheet **9•4**
square **9•2**
square root **9•1**
tangent **9•2**
vertex **9•3**
vertical **9•4**

9•1 Four-Function Calculator

People use calculators to make mathematical tasks easier. You might have seen your parents balance their checkbooks using a calculator. But a calculator is not always the fastest way to do a mathematical task. If your answer does not need to be exact, it might be faster to estimate. Sometimes you can do the problem in your head quickly, or a pencil and paper might be a better method. Calculators are particularly helpful for problems with many numbers or with numbers that have many digits.

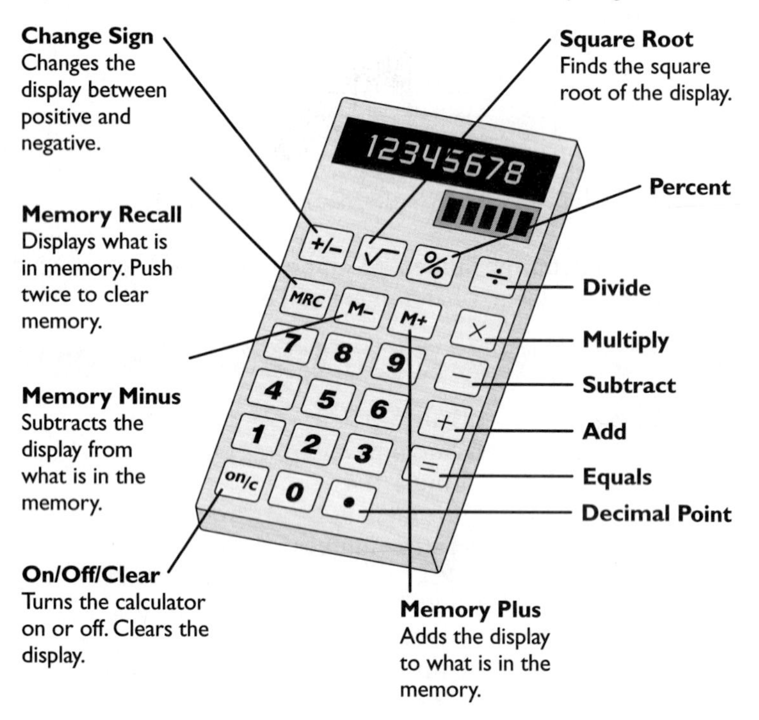

A calculator only gives you the answer to the problem you enter. Always have an estimate of the answer you expect. Then you can compare the calculator answer to your estimate to be sure you entered the problem correctly.

Basic Operations

Adding, subtracting, multiplying, and dividing are fairly straightforward.

Operation	Problem	Calculator Keys	Display
Addition	55 + 49.7	55 [+] 49.7 [=]	104.7
Subtraction	30 − 89	30 [−] 89 [=]	-59.
Multiplication	7.4 × 31.6	7.4 [×] 31.6 [=]	233.84
Division	4 ÷ 30	4 [÷] 30 [=]	0.1333333

Negative Numbers

To enter a **negative number** into your calculator, you press [+/−] after you enter the number.

Problem	Calculator Keys	Display
−35 + 24	35 [+/−] [+] 24 [=]	-11.
56 −(−.5)	56 [−] .5 [+/−] [=]	56.5
−42 × 13	42 [+/−] [×] 13 [=]	-546.
−12 ÷ (−3)	12 [+/−] [÷] 3 [+/−] [=]	4.

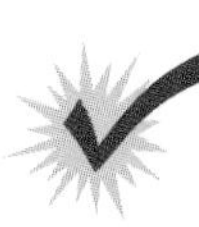

Check It Out

Find each answer on a calculator.

1. 16.1 + 28.9 − 43.7
2. 21 × (−0.75)
3. −7 ÷ 14.8

Memory

For complex or multi-step problems, use memory. You operate memory with three special keys. Here is the way many calculators operate. If yours does not work this way, check the instructions which came with your calculator.

Key	Function
MRC	One push displays (recalls) what is in memory. Push twice to clear memory.
M+	Adds display to what is in memory.
M−	Subtracts display from what is in memory.

When calculator memory contains something other than zero, the display will show [M] along with whatever number the display currently shows. What you do on your calculator does not change memory unless you use the special memory keys.

To solve $35 + 82 + 72 \times 4 + 35 - 16^2$, you could use the following keystrokes to do the problem with your calculator.

Keystrokes	Display
MRC MRC C	0.
16 × 16 M−	M 256.
72 × 4 M+	M 288.
35 + 82 M+	M 117.
35 M+	M 35.
MRC	M 184.

Notice the use of order of operations (p. 82).

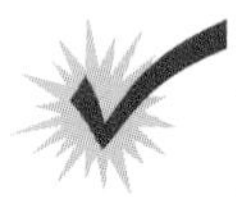

Check It Out

Use memory to find each answer.

4. $7 + 14 \times 5 - 73^3$
5. $8 + 42^4 - (-8) \times 35$

Special Keys

Some calculators have keys with special functions to save time.

Key	Function
$\boxed{\sqrt{x}}$	Finds the **square root** of the display.
$\boxed{\%}$	Changes display to the decimal expression of a **percent.**
$\boxed{\pi}$	Automatically enters **pi** to as many places as your calculator holds.

The Mystery of Memory

Play this memory-calculator game with a friend. Turn on a four-function calculator and clear its memory. Take turns entering numbers less than 50 and pressing the M+ key. When one player thinks the total in the calculator's memory is 200 or greater, check by pressing the MR key. Were you able to add the numbers correctly in your head?

The [%] and [π] keys save you time by saving you keystrokes. The [√] key allows you to find square roots precisely, something difficult to do by hand. See how they work in the examples below.

Problem: $7 + \sqrt{21}$
Keystrokes: 7 [+] 21 [√] [=]
Final display: 11.582575

If you try to find the square root of a negative number, your calculator will display an error message, such as 9 [+/−] [√] [E 3.]. There is no square root of −9, because no number times itself can give a negative number.

Problem: Find 5% of 30.
Keystrokes: 30 [×] 5 [%]
Final display: 1.5

The [%] key only changes a percent to its decimal form. If you know how to convert percents to decimals, you probably will not use the [%] key much.

Problem: Find the area of a **circle** with **radius** 3. (Use formula $A = \pi r^2$.)
Keystrokes: [π] [×] 3 [×] 3 [=]
Final display: 28.274333

If your calculator does not have the [π] key, you can use 3.14 or 3.1416 as an approximation for π.

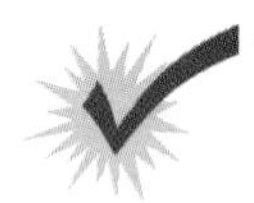

Check It Out

6. Without using the calculator, tell what the display would be if you entered 20 [M+] 3 [×] 5 [+] [MRC] [=].
7. Use memory functions to find the answer to $458 - 5^3 + 8 \times (-56)$.
8. Find the square root of 7,225.
9. Find 85% of 125.
10. Find 25% of the square root of 34.

9•1 EXERCISES

Find the value of each expression, using your calculator.

1. $29.75 + 88.4$
2. $26.44 - 11.93$
3. $-14.9 - 17.684$
4. $28 + 47 \times 50$
5. $17 + 25 \times (-22)$
6. $35 - 15 \times 78$
7. $-225 - 17 \times 33$
8. $17 + \sqrt{8100}$
9. $15 \div 50 + 13$
10. $1 \div (-400)$
11. 7% of 200
12. 140% of 800
13. $125 - \sqrt{47}$
14. $\sqrt{804} \div 17.35 + 620$
15. $\sqrt{68} \times 7 + 4$
16. $210 - \sqrt{5} + 16.8$

Use a calculator to answer items 17–25.

17. Find the perimeter if $x = 11.9$ cm.
18. Find the area if $x = 9.68$ cm.

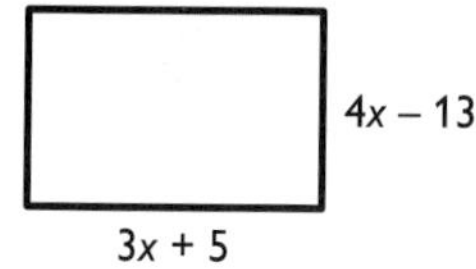

19. Find the circumference if $a = 3.7$ in.
20. Find the area if $a = 2$ in.

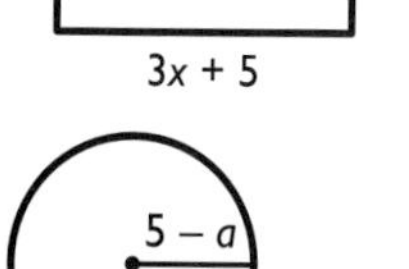

21. Find the area of $\triangle PQR$.
22. Find the perimeter of $\triangle PQR$. (*Remember:* $a^2 + b^2 = c^2$, p. 395.)
23. Find the circumference of circle Q.
24. Find the area of circle Q.
25. Find the shaded area of circle Q.

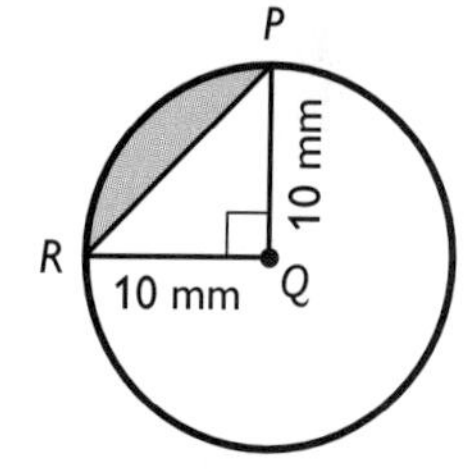

9•2 Scientific Calculator

Every mathematician and scientist has a scientific calculator to help quickly and accurately solve complex equations. Scientific calculators vary widely, some with a few functions and others with many functions. Some calculators can be programmed with functions of your choosing. The calculator below shows functions you might find on your scientific calculator.

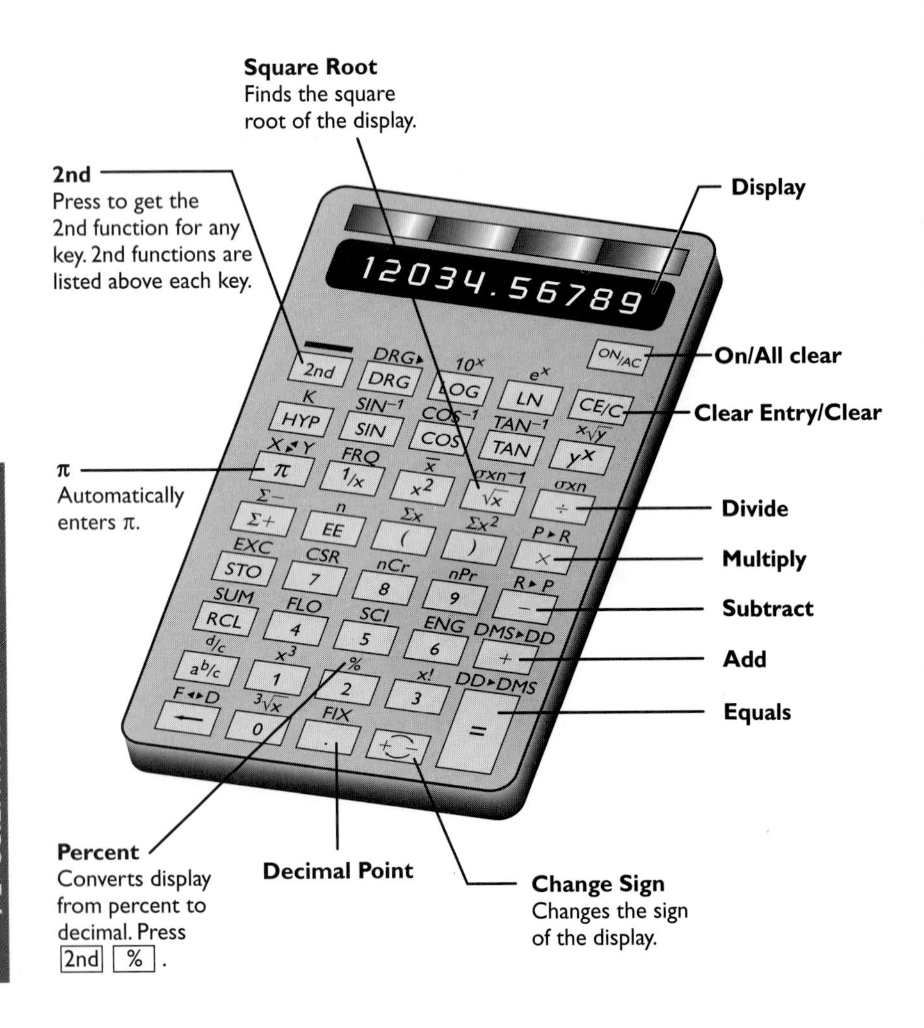

Frequently Used Functions

Since each scientific calculator is set up differently, your calculator may not work exactly as below. These key strokes work with the calculator illustrated on page 440. Use the reference book or card that came with your calculator to perform similar functions. See the index to find more information about the mathematics here.

Function	Problem	Key Strokes
Cube Root [$\sqrt[3]{x}$] Finds the cube root of the display.	$\sqrt[3]{343}$	343 [2nd] [$\sqrt[3]{x}$] [7.]
Cube [x^3] Finds the cube of the display.	17^3	17 [2nd] [x^3] [4913.]
Factorial [$x!$] Finds the factorial of the display.	7!	7 [2nd] [$x!$] [5040.]
Fix number of **decimal places** [FIX] Rounds display to number of places you determine.	Round 3.046 to the tenths place.	3.046 [2nd] [FIX] 2 [3.05]
Parentheses [(] [)] Use to group calculations.	$12 \times (7 + 8)$	12 [×] [(] 7 [+] 8 [)] [=] [180.]
Powers [y^x] Finds the x power of the display.	56^5	56 [y^x] 5 [=] [550731776.]
Powers of ten [10^x] Raises ten to the power displayed.	10^5	5 [2nd] [10^x] [100000.]

Function	Problem	Key Strokes
Reciprocal [1/x] Finds the reciprocal of the display.	Find the reciprocal of 8.	8 [1/x] [0.125]
Roots [$\sqrt[x]{y}$] Finds the x root of the display.	$\sqrt[4]{852}$	852 [2nd] [$\sqrt[x]{y}$] 4 [=] [5.402688131]
Square [x^2] Finds the square of the display.	17^2	17 [x^2] [289.]

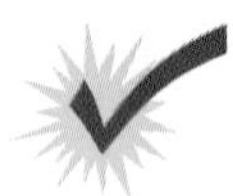

Check It Out

Use your calculator to find the following.

1. 12!
2. 14^4

Use your calculator to find the following to the nearest thousandth.

3. the reciprocal of 27
4. $(10^3 + 56^5 - \sqrt[3]{512}) \div 7!$

Tangent

Use [TAN] to find the **tangent** of an **angle** (pp. 398–400). Angles can be expressed as **degrees** or radians. Use the "Degree" or "DRG" or "DR" key to put your calculator in *degree* mode.

Find x to the nearest tenth.

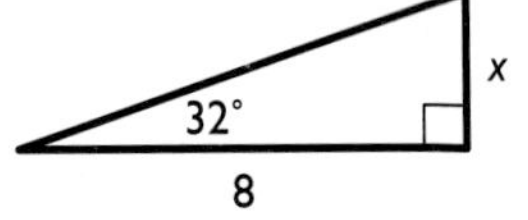

$\tan 32° = \frac{x}{8}$ $\quad x = 8 \times \tan 32°$

To find x, enter 8 [×] 32 [TAN] [=]. $\quad x \approx 5.0$

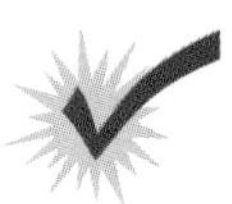

Check It Out

5. What is the tangent of 48°, expressed to the nearest tenth?
6. What is the tangent of 69°, expressed to the nearest tenth?

9•2 EXERCISES

Use a scientific calculator to find the following.

1. 69^2
2. 44^2
3. 13^3
4. 0.1^5

Give your answer to the nearest hundredth.

5. $\frac{60}{\pi}$
6. $9(\pi)$
7. $\frac{1}{9}$
8. $\frac{1}{\pi}$
9. $(15 - 4.4)^3 + 6$
10. $25 + (8 \div 6.2)$
11. $5! \times 4!$
12. $9! \div 4!$
13. $11! + 6!$
14. 5^{-3}
15. $\sqrt[4]{1{,}336{,}336}$
16. reciprocal of 0.0625
17. reciprocal of 25

Give the tangent of the angle to the nearest hundredth.

18. 55°
19. 88°
20. At a distance 72 ft from a building, the angle of the line of sight to the top of the building is 41°. How tall is the building to the nearest foot?

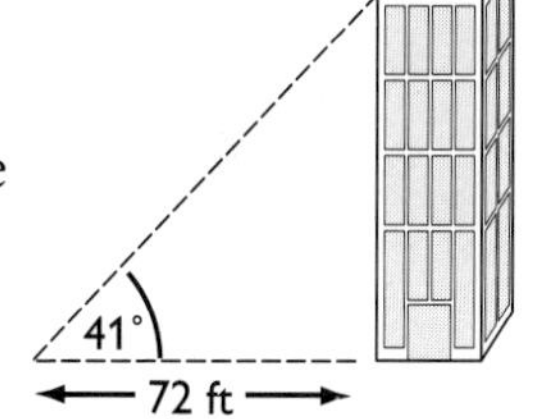

Magic Numbers

On a calculator, press the same number three times to display a three-digit number, for example 333. Then divide the number by the sum of the three digits and press the = key. Do you get 37?

Try this with other three-digit numbers. Write an algebraic expression that shows why the answer is always the same. See Hot Solutions for answer.

9•3 Geometry Tools

The Ruler

If you need to measure the dimensions of an object, or if you need to measure reasonably short **distances,** use a ruler.

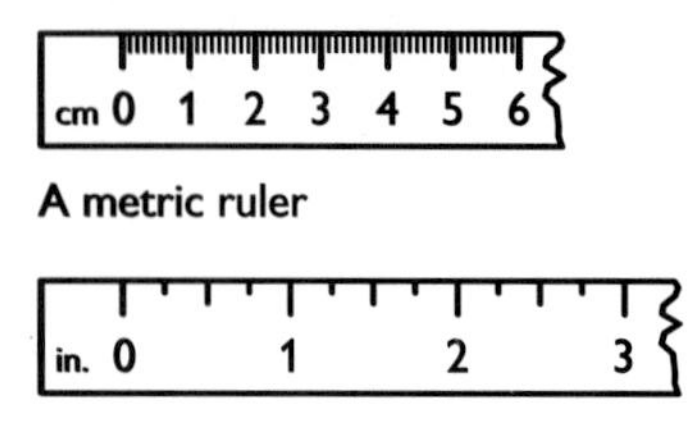

A metric ruler

A customary ruler

To get an accurate measure, be sure one end of the item being measured lines up with zero on your ruler.

The pencil below is measured first to the nearest tenth of a centimeter and then to the nearest eighth of an inch.

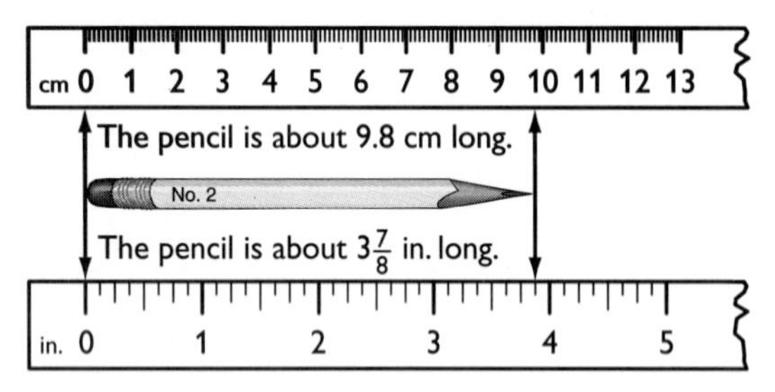

Check It Out

Use your ruler. Measure each line segment to the nearest tenth of a centimeter or the nearest eighth of an inch.

1. •————————————•
2. •——————————————————•

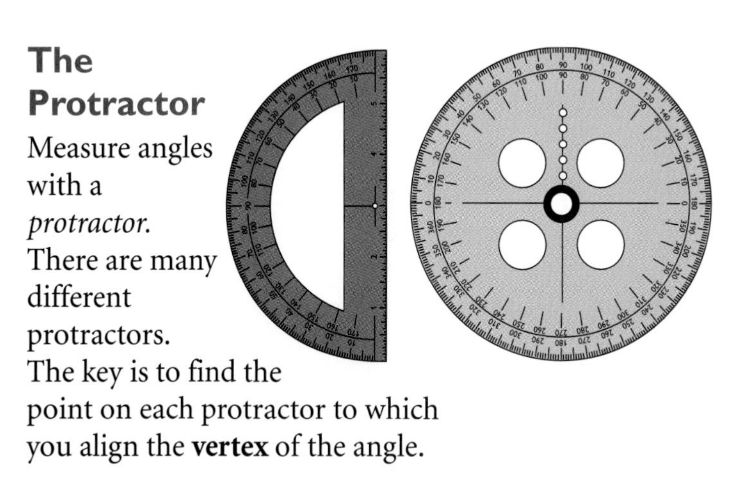

The Protractor

Measure angles with a *protractor.* There are many different protractors. The key is to find the point on each protractor to which you align the **vertex** of the angle.

MEASURING ANGLES WITH A PROTRACTOR

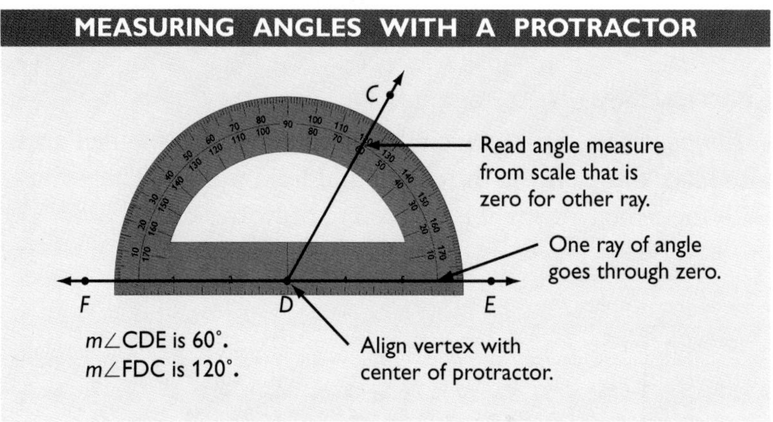

To draw an angle of a particular measure, draw one **ray** first, and position the center of the protractor at the endpoint. Then make a dot at the desired measure (45°, in this example).

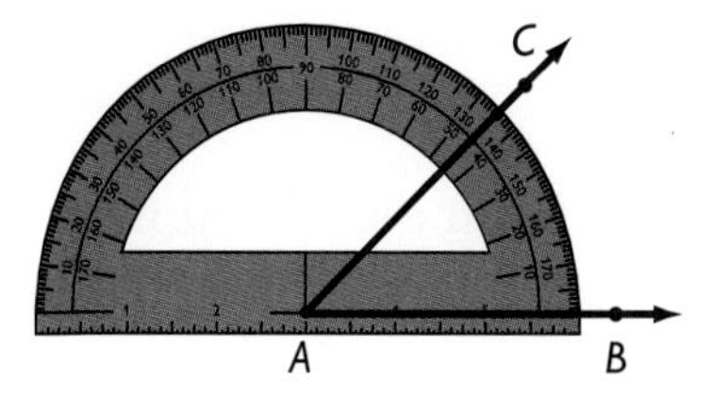

Connect A and C. Then $\angle BAC$ is a 45° angle.

Check It Out

Measure each angle to the nearest degree using your protractor.

3.

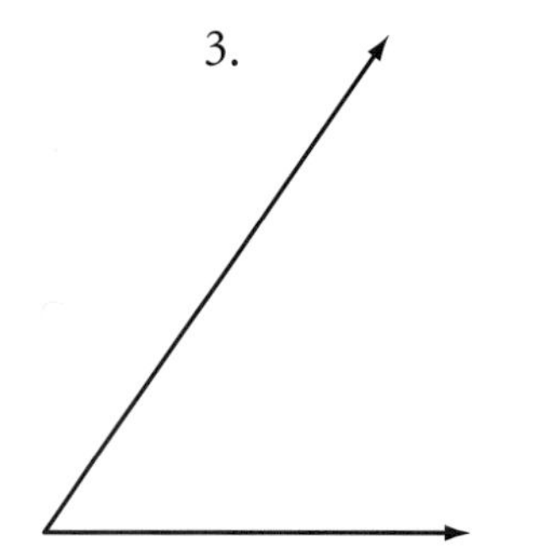

4.

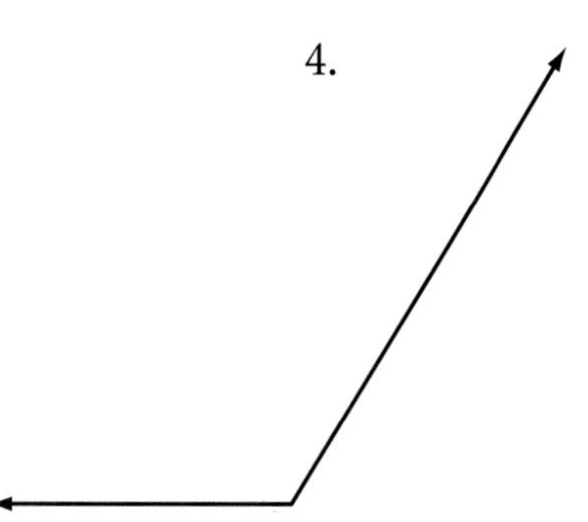

Compass

A *compass* is used to draw circles or parts of circles, called **arcs**. You place one **point** at the center and hold it there. The point with the pencil attached is pivoted to draw the arc or circle.

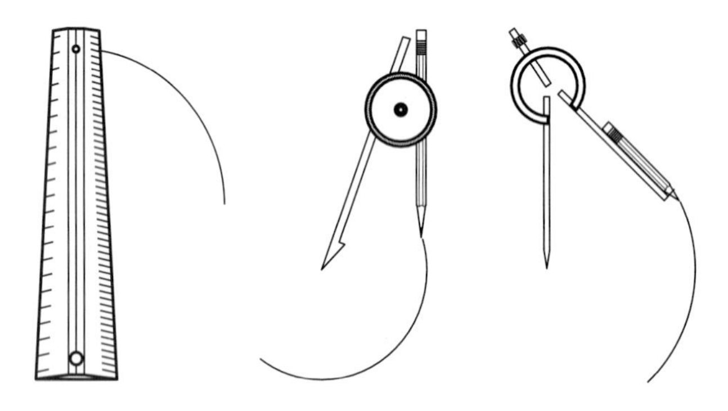

The distance between the point that is stationary (the center) and the pencil is the radius. Some compasses allow you to set the radius exactly.

For a review of *circles,* see page 388.

To draw a circle with a radius of $1\frac{1}{2}$ in., set the distance between the stationary point of your compass and the pencil at $1\frac{1}{2}$ in. Draw a circle.

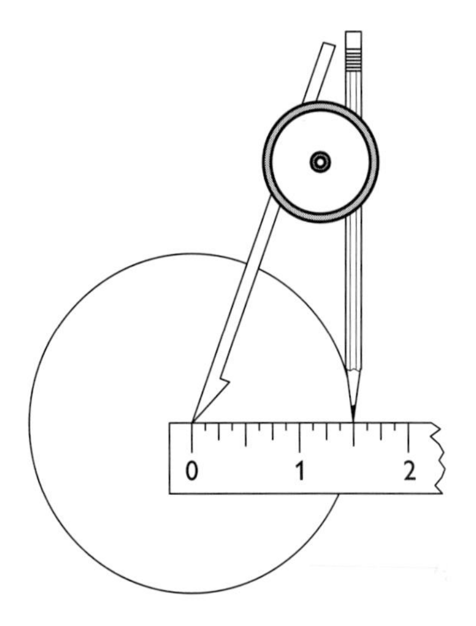

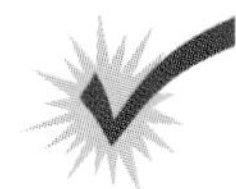

Check It Out

5. Draw a circle with a radius of 1 in. or 2.5 cm.
6. Draw a circle with a radius of 2 in. or 5.1 cm.

Construction Problem

A construction is a drawing problem in geometry that permits the use of only the straightedge and the compass. When you make a construction using straightedge and compass, you have to use what you know about geometry.

Follow the step-by-step directions below to inscribe an equilateral triangle in a circle.

- Draw a circle with center K.
- Draw a diameter ($\overline{SJ}$).
- Using S as a center and $\overline{SK}$ as a radius, draw an arc intersecting the circle at L and P.
- Connect L, P, and J to form the triangle.

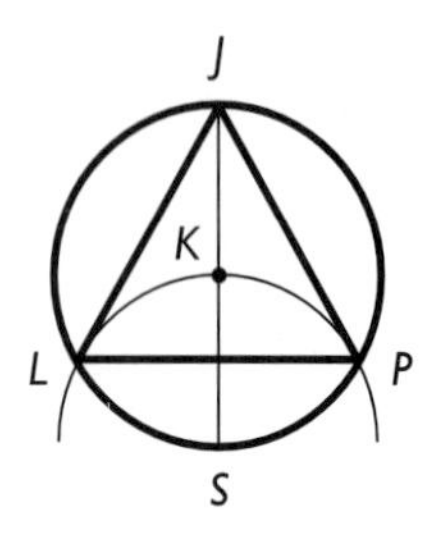

You can create a more complex design by inscribing another triangle in your circle using J as a center for drawing another intersecting arc.

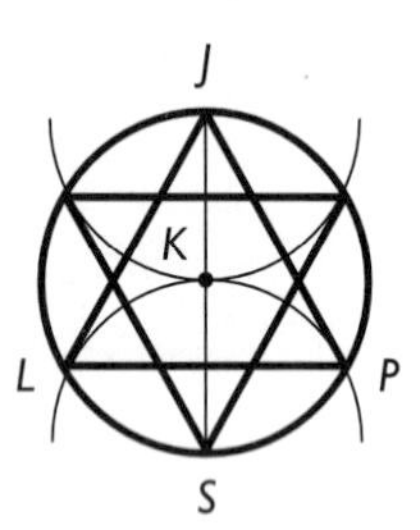

Once you have the framework, you can fill in different sections to come up with a variety of designs based on constructions.

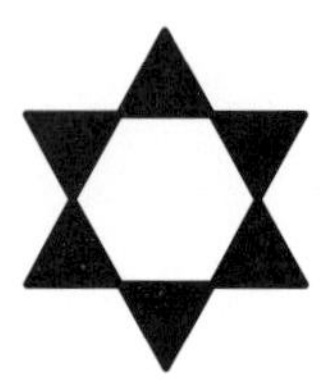

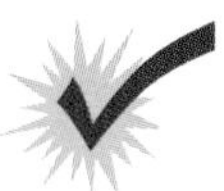

Check It Out

7. Draw the framework based on two triangles inscribed in a circle. Fill in sections to copy the design below.

8. Create your own design based on one or two triangles inscribed in a circle.

Mandalas

A mandala is a design which consists of geometric shapes and symbols that are meaningful to the artist. The word *mandala* means "circle" in Sanskrit, and the mandala design is usually contained within a circle.

In Hinduism and Buddhism, mandalas are used as aids to meditation and often incorporate symbols for the gods or the universe. Western artists create mandalas to symbolize their own lives or the lives of famous people. Within the geometric patterns, symbols for animals, the elements (earth, wind, fire, and water), the sun and stars, as well as personal symbols, appear frequently.

9·3 EXERCISES

Using a ruler, measure the length of each side of $\triangle ABC$. Give your answer in inches or centimeters, rounded to the nearest $\frac{1}{8}$ in. or $\frac{1}{10}$ cm.

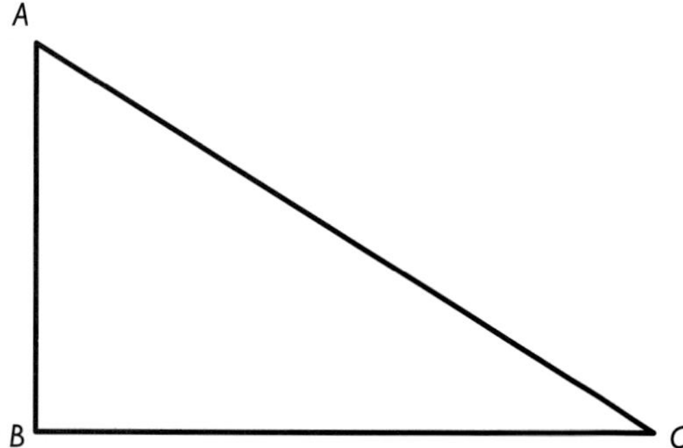

1. AB
2. BC
3. AC

Using a protractor, measure each angle in $\triangle ABC$.

4. $\angle A$
5. $\angle B$
6. $\angle C$
7. When you use a protractor to measure an angle, how do you know which of the two scales to read?

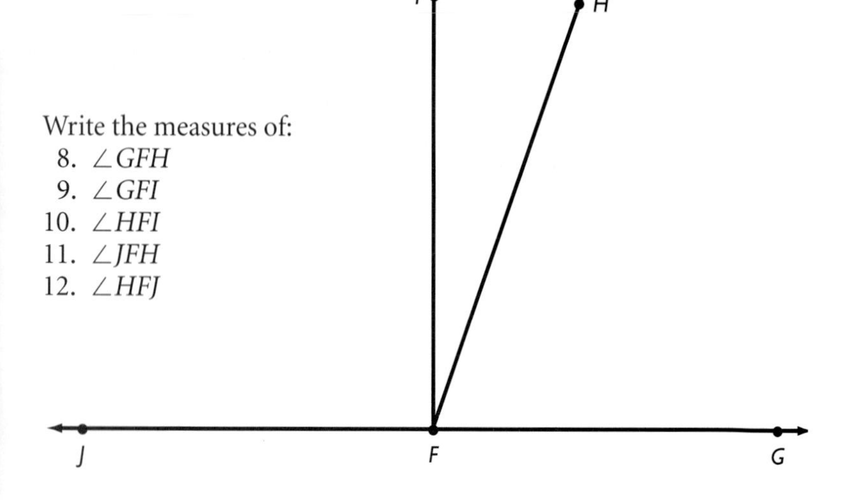

Write the measures of:

8. $\angle GFH$
9. $\angle GFI$
10. $\angle HFI$
11. $\angle JFH$
12. $\angle HFJ$

Match each tool with the function.

Tool	Function
13. compass	A. measure distance
14. protractor	B. measure angles
15. ruler	C. draw circles or arcs

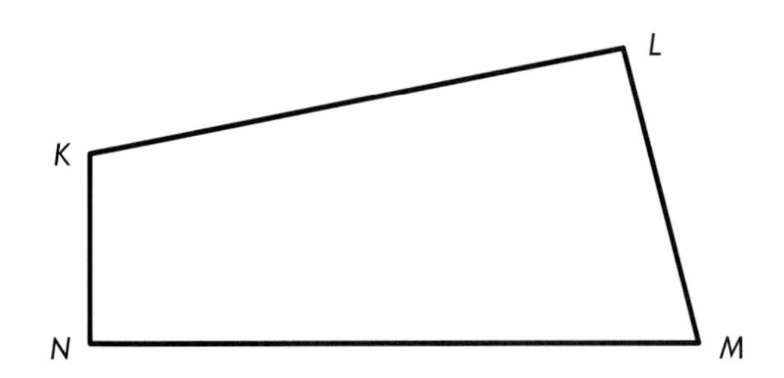

Tell the measures of:

16. $\angle KLM$
17. $\angle LMN$
18. $\angle MNK$
19. $\angle NKL$
20. Use a protractor to copy $\angle LMN$

Using a ruler, protractor, and compass, copy the figures below.

21.

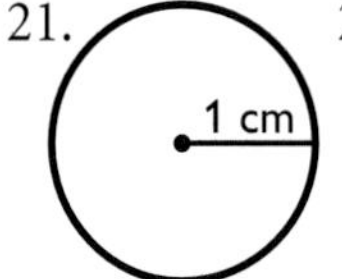

22.

23.

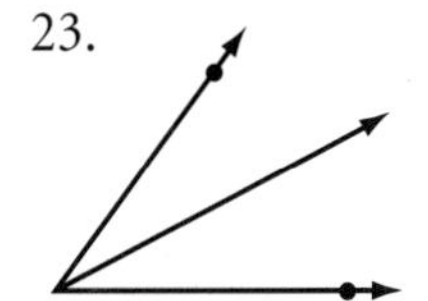

24.

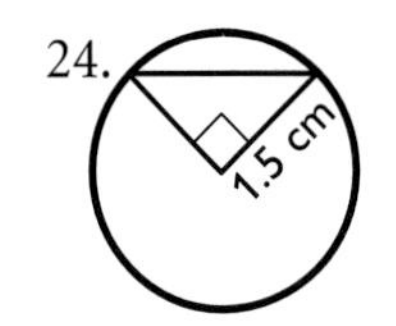

25.

9·4 Spreadsheets

What Is a Spreadsheet?

People have used **spreadsheets** as tools to keep track of information, such as finances, for a long time. Spreadsheets were paper-and-pencil math tools before becoming computerized. You may be familiar with computer spreadsheet programs.

A spreadsheet is a computer tool where information is arranged into **cells** within a grid and calculations are performed within the cells. When one cell is changed, all other cells that depend on it automatically change.

Spreadsheets are organized into **rows** and **columns.** Rows are **horizontal** and are numbered. Columns are **vertical** and are named by capital letters. The cells are named for their rows and columns.

File Edit

	A	B	C	D
1	1	3	1	
2	2	6	4	
3	3	9	9	
4	4	12	16	
5	5	15	25	
6				
7				
8				

The cell A3 is in Column A, Row 3. In this spreadsheet, there is a 3 in cell A3.

Check It Out

In the spreadsheet above, what number appears in each cell?

1. A2
2. B1
3. C5

Answer the following statements with true or false.

4. A row is labeled with numbers.
5. A column is horizontal.

Spreadsheet Formulas

A cell can contain a number, or it may contain the information it needs to generate a number. A **formula** generates a number dependent on other cells in the spreadsheet. The way the formulas are written depends on the particular spreadsheet computer software you are using. You enter a formula and the value generated shows, not the formula.

CREATING A SPREADSHEET FORMULA

	A	B	C	D
1	Item	Price	Qty	Total
2	sweater	$25	2	$50
3	pants	$20	3	
4	shirt	$15	2	
5				
6				

Express the value of the cell in relationship to other cells.

Total = Price × Qty

D2 = B2 × C2

If you change the value of a cell and a formula depends on it, the result of the formula will change.

In the spreadsheet above, if you entered 3 sweaters instead of 2 (C2 = 3), the Total column would automatically change to $75.

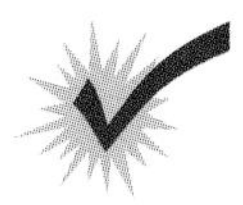

Check It Out

Use the spreadsheet in the box above. Write the formula for:

6. D3
7. D4
8. If D5 is to be the total of column D, write the formula for D5.

Fill Down and Fill Right

Now that you know the basics, let's look at some ways to make spreadsheets do even more of the work for you. *Fill down* and *fill right* are two spreadsheet commands that can save you a lot of time and effort.

To use *fill down,* select a portion of a column. *Fill down* will take the top cell that has been selected and copy it into the lower cells. If the top cell in the selected range contains a number, such as 5, *fill down* will generate a column containing all 5s.

If the top cell of the selected range contains a formula, the *fill down* feature will automatically adjust the formula as you go from cell to cell.

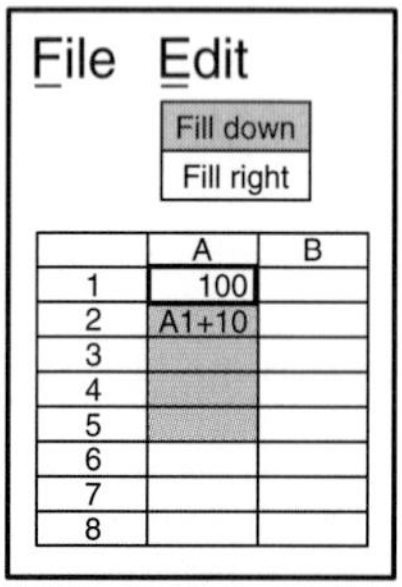

File Edit
Fill down
Fill right

	A	B
1	100	
2	A1+10	
3		
4		
5		
6		
7		
8		

The selected column is highlighted.

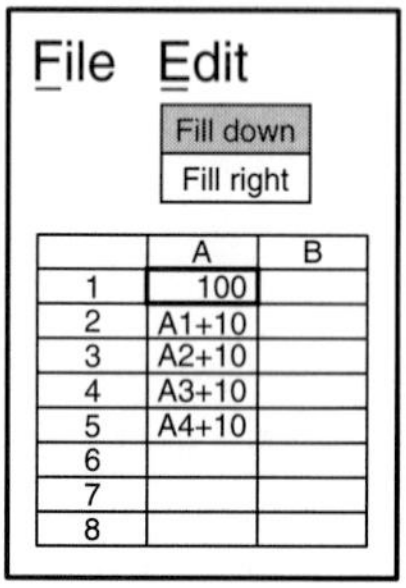

File Edit
Fill down
Fill right

	A	B
1	100	
2	A1+10	
3	A2+10	
4	A3+10	
5	A4+10	
6		
7		
8		

The spreadsheet fills the column and adjusts the formula.

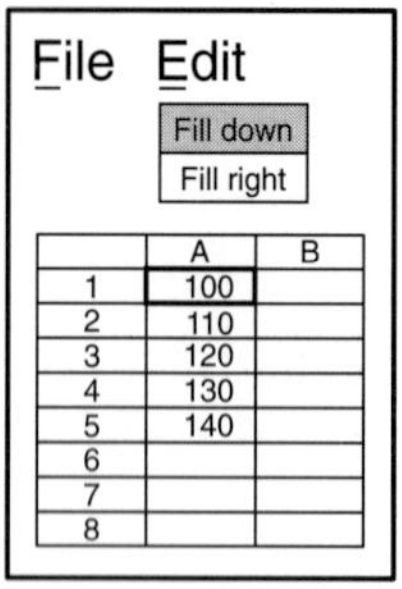

File Edit
Fill down
Fill right

	A	B
1	100	
2	110	
3	120	
4	130	
5	140	
6		
7		
8		

These are the values that actually appear.

Fill right works in a similar manner, except it goes across, copying the leftmost cell of the selected range in a row.

File Edit
Fill down
Fill right

	A	B	C	D	E
1	100				
2	A1+10				
3	A2+10				
4	A3+10				
5	A4+10				
6					
7					
8					

Row 1 is selected.

File Edit
Fill down
Fill right

	A	B	C	D	E
1	100	100	100	100	100
2	A1+10				
3	A2+10				
4	A3+10				
5	A4+10				
6					
7					
8					

The 100 fills to the right.

If you select A1 to E1 and *fill right*, you will get all 100s.
If you select A2 to E2 and *fill right*, you will "copy" the formula A1 + 10 as shown.

File Edit
Fill down
Fill right

	A	B	C	D	E
1	100	100	100	100	100
2	A1+10				
3					
4					
5					
6					
7					
8					

Row 2 is selected.

File Edit
Fill down
Fill right

	A	B	C	D	E
1	100	100	100	100	100
2	A1+10	B1+10	C1+10	D1+10	E1+10
3	A2+10				
4	A3+10				
5	A4+10				
6					
7					
8					

The spreadsheet fills the row and adjusts the formula.

Check It Out

Use the spreadsheet above.

9. Select A2 to A8 and *fill down*. What formula will be in A7? What number?
10. Select A3 to E3 and *fill right*. What formula will be in D3? What number?

Spreadsheet Graphs

You can graph from a spreadsheet. As an example, let's use a spreadsheet to compare the **perimeter** of a square to the length of a side.

File Edit

	A	B	C	D	E
1	side	perimeter			
2	1	4			
3	2	8			
4	3	12			
5	4	16			
6	5	20			
7	6	24			
8	7	28			
9	8	32			
10	9	36			
11	10	40			

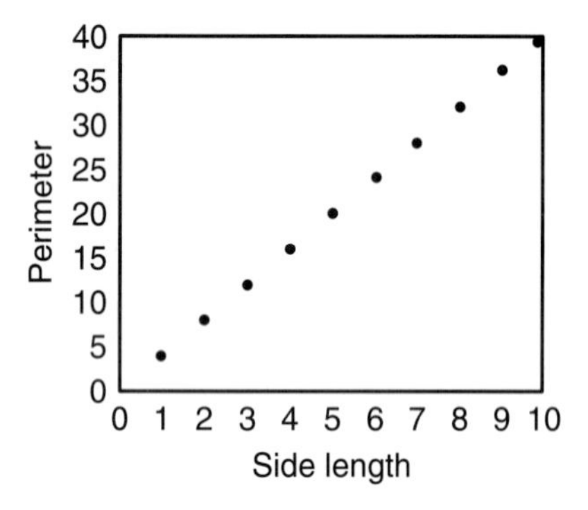

Most spreadsheets have a function that displays tables as graphs. See your spreadsheet reference for more information.

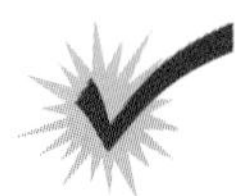

Check It Out

11. What cells gave the point (4, 16)?
12. What cells gave the point (8, 32)?

9·4 EXERCISES

For the spreadsheet shown to the right, what number appears in each of the following cells?

1. B3 2. C1 3. A2

In which cell does each number appear?

4. 15 5. 7 6. 800

File Edit

	A	B	C	D
1	1	1	200	1
2	2	3	500	6
3	3	5	800	15
4	4	7	1100	28
5				
6				
7				
8				

7. If the formula behind cell A2 is A1 + 1 and the formulas were copied down, what formula is behind cell A3?
8. Column B holds formulas dependent on the cell above. What formula might be behind cell B4?
9. The formula behind cell D2 is A2 × B2. What formula might be behind cell D4?
10. If row 5 were included in the spreadsheet, what numbers would be in that row?

Use the spreadsheet below to answer items 11–15.

File Edit

Fill down

Fill right

	A	B	C	D
1	5	10		
2	A1+6	B1×2		
3				
4				
5				
6				
7				
8				

11. If you select A2 to A5 and fill down, what formula will appear in A3?
12. If you select A3 to A5 and fill down, what numbers will appear in A3 to A5?
13. If you select B1 to E1 and fill right, what will appear in C1, D1, and E1?
14. If you select B2 to E2 and fill right, what formula will appear in D2?
15. If you fill right as indicated in items 13 and 14 above, what numbers will appear in B2, C2, and D2?

What have you learned?

You can use the problems and the list of words that follow to see what you have learned in this chapter. You can find out more about a particular problem or word by referring to the boldfaced topic number (for example, **9•2**).

Problem Set

Use your calculator for items 1–6. **9•1**

1. $25 + 37 \times 12$
2. 3,425% of 2,300

Round answers to the nearest tenth.

3. $12 \times \sqrt{225} + 17.25$
4. $3 \times \pi - \sqrt{.49}$

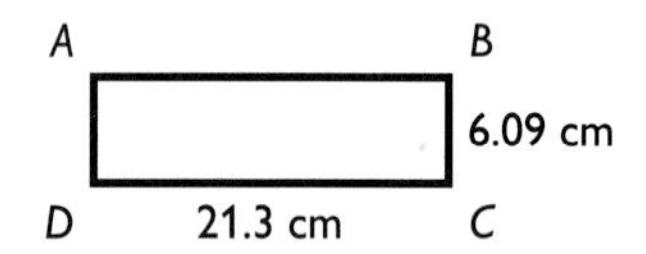

5. Find the perimeter of rectangle *ABCD.*
6. Find the area of rectangle *ABCD.*

Use a scientific calculator for items 7–12. Round decimal answers to the nearest hundredth. **9•2**

7. 2.027^5
8. Find the reciprocal of 4.5.
9. Find the square of 4.5.
10. Find the square root of 5.4.
11. $(4 \times 10^3) \times (7 \times 10^6)$
12. $0.6 \times (3.6 + 13)$

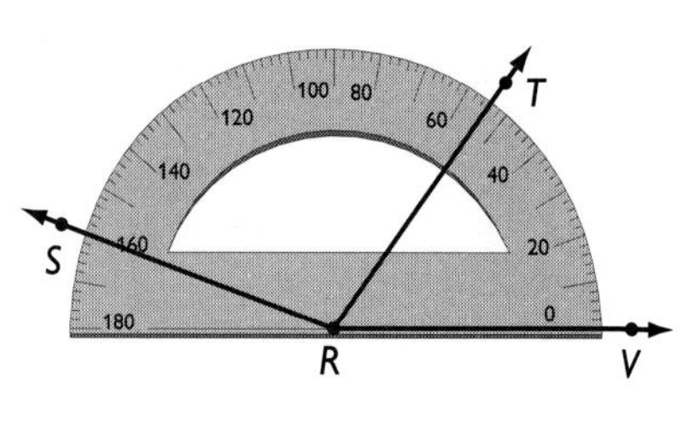

13. What is the measure of $\angle VRT$? **9•3**
14. What is the measure of $\angle VRS$? **9•3**
15. What is the measure of $\angle SRT$? **9•3**
16. Does *RT* divide $\angle VRS$ into 2 equal angles? **9•3**

17. What part of the ruler do you always line up with the end of the item measured? **9•3**

For items 18–20, refer to the spreadsheet below. **9•4**

File Edit

Fill down
Fill right

	A	B	C	D
1	6	26	100	77
2	8	28	200	
3	10	30	300	
4				

18. Name the cell holding 28.
19. A formula for cell C3 is 3 × C1. Name another formula for cell C3.
20. Cell D1 contains the number 77 and no formula. After using the command fill down, what number will be in cell D10?

hot words

WRITE DEFINITIONS FOR THE FOLLOWING WORDS.

angle **9•2**
arc **9•3**
cell **9•4**
circle **9•1**
column **9•4**
cube **9•2**
cube root **9•2**
decimal **9.1**
degree **9•2**
distance **9•3**
factorial **9•2**
formula **9•4**
horizontal **9•4**
negative number **9•1**
parentheses **9•2**
percent **9•1**
perimeter **9•4**
pi **9•1**
point **9•3**
power **9•2**
radius **9•1**
ray **9•3**
reciprocal **9•2**
root **9•2**
row **9•4**
spreadsheet **9•4**
square **9•2**
square root **9•1**
tangent **9•2**
vertex **9•3**
vertical **9•4**

PART **THREE**

hot solutions

Solutions

1 **Numbers and Computation** 462
2 **Fractions, Decimals, and Percents** 464
3 **Powers and Roots** 467
4 **Data, Statistics, and Probability** 469
5 **Logic** 473
6 **Algebra** 475
7 **Geometry** 483
8 **Measurement** 486
9 **Tools** 487

Index

Chapter 1: Numbers and Computation

p. 72

1. 600 **2.** 60,000,000

3. $(2 \times 10{,}000) + (4 \times 1{,}000) + (7 \times 100) + (3 \times 10) + (5 \times 1)$

4. 406,758; 396,758; 46,758; 4,678

5. 52,534,880; 52,535,000; 53,000,000

6. 0 **7.** 12 **8.** 5,889 **9.** 0

10. 600 **11.** 1,700

12. $(4 + 7) \times 3 = 33$ **13.** $(30 + 15) \div 5 + 5 = 14$

14. No **15.** No **16.** Yes **17.** No

18. $2^3 \times 5$ **19.** $2 \times 5 \times 11$ **20.** $2 \times 5 \times 23$

21. 4 **22.** 5 **23.** 9

24. 60 **25.** 120 **26.** 90

p. 73

27. 60

28. 7, 7 **29.** 15, −15 **30.** 12, 12 **31.** 10, −10

32. 2 **33.** −4 **34.** −11 **35.** 16 **36.** 0 **37.** 6

38. 42 **39.** −4 **40.** 7 **41.** 24 **42.** −36 **43.** −50

44. It will be a negative integer.

45. It will be a positive integer.

1•1 Place Value of Whole Numbers

p. 74

1. 40,000 **2.** 4,000,000 **3.** Forty million, three hundred seventy-six thousand, five hundred

4. Fifty-seven trillion, three hundred twenty billion, one hundred million

p. 75

5. $(9 \times 10{,}000) + (8 \times 1{,}000) + (2 \times 10) + (5 \times 1)$

6. $(4 \times 100{,}000) + (6 \times 100) + (3 \times 10) + (7 \times 1)$

p. 76

7. $<$ **8.** $>$

p. 76 **9.** 7,520; 72,617; 77,302; 740,009

10. 37,300 **11.** 490,000 **12.** 2,000,000
13. 800,000

1•2 Properties

p. 78 **1.** Yes **2.** No **3.** No **4.** Yes

p. 79 **5.** 28,407 **6.** 299 **7.** 0 **8.** 4.8

9. $(3 \times 2) + (3 \times 5)$ **10.** $6 \times (8 + 4)$

1•3 Order of Operations

p. 82 **1.** 12 **2.** 87

1•4 Factors and Multiples

p. 84 **1.** 1, 2, 4, 8 **2.** 1, 2, 3, 4, 6, 8, 12, 16, 24, 48

p. 85 **3.** 1, 2 **4.** 1,5

5. 2 **6.** 6

p. 86 **7.** Yes **8.** No **9.** Yes **10.** Yes

p. 88 **11.** Yes **12.** No **13.** Yes **14.** No **15.** 17 and 19 or 29 and 31 or 41 and 43 or...

16. $2^4 \times 5$ **17.** $2^3 \times 3 \times 5$

p. 89 **18.** 6 **19.** 8

p. 90 **20.** 18 **21.** 140

1•5 Integer Operations

p. 92 **1.** -6 **2.** $+200$

p. 93 **3.** 12, 12 **4.** 5, -5 **5.** 9, 9 **6.** 0, 0

7. -2 **8.** 0 **9.** -5 **10.** -3

p. 94 **11.** 10 **12.** -3 **13.** 3 **14.** -48

Oops! If a is only positive, $2 + a > 2$ is always true. If a can be positive, zero, or negative, $2 + a > 2$ is sometimes true, but $2 + a$ can also be equal to or less than 2.

HOT SOLUTIONS

Chapter 2: Fractions, Decimals, and Percents

p. 100

1. The second, because $\frac{1}{2} > \frac{3}{7}$.
2. About 9 days **3.** 10 servings **4.** 85%
5. C. $\frac{8}{11}$
6. $1\frac{1}{6}$ **7.** $1\frac{3}{4}$ **8.** $3\frac{1}{4}$ **9.** $8\frac{3}{10}$
10. B. $\frac{4}{3} = 1\frac{1}{3}$
11. $\frac{2}{5}$ **12.** $\frac{1}{2}$ **13.** $\frac{3}{4}$ **14.** 3
15. Hundredths **16.** 3.0 + 0.003

p. 101

17. 400.404 **18.** 0.165; 1.065; 1.605; 1.650
19. 16.154 **20.** 1.32 **21.** 30.855 **22.** 7.02
23. 30% **24.** 27.8 **25.** 55

2•1 Fractions and Equivalent Fractions

p. 103 **1.** $\frac{5}{8}$ **2.** $\frac{3}{8}$ **3.** Answers will vary.
p. 105 **4–7.** Answers will vary.
p. 106 **8.** $\neq$ **9.** $=$ **10.** $\neq$
p. 108 **11.** $\frac{4}{5}$ **12.** $\frac{3}{4}$ **13.** $\frac{2}{5}$
p. 110 **14.** $7\frac{1}{6}$ **15.** $11\frac{1}{3}$ **16.** $6\frac{2}{5}$ **17.** $9\frac{1}{4}$
18. $\frac{37}{8}$ **19.** $\frac{77}{6}$ **20.** $\frac{49}{2}$ **21.** $\frac{98}{3}$

2•2 Comparing and Ordering Fractions

p. 113 **1.** $>$ **2.** $>$ **3.** $=$ **4.** $<$
5. $>$ **6.** $<$ **7.** $<$
p. 114 **8.** $\frac{1}{4}; \frac{2}{5}; \frac{1}{2}; \frac{3}{5}$ **9.** $\frac{2}{3}; \frac{13}{18}; \frac{7}{9}; \frac{5}{6}$ **10.** $\frac{1}{2}; \frac{4}{7}; \frac{5}{8}; \frac{2}{3}; \frac{11}{12}$

2•3 Addition and Subtraction of Fractions

p. 116 **1.** $1\frac{1}{5}$ **2.** $1\frac{3}{34}$ **3.** $\frac{1}{2}$ **4.** $\frac{1}{2}$
p. 118 **5.** $1\frac{2}{5}$ **6.** $1\frac{3}{14}$ **7.** $\frac{1}{20}$ **8.** $\frac{11}{24}$
9. $9\frac{5}{6}$ **10.** $34\frac{5}{8}$ **11.** 61

p. 119 **12.** $23\frac{39}{40}$ **13.** $20\frac{1}{24}$ **14.** $22\frac{7}{15}$

p. 120 **15.** $7\frac{1}{2}$ **16.** $3\frac{37}{70}$ **17.** $11\frac{1}{8}$

2•4 Multiplication and Division of Fractions

p. 123 **1.** $\frac{1}{3}$ **2.** $\frac{1}{12}$ **3.** 2

4. $\frac{1}{10}$ **5.** $\frac{8}{15}$ **6.** 2

7. $\frac{7}{3}$ **8.** $\frac{1}{3}$ **9.** $\frac{5}{22}$

p. 124 **10.** $1\frac{1}{2}$ **11.** $\frac{1}{14}$ **12.** $\frac{1}{4}$

2•5 Naming and Ordering Decimals

p. 127 **1.** 1.50 **2.** 0.32 **3.** 16.63 **4.** 0.03

5. Three hundred sixty-five thousandths

6. One and one hundred two thousandths

7. Fifty-four thousandths

p. 129 **8.** Five ones; five and three hundred six thousandths

9. Eight thousandths; fifty-eight thousandths

10. One thousandth; six and fifteen ten thousandths

11. Six hundred thousandths; two hundred six hundred thousandths

12. < **13.** > **14.** <

p. 130 **15.** 0.753; 0.7539; 0.754; 0.759

16. 12.00427; 12.0427; 12.427; 12.4273

17. 2.12 **18.** 38.41

2•6 Decimal Operations

p. 132 **1.** 7.1814 **2.** 96.674 **3.** 38.54 **4.** 802.0556

p. 133 **5.** 13 **6.** 1 **7.** 15 **8.** 280

p. 134 **9.** 59.481 **10.** 80.42615

p. 135 **11.** 900 **12.** 4

13. 0.072 **14.** 0.0028231

p. 136 **15.** 21.6 **16.** 5.23 **17.** 92 **18.** 25.8
p. 137 **19.** 10.06 **20.** 24.8
p. 138 **21.** 0.07 **22.** 0.65
Luxuries or Necessities? About 923,000,000

2•7 Meaning of Percent

p. 140 **1.** 44% **2.** 33%
p. 141 **3.** 150 **4.** 100 **5.** 150 **6.** 250
p. 142 **7.** \$1.00 **8.** \$6
Honesty Pays 20%

2•8 Using and Finding Percents

p. 144 **1.** 60 **2.** 665 **3.** 11.34 **4.** 27
p. 145 **5.** 665 **6.** 72 **7.** 130 **8.** 340
p. 146 **9.** $33\frac{1}{3}$% **10.** 450% **11.** 400% **12.** 60%
p. 147 **13.** 104 **14.** 20 **15.** 25 **16.** 1,200
p. 148 **17.** 25% **18.** 95% **19.** 120% **20.** 20%
p. 149 **21.** 11% **22.** 50% **23.** 16% **24.** 30%
p. 150 **25.** Discount: \$162.65, Sale Price: \$650.60
26. Discount: \$5.67, Sale Price: \$13.23
p. 151 **27.** 100 **28.** 2 **29.** 15 **30.** 30
p. 152 **31.** I = \$1,800, A = \$6,600
32. I = \$131.25, A = \$2,631.25

2•9 Fraction, Decimal, and Percent Relationships

p. 155 **1.** 80% **2.** 65% **3.** 45% **4.** 38%
5. $\frac{11}{20}$ **6.** $\frac{29}{100}$ **7.** $\frac{17}{20}$ **8.** $\frac{23}{25}$

HOT SOLUTIONS

p. 156 **9.** $\frac{89}{200}$ **10.** $\frac{43}{125}$

p. 157 **11.** 8% **12.** 66% **13.** 39.8% **14.** 74%

15. 0.145 **16.** 0.0001 **17.** 0.23 **18.** 0.35

p. 158 **The Ups and Downs of Stocks** 1%

p. 160 **19.** 0.8 **20.** 0.55 **21.** 0.875 **22.** $0.41\overline{6}$

23. $2\frac{2}{5}$ or $\frac{12}{5}$ **24.** $\frac{7}{125}$ **25.** $\frac{7}{50}$ **26.** $1\frac{1}{5}$ or $\frac{6}{5}$

Chapter 3: Powers and Roots

p. 166 **1.** 5^7 **2.** a^5

3. 4 **4.** 81 **5.** 36

6. 8 **7.** 125 **8.** 343

9. 1,296 **10.** 2,187 **11.** 512

12. 1,000 **13.** 10,000,000 **14.** 100,000,000,000

15. 4 **16.** 7 **17.** 11

18. 5 and 6 **19.** 3 and 4 **20.** 8 and 9

p. 167 **21.** 3.873 **22.** 6.164

23. 2 **24.** 4 **25.** 7

26. Very small **27.** Very large

28. 7.8×10^7 **29.** 2×10^5 **30.** 2.8×10^{-3}

31. 3.02×10^{-5}

32. 8,100,000 **33.** 200,700,000

34. 4,000 **35.** 0.00085 **36.** 0.00000906

37. 0.0000007

38. 12 **39.** 13 **40.** 18

3•1 Powers and Exponents

p. 168 **1.** 4^3 **2.** 6^9 **3.** x^4 **4.** y^6

p. 169 **5.** 25 **6.** 100 **7.** 9 **8.** 49

p. 170 **9.** 64 **10.** 1,000 **11.** 27 **12.** 512

p. 171 **The Future of the Universe** $10^4 \times (10^{12})^8 = 10^{100}$

p. 172 **13.** 128 **14.** 59,049 **15.** 81 **16.** 390,625

HOT SOLUTIONS

p. 173 **17.** 1,000 **18.** 1,000,000 **19.** 1,000,000,000
20. 100,000,000,000,000

p. 174 **21.** 324 **22.** 9,765,625 **23.** 33,554,432
24. 20,511,149

3·2 Square and Cube Roots

p. 176 **1.** 4 **2.** 7 **3.** 10 **4.** 12

p. 177 **5.** Between 7 and 8 **6.** Between 4 and 5
7. Between 2 and 3 **8.** Between 9 and 10

p. 179 **9.** 1.414 **10.** 7.071 **11.** 8.660 **12.** 9.950
13. 4 **14.** 7 **15.** 10 **16.** 5

p. 180 **Squaring Triangles** 21, 28; it is the sequence of squares.

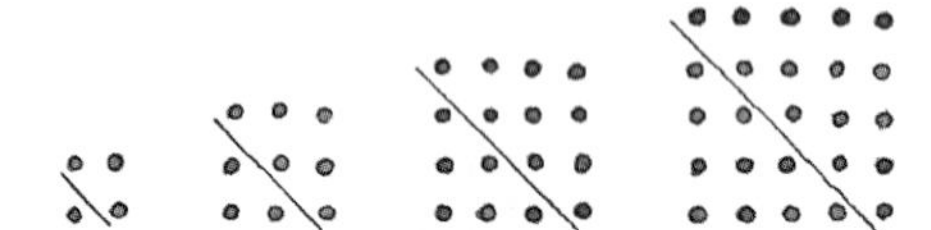

3·3 Scientific Notation

p. 182 **1.** Very small **2.** Very large
Bugs 1.2×10^{18}

p. 183 **3.** 6.8×10^4 **4.** 7×10^6 **5.** 3.05×10^{10}
6. 7.328×10^7

p. 184 **7.** 3.8×10^{-3} **8.** 4×10^{-7} **9.** 6.03×10^{-11}
10. 7.124×10^{-4}

p. 185 **11.** 53,000 **12.** 924,000,000 **13.** 120,500
14. 8,840,730,000,000

p. 186 **15.** 0.00071 **16.** 0.000005704 **17.** 0.0865
18. 0.000000000030904

3·4 Laws of Exponents

p. 188 **1.** 23 **2.** 17 **3.** 18 **4.** 19

Chapter 4: Data, Statistics, and Probability

p. 194
1. Late morning **2.** Seventh **3.** No
4. 75% **5.** No, percents do not show quantities.
6. Histogram

p. 195
7. Positive **8.** 34 **9.** Mode
10. 4 **11.** 21
12. 3, 5; 3, 7; 5, 7
13. $\frac{3}{5}$ **14.** 0 **15.** $\frac{2}{15}$

4•1 Collecting Data

p. 197
1. Adults over the age of 45; 150,000
2. Elk in Roosevelt National Forest; 200

p. 198
3. No, it is limited to people who are friends of her parents, and they may have similar beliefs.
4. Yes, if the population is the class. Each student has the same chance of being picked.

p. 199
5. It assumes you like pizza.
6. It does not assume you watch TV after school.
7. Do you recycle newspapers?
8. 2 **9.** Bagels **10.** Pizza; students chose pizza more than any other food.

4•2 Displaying Data

p. 202
1. A word has 11 letters.

2. 1994 WINTER OLYMPICS

No. of Gold Medals	0	1	2	3	4	5	6	7	8	9	10	11
No. of Countries	8	4	2	2	1	0	1	1	0	1	1	1

p. 203 **3.** 25 g **4.** 11.5 g **5.** 50%

p. 205 **6.** About half **7.** About a quarter

8.

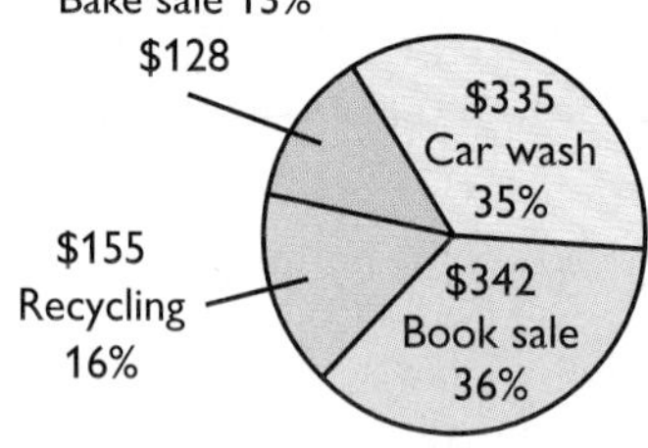

And the Winner Is... Improved sales; weekly; bar graph

p. 206 **9.** 8:00 A.M. **10.** 6

11.

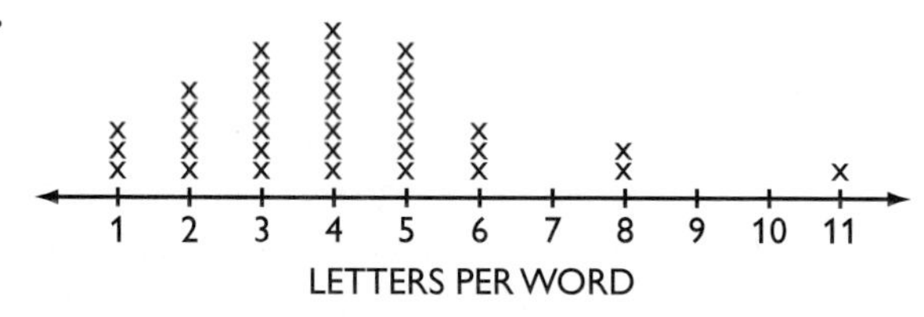

p. 207 **12.** Gabe **13.** Gabe

p. 208 **14.** 7 **15.** 37.1; 27.2

p. 210 **16.** September **17.** Answers will vary.

18.

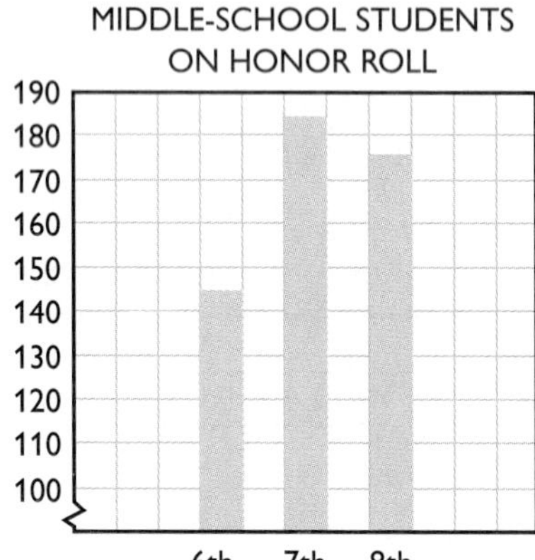

p. 211 **19.** $120,000,000,000 **20.** Answers will vary. Possible answer: It increased slightly for the first three years and then remained constant.

p. 212 **21.** 16

22.

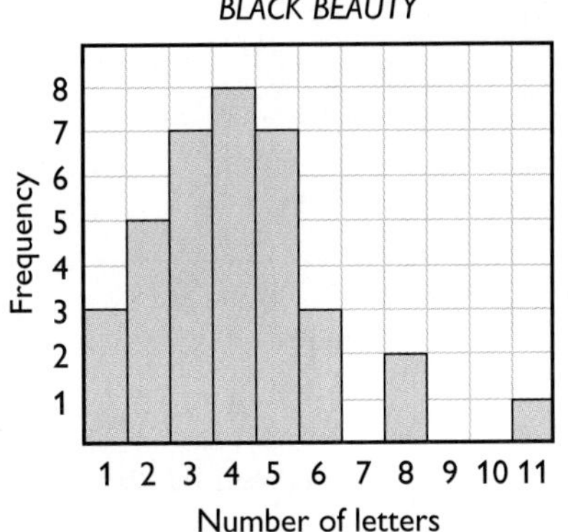

4•3 Analyzing Data

p. 215 **1.**

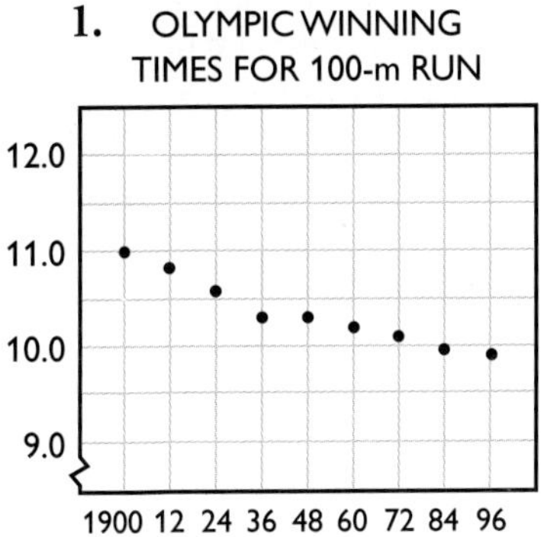

2.

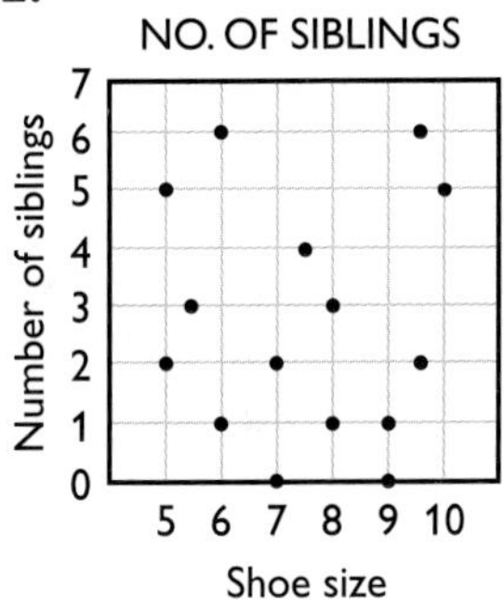

p. 217 **3.** Age and letters in name **4.** Negative
5. Miles cycled and hours

p. 218 **6.**

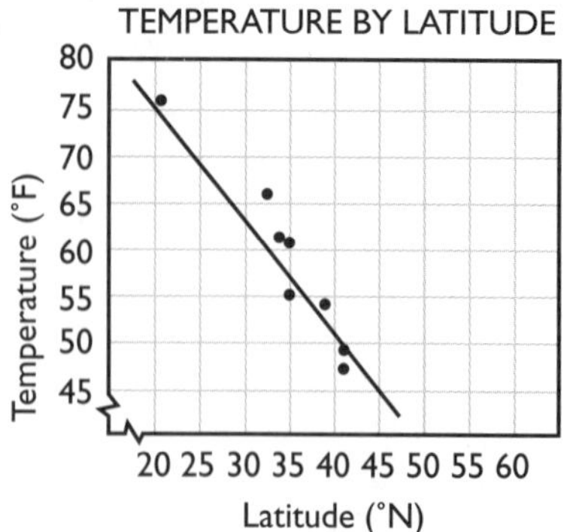

7. About 32°F

p. 220 **8.** Flat **9.** Normal **10.** Skewed to right
11. Bimodal **12.** Skewed to left

4•4 Statistics

p. 223 **1.** 8 **2.** 84 **3.** 197 **4.** 27° **5.** 92 points

Graphic Impressions Answers may vary. Possible answers: that a spider is 4 times larger than a guppy and a crocodile is 3 times longer than a spider, instead of their life spans having that relationship; the bar graph is a more accurate portrayal.

p. 225 **6.** 11 **7.** 2.1 **8.** 18 **9.** 23,916

p. 227 **10.** 7 **11.** 1.6 **12.** 10 **13.** 49

Olympic Decimals Technical merit 9.52, composition and style 9.7

p. 228 **14.** 750 **15.** 5.1 **16.** 52° **17.** 34 points

p. 229 **18.** 37 **19.** 6.8

p. 230 **How Mighty Is the Mississippi?** 2,949.5 mi; 2,620 mi; 1,830 mi

4•5 Combinations and Permutations

p. 234 **1.** 216 three-digit numbers

2. 36 routes

Monograms 17,576 monograms

p. 236 **3.** 210 **4.** 720 **5.** 40,320 **6.** 1,190 **7.** 362,880

p. 238 **8.** 84 **9.** 91 **10.** 220 **11.** Twice as many permutations as combinations

4•6 Probability

p. 241 **1.** $\frac{1}{2}$ **2.** $\frac{1}{20}$ **3.** Answers will vary.

p. 242 **4.** $\frac{3}{4}$ **5.** 0 **6.** $\frac{1}{6}$ **7.** $\frac{4}{11}$

p. 243 **8.** $\frac{1}{4}$, 0.25, 1:4, 25% **9.** $\frac{1}{8}$, 0.125, 1:8, 12.5%
10. $\frac{1}{8}$, 0.125, 1:8, 12.5% **11.** $\frac{1}{25}$, 0.04, 1:25, 4%

p. 244 **12.** The numbers 2, 5, 4, 2, 3, 1, 6, 3
13–14. Answers will vary.

p. 245 **Lottery Fever** Struck by lightning; $\frac{260}{260{,}000{,}000}$ is about 1 in 1 million, compared to the 1-in-16-million chance of winning a 6-out-of-50 lottery.

p. 247 **15.**

Second Spin

First Spin	R	B	G	Y
R	RR	RB	RG	RY
B	BR	BB	BG	BY
G	GR	GB	GG	GY
Y	YR	YB	YG	YY

16. $\frac{7}{16}$

p. 248 **17.**

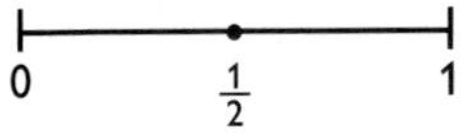

18.

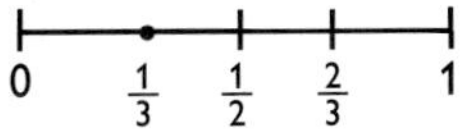

19.

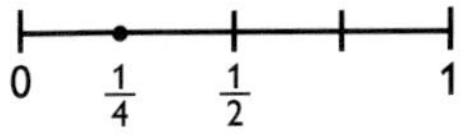

20.

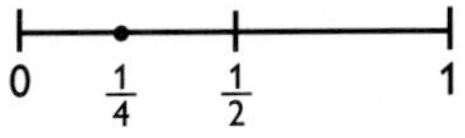

p. 250 **21.** $\frac{1}{4}$; independent **22.** $\frac{91}{190}$; dependent
23. $\frac{1}{16}$ **24.** $\frac{13}{204}$

Chapter 5: Logic

p. 256 **1.** True **2.** False **3.** False **4.** True **5.** True
6. True **7.** True **8.** False **9.** True

Continued

p. 256 (cont.)

10. If it is Tuesday, then the jet flies to Belgium.

11. If it is Sunday, then the bank is closed.

12. If $x^2 = 49$, then $x = 7$. **13.** If an angle is acute, then it has a measure less than 90°.

14. The playground will not close at sundown.

15. These two lines do not form an angle.

p. 257

16. If you do not pass all your courses, then you will not graduate. **17.** If two lines do not intersect, then they do not form four angles.

18. If you do not buy an adult ticket, then you are not over 12 years old. **19.** If a pentagon is not equilateral, then it does not have five equal sides.

20. Thursday **21.** Any nonisosceles trapezoid

22. $\{a, c, d, e, 3, 4\}$ **23.** $\{e, m, 2, 4, 5\}$

24. $\{a, c, d, e, m, 2, 3, 4, 5\}$ **25.** $\{e, 4\}$

5·1 If/Then Statements

p. 259

1. If lines are perpendicular, then they meet to form right angles. **2.** If an integer ends in 0 or 5, then it is a multiple of 5.
3. If an integer is odd, then it ends in 1, 3, 5, 7, or 9.
4. If Jacy is too young to vote, then he is 15 years old.

p. 260

5. A rectangle does not have four sides.
6. The donuts were not eaten before noon.
7. If an integer does not end with 0 or 5, then it is not a multiple of 5. **8.** If I am not in Seattle, then I am not in the state of Washington.

p. 261

9. If an angle is not a right angle, then it does not have a measure of 90°. **10.** If $2x = 6$, then $x = 3$.

p. 262

Rapunzel, Rapunzel, Let Down Your Hair If there are more than 150,001 people in your town, the same argument applies.

5·2 Counterexamples

p. 264

1. True; false; counterexample: skew lines **2.** True; true

5•3 Sets

p. 266

1. False **2.** True **3.** True

4. $\{1\}, \{4\}, \{1, 4\}, \emptyset$

5. $\{m\}, \emptyset$

6. $\{a\}, \{b\}, \{c\}, \{a, b\}, \{b, c\}, \{a, c\}, \{a, b, c\}, \emptyset$

p. 267

7. $\{1, 2, 9, 10\}$ **8.** $\{m, a, p, t, h\}$

9. $\{9\}$ **10.** $\emptyset$

p. 268

11. $[1, 2, 3, 4, 5, 6]$ **12.** $\{1, 2, 3, 4, 5, 6, 9, 12, 15\}$

13. $\{6, 12\}$ **14.** $\{6\}$

Chapter 6: Algebra

p. 274

1. $2x - 3 = x + 9$ **2.** $4(n + 2) = 2n - 4$

3. $6(x + 5)$ **4.** $3(4n - 5)$

5. $a + 3b$ **6.** $11n - 10$ **7.** 20 mi

8. $x = 7$ **9.** $y = -20$ **10.** $x = 9$ **11.** $y = 54$

12. $n = 3$ **13.** $y = -3$ **14.** $n = 4$ **15.** $x = 6$

16. 18 girls **17.** 6.5 cm **18.**

−3 −2 −1 0

$x < -2$

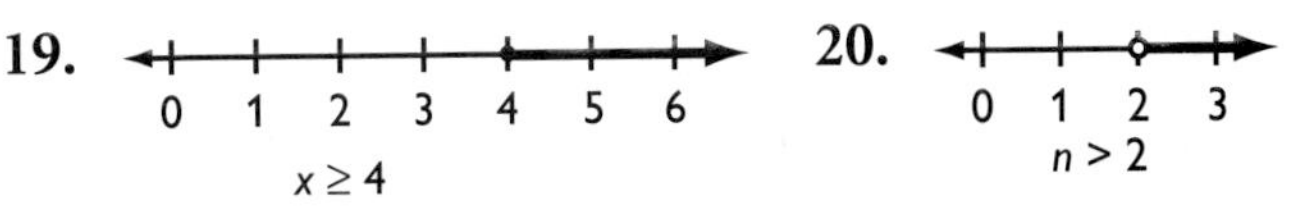

p. 275

21–24.

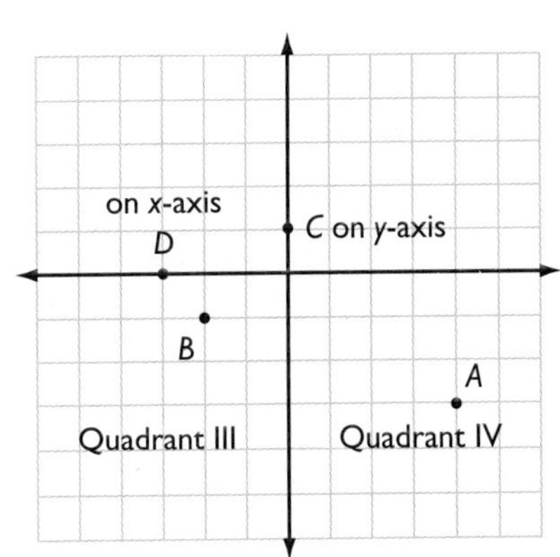

25. $-\frac{3}{5}$

Continued

HOT SOLUTIONS

p. 275 (cont.)

26.

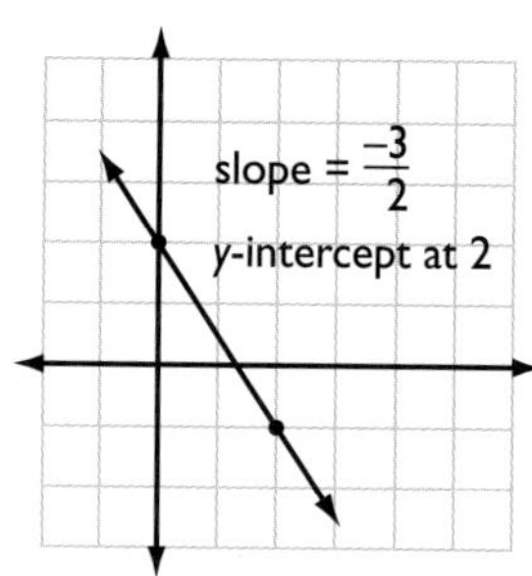

27.

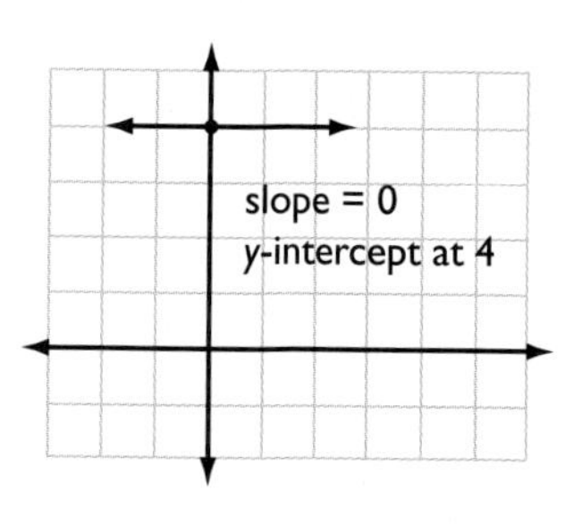

28.

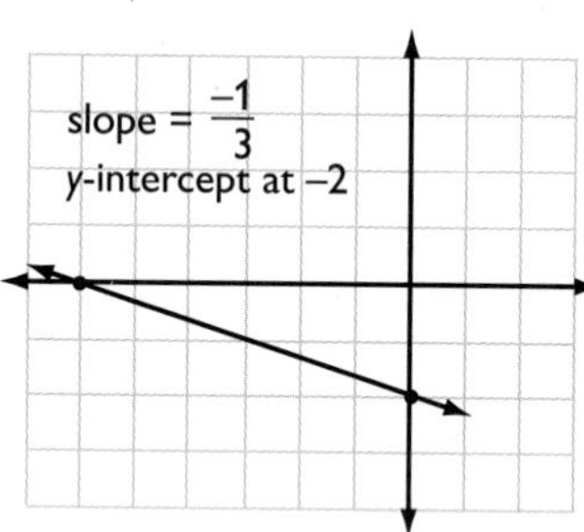

29.

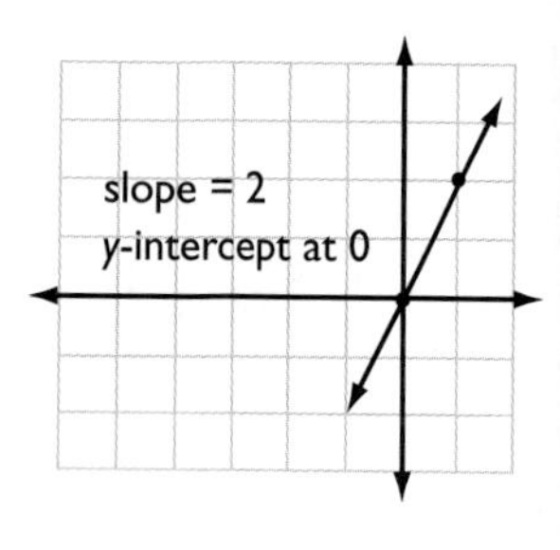

30. $y = x - 4$ **31.** $y = \frac{2}{3}x + 1$ **32.** $x = -4$
33. $y = 7$ **34.** $y = -\frac{1}{9}x + 2$ **35.** $y = \frac{1}{2}x + \frac{1}{2}$

6·1 Writing Expressions and Equations

p. 276 **1.** 2 **2.** 1 **3.** 3 **4.** 2

p. 277 **5.** $5 + x$ **6.** $n + 10$ **7.** $y + 8$ **8.** $n + 1$

p. 278 **9.** $14 - x$ **10.** $n - 2$ **11.** $y - 6$ **12.** $n - 4$

p. 279 **13.** $3x$ **14.** $7n$ **15.** $0.25y$ **16.** $12n$

p. 280 **17.** $\frac{x}{7}$ **18.** $\frac{16}{n}$ **19.** $\frac{40}{y}$ **20.** $\frac{a}{11}$

p. 281 **21.** $8n - 12$ **22.** $\frac{4}{x} - 1$ **23.** $2(n - 6)$

Sea World to the Rescue $2{,}378 + 25x = 9{,}000$

p. 282 **24.** $x - 8 = 5x$ **25.** $4n - 5 = 4 + 2n$
26. $\frac{x}{6} + 1 = x - 9$

6•2 Simplifying Expressions

p. 284 **1.** No **2.** Yes **3.** No **4.** Yes

5. $5 + 2x$ **6.** $7n$ **7.** $4y + 9$ **8.** $6 \cdot 5$

p. 285 **Prime Time** $23.3a$

p. 286 **9.** $4 + (8 + 11)$ **10.** $5 \cdot (2 \cdot 9)$ **11.** $2x + (5y + 4)$

12. $(7 \cdot 8)n$

13. $6(100 - 2) = 588$ **14.** $3(100 + 5) = 315$

15. $9(200 - 1) = 1{,}791$ **16.** $4(300 + 10 + 8) = 1{,}272$

p. 287 **17.** $14x + 8$ **18.** $24n - 16$ **19.** $-7y + 4$
20. $9x - 15$

p. 288 **21.** $7(x + 5)$ **22.** $3(6n - 5)$ **23.** $15(c + 4)$
24. $20(2a - 5)$

p. 290 **25.** $13x$ **26.** $4y$ **27.** $10n$ **28.** $-4a$

29. $3y + 8z$ **30.** $13x - 20$ **31.** $9a + 4$ **32.** $9n - 4$

6•3 Evaluating Expressions and Formulas

p. 292 **1.** 22 **2.** 1 **3.** 23 **4.** 20

p. 293 **5.** 34 cm **6.** 28 ft

Maglev $1\frac{1}{4}$ hr, $2\frac{1}{4}$ hr, $3\frac{3}{4}$ hr

p. 294 **7.** 36 mi **8.** 1,500 km **9.** 440 mi **10.** 8 ft

6•4 Solving Linear Equations

p. 296 **1.** -4 **2.** x **3.** 35 **4.** $-10y$

p. 297 **5.** True, false, false **6.** False, true, false

7. False, true, false **8.** False, false, true

9. Yes **10.** No **11.** No **12.** Yes

p. 298 **13.** $x + 3 = 12$ **14.** $x - 3 = 6$ **15.** $3x = 27$
16. $\frac{x}{3} = 3$

p. 299 **17.** $x = 9$ **18.** $n = 16$ **19.** $y = -7$ **20.** $a = 9$

p. 301 **21.** $x = 7$ **22.** $y = 32$ **23.** $n = -3$ **24.** $a = 36$

p. 302 **25.** $x = 3$ **26.** $y = 50$ **27.** $n = -7$ **28.** $a = -6$

p. 303 **29.** $n = 4$ **30.** $x = -2$

p. 305 **31.** $n = 5$ **32.** $x = -6$
33. $w = \frac{A}{l}$ **34.** $y = \frac{3x + 8}{2}$

p. 306 **How Risky Is It?** Yes; increasing life expectancy; less risky

6•5 Ratio and Proportion

p. 308 **1.** $\frac{3}{9} = \frac{1}{3}$ **2.** $\frac{9}{12} = \frac{3}{4}$ **3.** $\frac{12}{3} = \frac{4}{1} = 4$

p. 309 **4.** Yes **5.** No

p. 310 **6.** 5.5 gal **7.** \$450 **8.** 970,000 **9.** 22,601,000

6•6 Inequalities

p. 313 **1.** –2 –1 0 1 2 3 **2.** –5 –4 –3 –2 –1 0

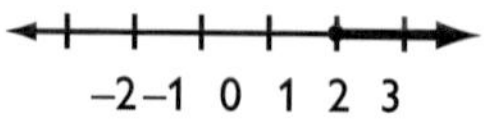

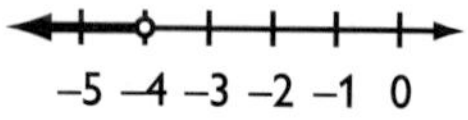

3. –4 –3 –2 –1 0 1 **4.** –2 –1 0 1 2 3

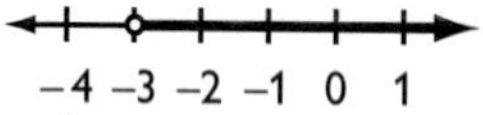

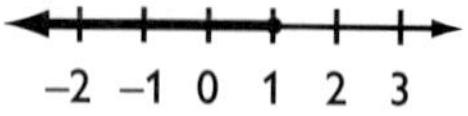

p. 314 **5.** $x > -3$ **6.** $n \leq -2$ **7.** $y < 3$ **8.** $x \geq 1$

6•7 Graphing on the Coordinate Plane

p. 316 **1.** y-axis **2.** Quadrant II **3.** Quadrant IV
4. x-axis

p. 317 **5.** $(-2, 4)$ **6.** $(1, -3)$ **7.** $(-4, 0)$ **8.** $(0, 1)$

p. 318

9. *H* is on Quadrant II.
10. *J* is on Quadrant IV.
11. *K* is on Quadrant III.
12. *L* is on the *x*-axis.

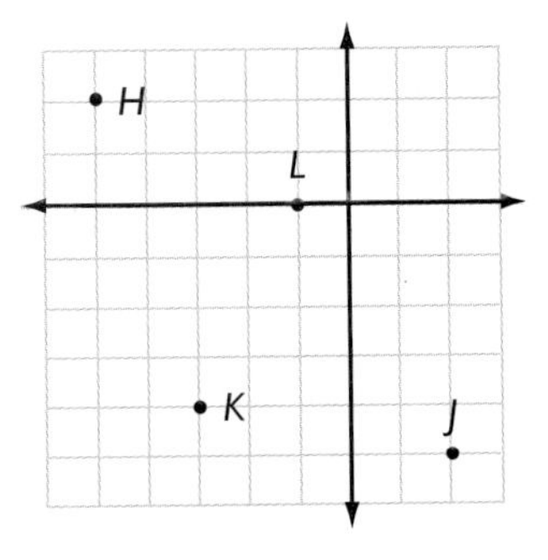

p. 320

13.

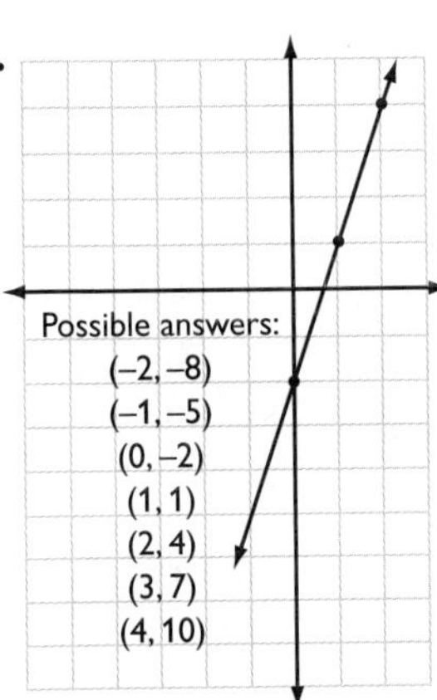

14.

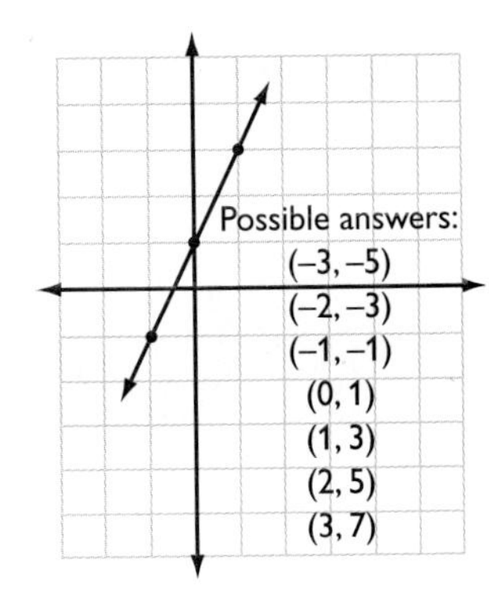

15.

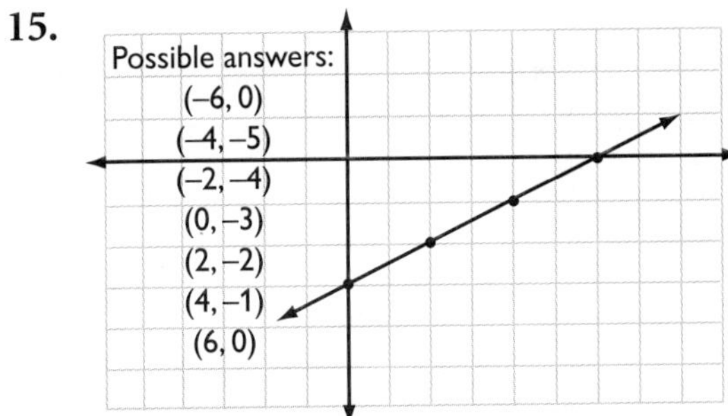

HOT SOLUTIONS

Continued

p. 320 (cont.)

16.

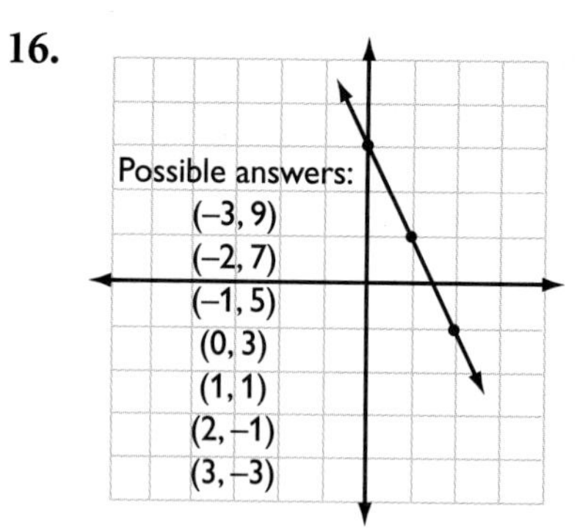

p. 321

17.

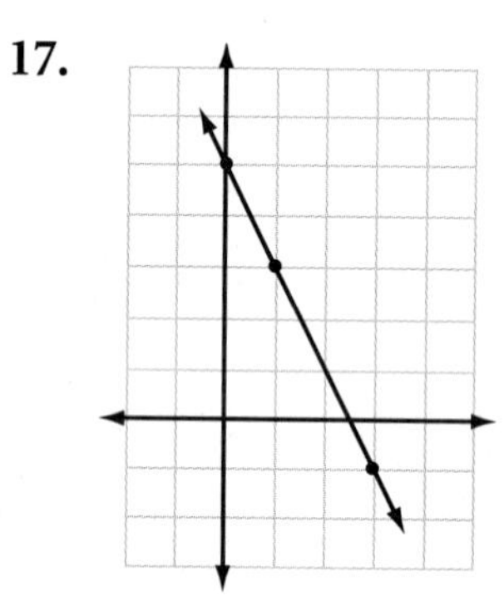

18.

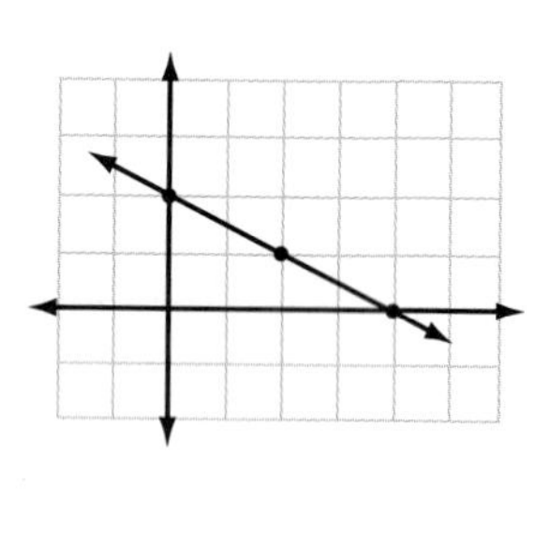

19.

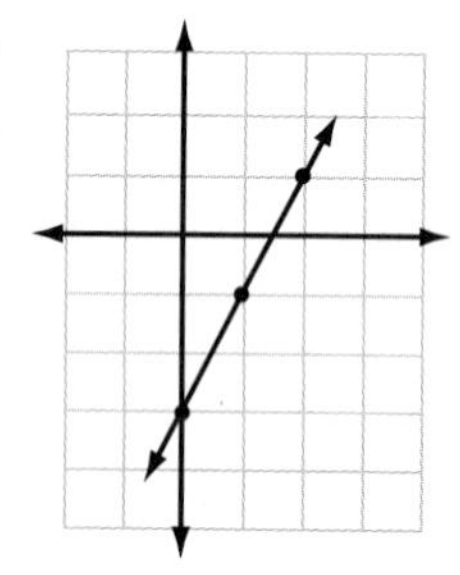

20.

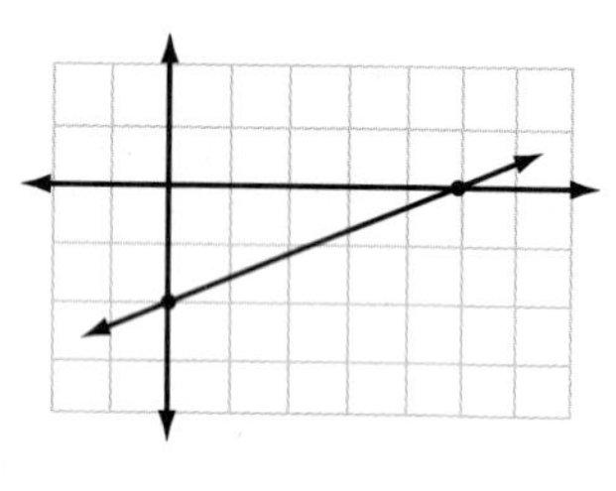

p. 322

21.

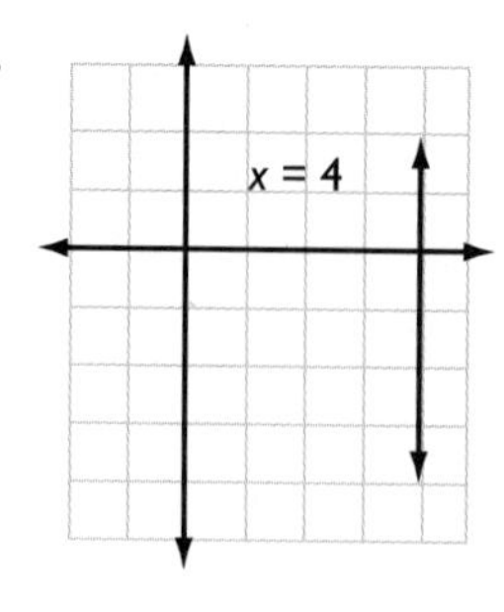

22.

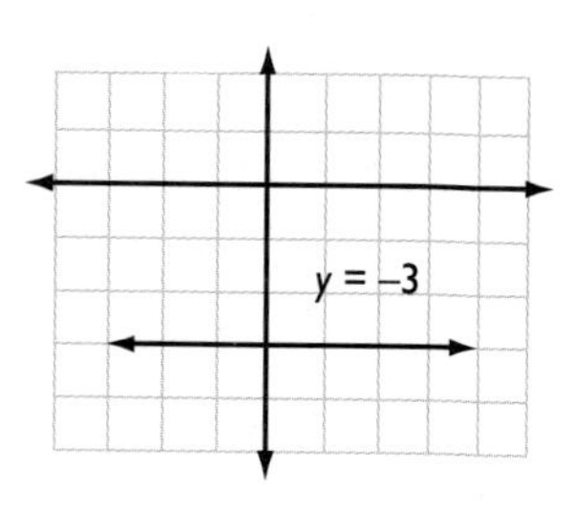

p. 322 **23.**

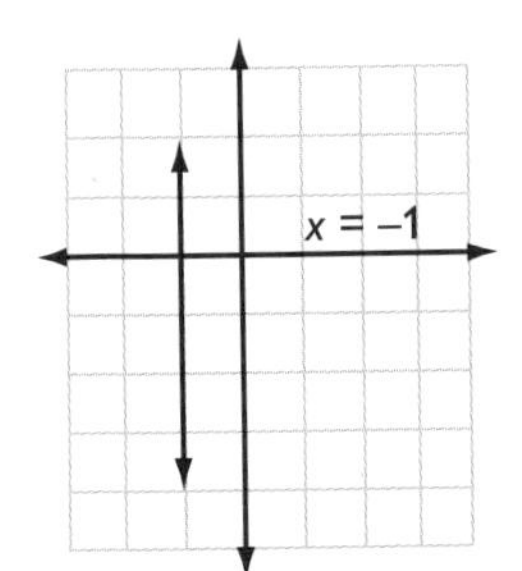

24.

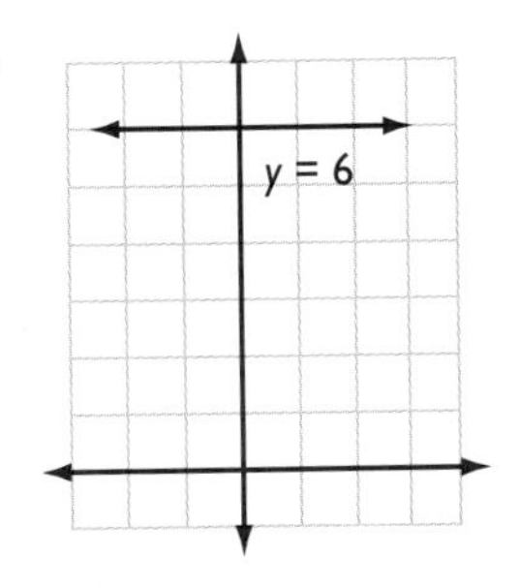

 Slope and Intercept

p. 325 **1.** $\frac{2}{3}$ **2.** $\frac{-5}{1} = -5$

p. 326 **3.** -1 **4.** $\frac{3}{2}$ **5.** $-\frac{1}{2}$ **6.** 5

p. 327 **7.** 0 **8.** No slope **9.** No slope **10.** 0

p. 328 **11.** 0 **12.** $-\frac{1}{3}$

p. 329 **13.**

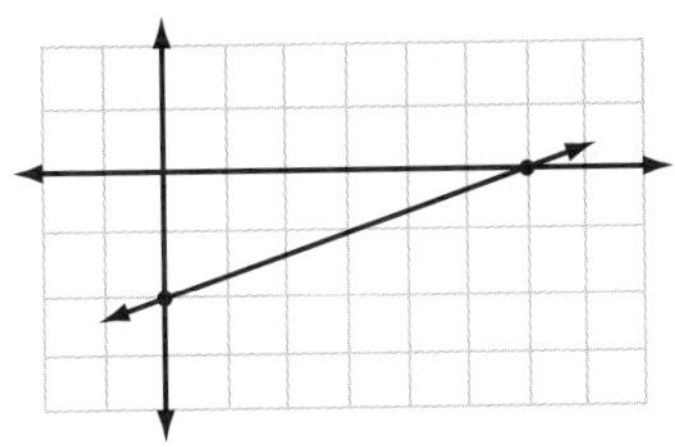

15.

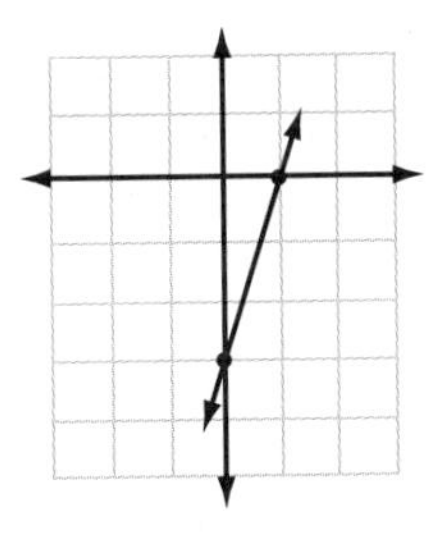

14.

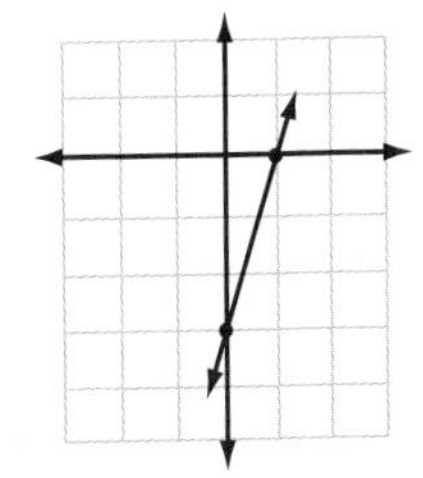

16.

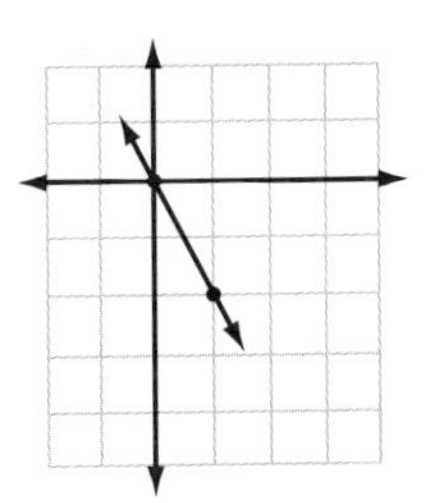

p. 330 **17.** Slope $= -2$, y-intercept at 3 **18.** Slope $= \frac{1}{5}$, y-intercept at -1 **19.** Slope $= -\frac{3}{4}$, y-intercept at 0 **20.** Slope $= 4$, y-intercept at -3

p. 331

21.

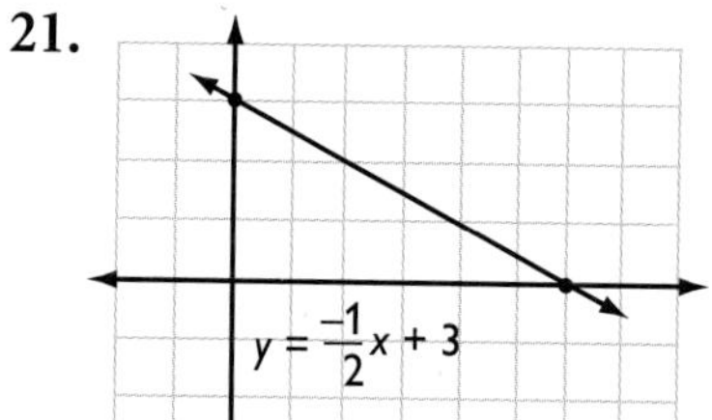

22.

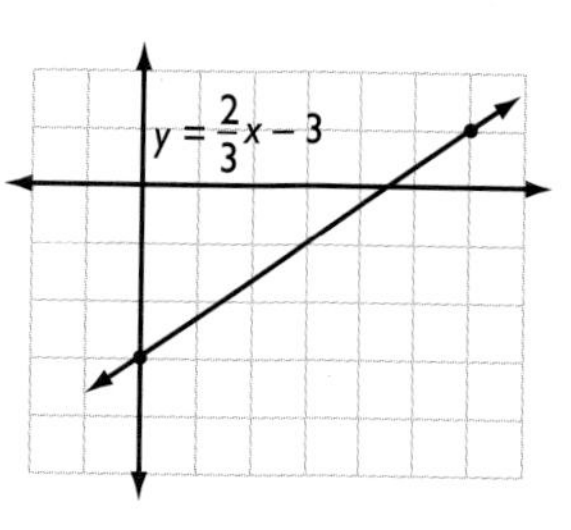

23.

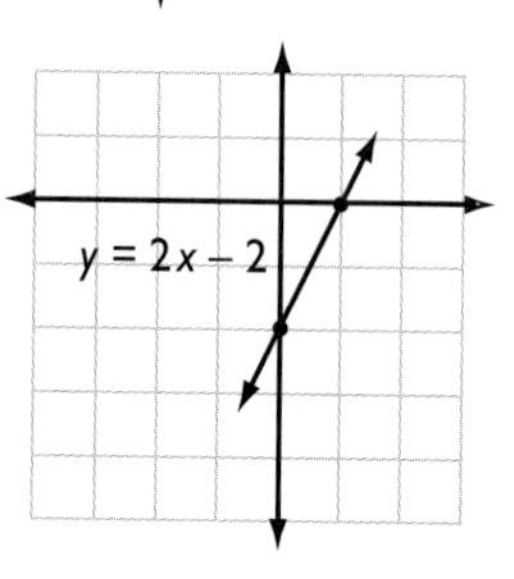

24.

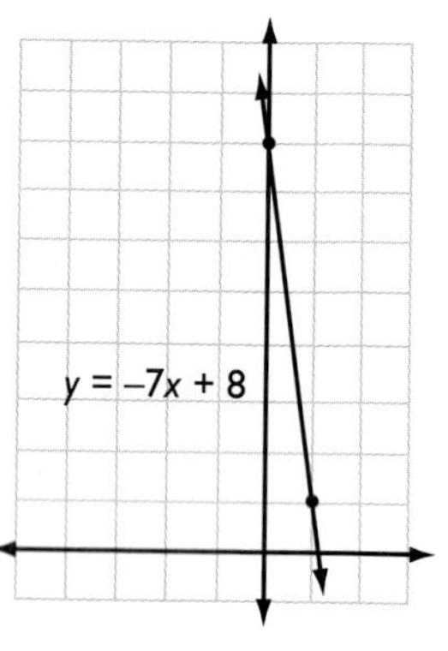

p. 332

25.

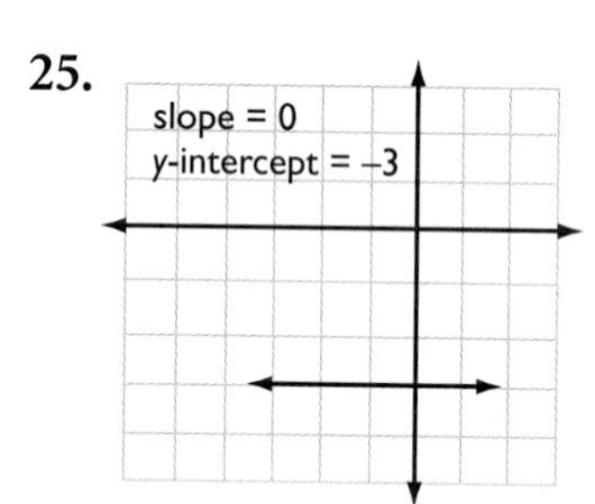

26.

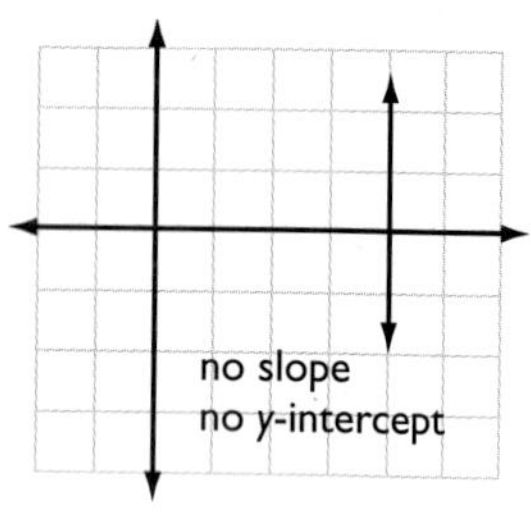

27.

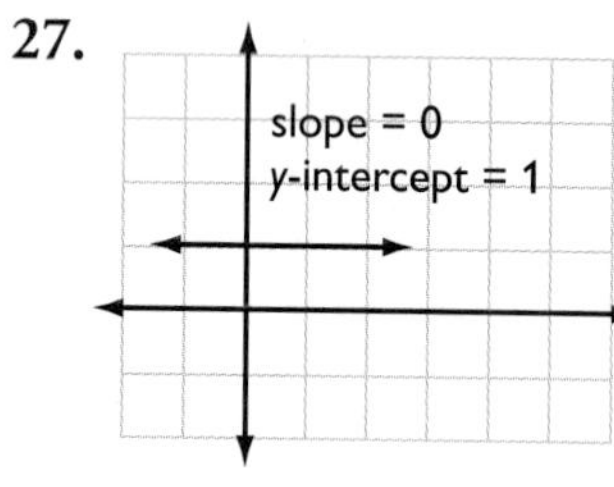

28.

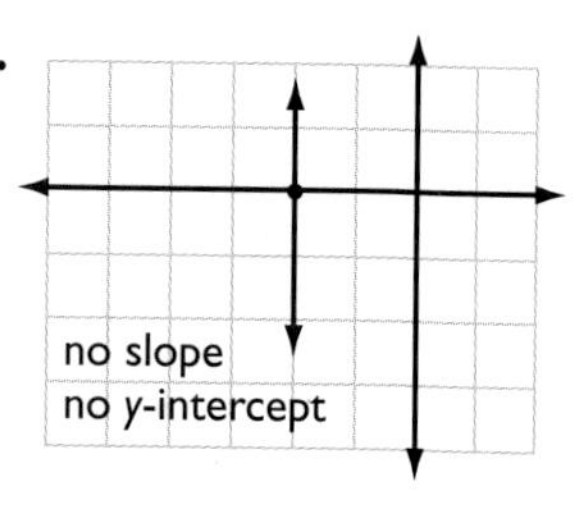

p. 333 **29.** $y = -2x + 4$ **30.** $y = \frac{2}{3}x - 2$ **31.** $y = 3x - 4$

p. 334 **32.** $y = x - 2$ **33.** $y = -2x + 5$ **34.** $y = \frac{3}{4}x - 3$ **35.** $y = 2$

Chapter 7: Geometry

p. 340 **1.** ∠CBA **2.** ∠ABD **3.** ∠CBD **4.** 52° **5.** 108° **6.** Square **7.** 30 cm **8.** 48 in. **9.** 48 m^2 **10.** 96 $in.^2$ **11.** 65 ft^2 **12.** 386.6 cm^2 **13.** 82 ft^2

p. 341 **14.** 118 $in.^3$ **15.** Prism and cylinder **16.** 157 ft and 1,963 ft^2 **17.** Not a right triangle **18.** 13.6 in. **19.** $\overline{BC}$, $\overline{BA}$, $\frac{BC}{BA}$ **20.** About 25°

7•1 Naming and Classifying Angles and Triangles

p. 342 **1.** $\overleftrightarrow{PQ}$, $\overleftrightarrow{QP}$ **2.** P

p. 344 **3.** G **4.** $\angle DGE$ or $\angle EGD$, $\angle EGF$ or $\angle FGE$, $\angle DGF$ or $\angle FGD$

p. 345 **5.** 20° **6.** 115° **7.** 135°

p. 346 **8.** $m\angle DBC = 120°$; obtuse angle

9. $m\angle ABC = 180°$; straight angle

10. $m\angle ABD = 60°$; acute angle

p. 348 **11.** $m\angle Z = 90°$ **12.** $m\angle M = 60°$

13. D **14.** B and D

7•2 Naming and Classifying Polygons and Polyhedrons

p. 351 **1.** Possible answers: *RSPQ; QPSR; SPQR; RQPS; PQRS; QRSP; PSRQ; SRQP* **2.** 360° **3.** 105°

p. 352 **4.** No; yes; no; yes; no **5.** Yes; it has four sides that are the same length and opposite sides are parallel.

p. 354 **6.** Yes; quadrilateral **7.** No **8.** Yes; hexagon

Oh, Obelisk! 50°

p. 356 **9.** 1,440° **10.** 108°

p. 357 **11.** Triangular prism
12. Triangular pyramid or tetrahedron

7•3 Symmetry and Transformations

p. 361

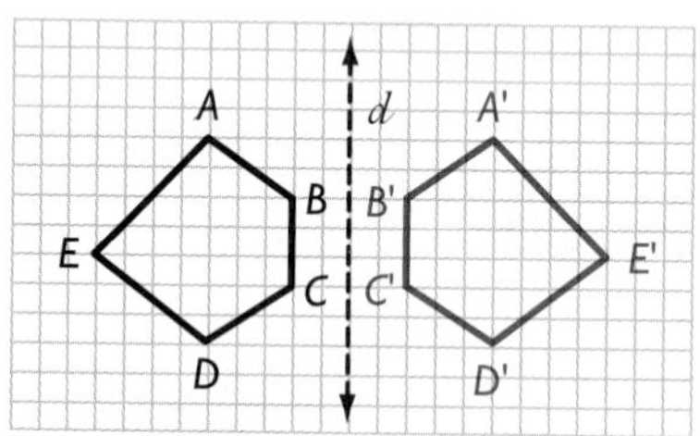

Fish Farming 15,000 m^2

p. 362 **2.** No **3.** Yes, two **4.** No **5.** Yes, one

p. 363 **6.** 180° **7.** 270°

p. 364 **8.** Yes **9.** No **10.** No

7•4 Perimeter

p. 367 **1.** 29 cm **2.** 39 in. **3.** 6 m **4.** 20 ft

p. 368 **5.** 60 cm **6.** 48 cm

The Pentagon 924 ft.

p. 369 **7.** 40 in. **8.** 48 m

7•5 Area

p. 372 **1.** About 40 cm^2

p. 374 **2.** $6\frac{2}{3}$ ft^2 or 960 $in.^2$ **3.** 36 cm^2

4. 54 m^2 or 54 square meters **5.** 8 m

p. 375 **6.** 60 $in.^2$ **7.** 540 cm^2

p. 376 **8.** 12 ft^2 **9.** 30 ft^2

7•6 Surface Area

p. 379 **1.** 126 m^2 **2.** 88 cm^2

p. 380 **3.** 560 cm^2 **4.** A **5.** 1,632.8 cm^2

7•7 Volume

p. 382 **1.** 3 cm^3 **2.** 6 ft^3

p. 383 **3.** 896 $in.^3$ **4.** 27 cm^3

p. 384 **5.** 113.04 $in.^3$ **6.** 56.52 cm^3

p. 386 **7.** 9.4 m^3 **8.** 2,406.7 in^3

Good Night, T. Rex About 1,175,000 mi^3

7•8 Circles

p. 388 **1.** 9 in. **2.** 1.5 m **3.** $\frac{x}{2}$ **4.** 12 cm
5. 32 m **6.** $2y$

p. 390 **7.** 5π in. **8.** 20.1 cm **9.** 7.96 m **10.** $5\frac{1}{2}$ in.

p. 391 **11.** $\angle ABC$ **12.** 90° **13.** 120°

Around the World $2\frac{2}{5}$ times

p. 392 **14.** 20.25π in.^{2}; 63.6 in.^{2} **15.** 177 cm^2

7•9 Pythagorean Theorem

p. 394 **1.** 9, 16, 25 **2.** Yes

p. 396 **3.** 14 cm **4.** 55 in.

7•10 Tangent Ratio

p. 399 **1.** 2.4 **2.** 1.33

Chapter 8: Measurement

p. 406 **1.** One hundredth **2.** One thousand

3. One thousandth

4. 0.8 **5.** 5.5 **6.** 15,840 **7.** 13

8. 108 in. **9.** 3 yd **10.** 274 cm **11.** 3 m

12. 684 in.^{2} **13.** 4,181 cm^2

p. 407 **14.** 50,000 **15.** 90 **16.** 5,184 **17.** 4,000

18. $\frac{6}{8}$ or $\frac{3}{4}$ **19.** 16 bottles **20.** About 13 cans

21. About 44 lb **22.** 3 lb **23.** 172,800 sec **24.** 9:4

25. 2.25 or $\frac{9}{4}$

8•1 Systems of Measurement

p. 409 **1.** Metric **2.** Customary

p. 410 **3.** Between 48.6 and 49, so we can't give the answer to the nearest tenth. We can say about 49 meters.

4. No; the answer is given to more decimal places than the approximate data. The actual answer can be anywhere between 17.6 and 17.7 mi/gal.

8·2 Length and Distance

p. 412 **1.** Answers will vary. **2.** Answers will vary.

p. 413 **3.** 800 **4.** 3.5 **5.** 4 **6.** 10,560

p. 414 **7.** 71.1 cm **8.** 89.7 yd **9.** B **10.** C **11.** B

8·3 Area, Volume, and Capacity

p. 416 **1.** 1,600 mm^2 **2.** 288 $in.^2$

p. 417 **3.** 12 yd^3 **4.** 512 cm^3 = 512,000 mm^3

p. 418 **5.** The juice

p. 419 **In the Soup!** 1,792 cartons; 912.6 ft^3

8·4 Mass and Weight

p. 420 **1.** 80 **2.** 3.75 **3.** 8,000,000 **4.** 0.375

p. 421 **Poor SID** No, the mass is always the same.

8·5 Time

p. 422 **1.** 192 months **2.** September 10, 2013

p. 423 **The World's Largest Reptile** 5 times as long

8·6 Size and Scale

p. 424 **1.** *A* and *E* are similar. *B* and *D* are similar.

p. 425 **2.** 2 **3.** $\frac{1}{3}$

p. 426 **4.** $\frac{9}{4}$ **5.** 16 ft^2

Chapter 9: Tools

p. 432 **1.** 1,020 **2.** 7,875

3. 268.5 **4.** 6.1

5. 114.46 cm **6.** 581.196 cm^2

7. 55,840.59 **8.** 0.29 **9.** 20.25 **10.** 2.12

11. 3.2×10^{13} **12.** 29.12

p. 433 **13.** 74° **14.** 148° **15.** 74° **16.** Yes
17. Straightedge and compass
18. A2 **19.** C1 + C2 **20.** 66

9•1 Four-Function Calculator

p. 435 **1.** 1.3 **2.** −15.75 **3.** −0.4729729
p. 436 **4.** −388,940 **5.** 3,111,984
p. 438 **6.** 35 **7.** −115 **8.** 85 **9.** 106.25 **10.** 1.4577379

9•2 Scientific Calculator

p. 442 **1.** 479,001,600 **2.** 38,416 **3.** 0.037
4. 109,272.375 **5.** 1.1 **6.** 2.6
p. 443 **Magic Numbers** $\frac{100a + 10a + 1a}{a + a + a} = \frac{111a}{3a} = 37$

9•3 Geometry Tools

p. 444 **1.** 2 in. or 5.1 cm **2.** $2\frac{3}{4}$ in. or 7 cm
p. 446 **3.** 54° **4.** 122°
p. 447 **5.**

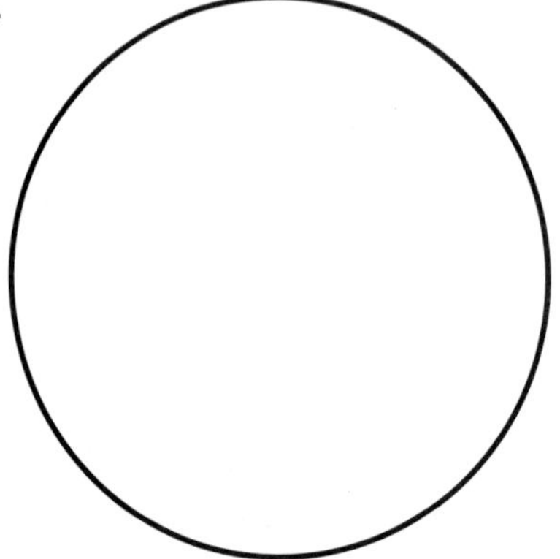

p. 447 **6.**

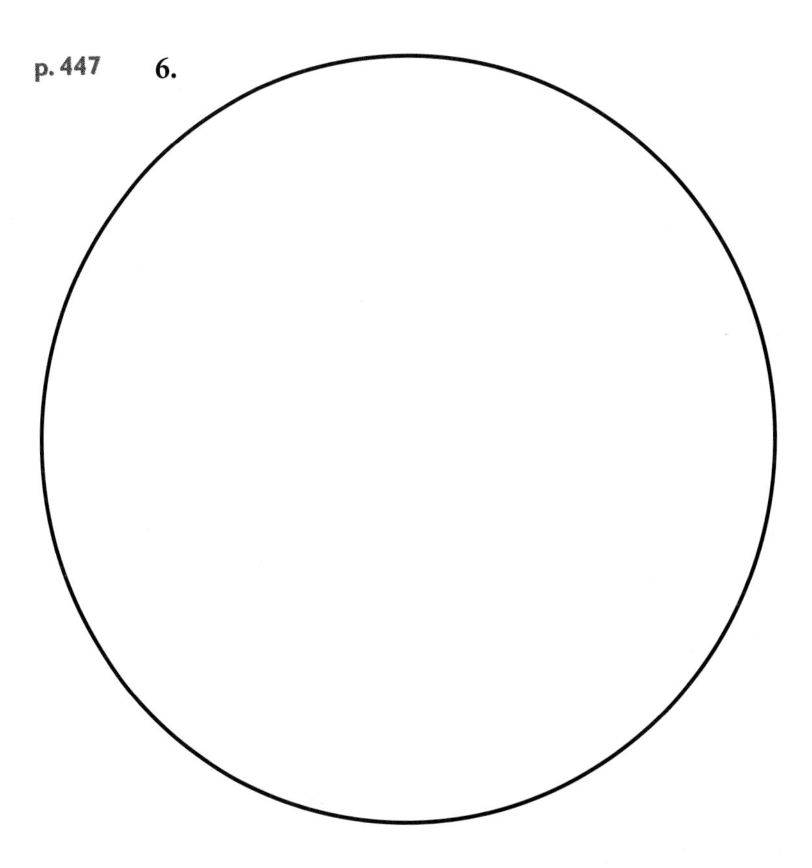

p. 449 **7.**

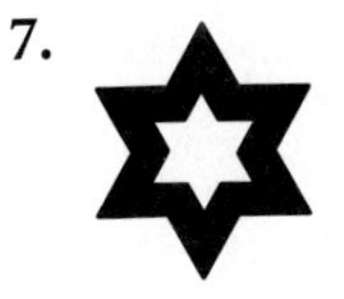

Compare drawing to picture in book.

8. Answers will vary.

9•4 Spreadsheets

p. 452 **1.** 2 **2.** 3 **3.** 25 **4.** True **5.** False

p. 453 **6.** B3 × C3 **7.** B4 × C4 **8.** D2 + D3 + D4

p. 455 **9.** A6 + 10; 160 **10.** D2 + 10; 120

p. 456 **11.** A5, B5 **12.** A9, B9

INDEX

A

Absolute values of integers, 92
Accuracy, 410
Acronym PEMDAS, 83, 189
Acute angle, 345
Acute triangle, 347
Addition
  associative property of, 78, 285
  calculator keys for, 435
  commutative property of, 78, 284
  of decimals, 132–133
  distributive property of, 79, 286
  of fractions, 116–117
  identity property of, 79
  integer operations with, 92
  of integers, 93
  of mixed numbers, 118–120
  order of operations with, 82
  shortcuts for, 80
  writing expressions involving, 276–277
  zero property of, 79
Additive inverses, 296
Adjacent angles, 346
Analyzing data, 214–220
Angles, 343–346
  central, 390–391
  classifying, 345–346
  measuring, 344
  naming, 343
  of polygons, 355–356
  of quadrilaterals, 350
  tangents of, 398–399
Approximation, 76
Arc, 391
Area, 372–376, 416
  of circle, 392
  estimating, 372
  of parallelogram, 374
  of rectangle, 373
  of trapezoid, 376
  of triangle, 375
Associative property of addition and multiplication, 78, 285
Averages, 222
  weighted, 228–229
Axes, 316

B

Bar graph, 209
  double, 210
Bases, 168
  finding, 147
Benchmarks in percents, 141
Bimodal data, 226
Bimodal distribution, 220
Box plot, 203

C

Calculator memory, 436
Calculators
  estimating square roots, 178
  evaluating powers, 174
  finding decimal quotients, 138
  finding discounts and sale prices, 150
  finding percents of decrease, 149
  finding percents of increase, 148
  FIX number of decimal places, 138
  four-function, 434–438
  scientific, 440–442
  special keys on, 437–438
Canceling factors, 123
Capacity, 418
Cells in spreadsheets, 452
Center of circle, 388
Center of rotation, 363
Central angles, 390–391
Circle graph, 204
Circles
  area of, 392
  parts of, 388
Circumference, 389–390
Classifying angles, 345–346
Classifying triangles, 346–347
Collecting data, 196–200
Columns in spreadsheets, 452
Combinations, 236–238
Commas separating periods, 74
Common denominators, 113, 117
Common factors, 85
  distributive property with, 288
  greatest, *see* Greatest common factor
Commutative property of addition and multiplication, 78, 284
Comparing
  decimals, 129–130
  fractions, 112
  mixed numbers, 113
  numbers, 75–76
Compass, 446–447

Compiling data, 199
Composite numbers, 87
Conclusion, 258
Conditional, 258
  contrapositive of, 261
  converse of, 259
  inverse of, 260
Cone, 356
  volume of, 384–385
Congruent sides, 346–347
Constant terms, 284
Construction problems, 447–448
Contrapositive of conditional, 261
Converse of conditional, 259
Conversion factors, 413
Conversion table, 414
Converting
  decimals to fractions, 160
  decimals to percents, 156
  fractions to decimals, 159
  fractions to percents, 154
  improper fractions to mixed numbers, 110
  mixed numbers to improper fractions, 110
  percents to decimals, 157
  percents to fractions, 155
  scientific notation to standard form, 185–186
Coordinate plane
  graphing on, 316–322
  locating points on, 318
Correlation, 216–217
Counterexamples, 264
Cross products
  of fractions, 106
  in proportions, 309
Cube, 356–357
Cube function key, 441
Cube root function key, 441
Cube roots, 179
Cubes of numbers, 170
Cubic inch, 382
Cubic meter, 382
Customary equivalents, 413
Customary system, 408, 409
Cylinder, 356
  surface area of, 380
  volume of, 384

D

Data
  analyzing, 214–220
  collecting, 196–200
  compiling, 199
  displaying, 202–212
  distribution of, 219–220
Decagon, 353
Decimal method for finding percents of numbers, 144
Decimal points, 126
  in products, 134
Decimal products
  decimal points in, 134
  estimating, 135
Decimal sums and differences, estimating, 133
Decimals
  addition of, 132–133
  comparing, 129–130
  converting fractions to, 159
  converting percents to, 157
  converting to fractions, 160
  converting to percents, 156
  division of, 136–138
    zeros in, 137
  fractions and, 159–161
  fractions and percents and, 154
  hundredths in, 126
  multiplication of, 134
    with zeros in products, 135
  naming, 126–129
  operations with, 132–138
  ordering, 130
  percents and, 156–158
  place-value chart for, 126, 128
  repeating, 159
  rounding, 130
  rounding quotients in, 138
  subtraction of, 132–133
  tenths in, 126
  terminating, 159
  thousandths in, 127
  zeros to right of, 129
Decrease, percents of, 149
Degrees, 391
Denominators, 103
  common or like, 113, 117
    adding mixed numbers with, 118
    addition and subtraction of fractions with, 116
    subtracting mixed numbers with, 120
  in reciprocals of numbers, 123
  unlike
    adding mixed numbers with, 119
    addition and subtraction of fractions with, 117
    ordering fractions with, 114
    subtracting mixed numbers with, 120

Dependent events, 249
Diagonal of polygon, 353
Diameter, 388
Differences, 277
  decimal, estimating, 133
Discounts, 150
Displaying data, 202–212
Distance, 413
Distance traveled, formula for, 294
Distribution of data, 219–220
Distributive property, 79, 286
  with common factors, 288
  solving equations involving, 304
Divisibility rules for numbers, 86
Division
  calculator keys for, 435
  of decimals, 136–138
    zeros in, 137
  of fractions, 124
  of integers, 94
  order of operations with, 82
  writing expressions involving, 279–280
Double-bar graph, 210

E

Element of set, 266
Empty set, 266
Equal sign (=), 75, 282
Equations, 282
  equivalent, 298
  false, 296
  graphing with two variables, 319–321
  linear, 296–305
  solutions of, 297
  solving, *see* Solving equations
  true, 296
  writing, *see* Writing equations
Equilateral triangle, 347
Equivalent equations, 298
Equivalent expressions, 282, 287
Equivalent fractions, 104–106
Eratosthenes, sieve of, 87
Estimating
  area, 372
  decimal products, 135
  decimal sums and differences, 133
  percents of numbers, 141–142, 151
    mental math in, 141–142
  square roots, 177–178
Evaluating
  expressions, 292
  expressions with exponents, 188
  formulas, 292–294
  higher powers, 172
Expanded notation for numbers, 75
Experimental probability, 240
Exponents, 168–172
  evaluating expressions with, 188
  laws of, 188
  multiplication using, 168–169
  order of operations with, 82, 188
  in prime factorization, 88
Expressions, 276
  equivalent, 282, 287
  evaluating, 292
    with exponents, 188
  simplifying, 284–290
  writing, *see* Writing expressions

F

Factorial function key, 441
Factorial notation, 236
Factorization, prime, *see* Prime factorization
Factors, 84
  canceling, 123
False equations, 296
Fill down and fill right spreadsheet commands, 454–455
Finding
  bases, 147
  percents of numbers, 144–152
  simple interest, 152
FIX number of decimal places, 138
Flat distribution, 220
Flips, 360
Fluid ounces, 418
Formulas, 292
  evaluating, 292–294
  solving, for variables, 305
  spreadsheet, 453
Four-function calculators, 434–438
Fraction method for finding percents of numbers, 144
Fractional names for one, 104
Fractions, 102
  addition of, 116–117
  comparing, 112
  converting decimals to, 160
  converting percents to, 155
  converting to decimals, 159
  converting to percents, 154
  cross products of, 106
  decimals and, 159–161
  decimals and percents and, 154
  denominators of, 103
  division of, 124
  equivalent, 104–106

improper, *see* Improper fractions
multiplication of, 122–123
naming, 102
numerators of, 103
ordering, 114
percents and, 154–156
subtraction of, 116–117
writing, in lowest terms, 106–107
Frequency graphs, 206

G

GCF, *see* Greatest common factor
Geometry tools, 445–449
Graph
bar, 209
circle, 204
double bar, 210
line, 207
spreadsheet, 456
strip, 244
Graphing
on coordinate plane, 316–322
equations with two variables, 319–321
lines using slope and y-intercept, 329
Greater than (>), 75, 312
Greater than or equal to (≥), 312
Greatest common factor (GCF), 85, 107
prime factorization in finding, 89

H

Heptagon, 353
Hexagon, 353, 356
Higher powers, evaluating, 172
Histogram, 211
Horizontal lines, 322
slope-intercept form and, 332
slopes of, 327
Horizontal number line, 316
Horizontal rows in spreadsheets, 452
Hundredths in decimals, 126
Hypotenuse, 395
Hypothesis, 258

I

Identity property
of addition, 79
of multiplication, 79
If/then statements, 258–262
Improper fractions, 109
converting mixed numbers to, 110
converting to mixed numbers, 109
Increase, percents of, 148
Independent events, 249
Inequalities, 312–314
showing, 312–313
solving, 313–314
Integer operations, 92–94
Integers
absolute values of, 92
addition of, 93
division of, 94
multiplication of, 94
negative, 92
opposites of, 92
positive, 92
subtraction of, 93
Interest, 152
simple, finding, 152
Interest rate, 152
Intersection of sets, 267
Inverse of conditional, 260
Inverses, additive, 296
Isosceles triangle, 347

L

Large numbers, 182–183
LCM (least common multiple), 90
Leaf, 208
Leap years, 422
Least common multiple (LCM), 90
Length, 412
Less than (<), 75, 312
Less than or equal to (≤), 312
Like terms, 289
Line of best fit, 218
Line graphs, 207
Line segments, 353
Linear equations, solving, 296–305
Lines, 219–234, 342
graphing, using slope and y-intercept, 329
slopes of, 325–327
of symmetry, 360–361
writing equations of, 332–334
from two points, 333–334
Lotteries, 245
Lowest terms, 106
writing fractions in, 106–107

M

Magic numbers, 443
Mandalas, 449
Mass, 420
Mean, 222
Measurement, systems of, 408
Measuring angles, 344
Median, 224–225
Member of set, 266

Memory, calculator, 436
Mental math in estimating
  percents, 141–142
Metric equivalents, 413
Metric system, 408
Mixed numbers, 109
  addition of, 118–120
  comparing, 113
  converting improper fractions
    to, 110
  converting to improper fractions, 110
  converting percents to fractions, 156
  subtraction of, 118–120
Mode, 226
Multiples of numbers, 90
Multiplication
  associative property of, 78, 285
  calculator keys for, 435
  commutative property of, 78, 284
  of decimals, 134
    with zeros in products, 135
  distributive property of, 79, 286
  of fractions, 122–123
  integer operations with, 89
  of integers, 94
  one (or identity) property of, 79
  order of operations with, 82
  shortcuts for, 80
  using exponents, 168–169
  writing expressions involving,
    278–279
  zero property of, 79

N

Naming
  angles, 343
  decimals, 126–129
  decimals greater than and less than
    one, 128
  fractions, 102
  percents, 140–142
Negation, 260
Negative integers, 92
Negative numbers, 92
  entering into calculators, 435
Net, 378
Nonagon, 353
Normal distribution, 219
Notation
  expanded, for numbers, 75
  factorial, 236
  scientific, 182–186
Number palindromes, 80
Number system, 74
Numbers
  common factors of, 85
  comparing, 75–76
  composite, 87
  cubes of, 170
  divisibility rules for, 86
  expanded notation for, 75
  factors of, 84
  finding percents of, 144–152
  greatest common factors of, *see*
    Greatest common factor
  large, 182–183
  least common multiple of, 90
  magic, 443
  mixed, *see* Mixed numbers
  multiples of, 90
  negative, *see* Negative numbers
  ordering, 75–76
  percents of, estimating, 141, 151
  place-value chart for, 74
  positive, 92
  prime, 87
  prime factorization of, 88
  reciprocals of, 123
  relatively prime, 87
  rounding, 76
  small, 182, 184
  squares of, 169
  whole, place value of, 74–76
Numerators, 103
  in reciprocals of numbers, 123

O

Obtuse angle, 345
Obtuse triangle, 347
Octagon, 353
One
  fractional names for, 104
  naming decimals greater than and
    less than, 128
One (or identity) property of
  multiplication, 79
Operations, 82
  decimal, 132–138
  integer, 80, 92–94
  order of, 82
Opposites of integers, 92
Order of operations, 82
Ordered pairs, 317
  writing, 317
Ordering
  decimals, 130
  fractions, 114
  numbers, 75–76
Origin, 317
Ounce, 418
Outcome, 241
Outcome grid, 246

INDEX

P

Palindromes, number, 80
Parallelogram, 351–352
  area of, 374
Parentheses
  on calculator, 441
  order of operations with, 82
PEMDAS acronym, 83, 189
Pentagon, 353, 355
Percents, 140, 157
  benchmarks in, 141
  converting decimals to, 156
  converting fractions to, 154
  converting mixed number percents to fractions, 156
  converting to decimals, 157
  converting to fractions, 155
  decimals and, 156–158
  of decrease, 149
  estimating, 141–142, 151
    mental math in, 141–142
  finding, 144–152
  fractions and, 154–156
  fractions and decimals and, 154
  of increase, 148
  meaning of, 140
  naming, 140–142
Perfect squares, 177
Perimeter, 366–369
  of polygons, 366–367
  of rectangle, 367–368
    formula for, 292
  of right triangle, 369
Periods, commas separating, 74
Permutations, 235–236
Pi, 389, 392
Place value, 74
  of whole numbers, 74–76
Place-value chart
  for decimals, 126, 128
  for numbers, 74
Points, 317, 342
  locating, on coordinate plane, 318
  two, writing equations of lines from, 333–334
Polygons, 346, 353–356
  angles of, 355
  perimeter of, 366–367
Polyhedrons, 356–357
Population, 196
Positive integers, 92
Positive numbers, 92
Powers, 168, 188
  calculators evaluating, 174
  higher, evaluating, 172
  of ten, 173
Powers function key, 441
Prices, sale, 150
Prime factorization, 88
  in finding greatest common factor, 89
  in finding least common multiple, 90
Prime numbers, 87
Principal, 152
Prism, 356, 357
  volume of, 383
Probability, 240–250
  defined, 240
  experimental, 240
  expressing, 243
  theoretical, 241–242
Probability line, 247–248
Products, 279
  decimal, *see* Decimal products
  writing, 122
  zeros in, multiplication of decimals with, 135
Proportion method for finding percents of numbers, 145
Proportions, 308–309
  solving problems with, 310
Protractor, 344, 445
Pyramid, 356, 357
  surface area of, 379
  volume of, 384–385
Pythagorean Theorem, 369, 395
Pythagorean triple, 396

Q

Quadrants, 316
Quadrilaterals, 350–356, 353
  angles of, 350
  defined, 350
Questionnaires, 198
Quotients, 279
  rounding, in decimals, 138

R

Radius, 388
Random sample, 197
Range, 228
Rates, 308
  interest, 152
Ratio, 308
Rays, 342
Reciprocal function key, 442
Reciprocals of numbers, 123

Rectangle, 351, 353
area of, 373
perimeter of, 367–368
formula for, 292
Rectangular prism, 357
surface area of, 378
Reflection symmetry, 362
Reflections, 360–361
Reflex angle, 345
Regular polygons, 355, 367
Regular shapes, 351
Relatively prime numbers, 87
Repeating decimals, 159
Rhombus, 352
Right angle, 345
Right triangle, 347, 394
perimeter of, 369
Roots function key, 442
Rotations, 363
Rounding
decimals, 130
numbers, 76
quotients in decimals, 138
Rows in spreadsheets, 452
Ruler, 444

S

Sale prices, 150
Sample, 196
random, 197
Sampling with and without replacement, 250
Scale factors, 425–426
Scalene triangle, 347
Scatter plots, 214–215
Scientific calculators, 440–442
Scientific notation, 182–186
converting to standard form, 185–186
writing large numbers using, 183
writing small numbers using, 184
Sets, 266–268
defined, 266
intersection of, 267
union of, 266–267
Shortcuts for addition and multiplication, 79
Sides of angle, 343
Sieve of Eratosthenes, 87
Similar figures, 424
Simple interest, finding, 152
Simplifying expressions, 284–290
Skewed distribution, 219
Slides, 364
Slope-intercept form, 330
horizontal and vertical lines and, 332
writing equations in, 330–331
Slopes, 324–327
of horizontal and vertical lines, 327
of lines, 325–327
and *y*-intercept, graphing lines using, 329
Small numbers, 182, 184
Solid shapes, 356–357
Solutions of equations, 297
Solving equations
involving distributive property, 304
linear, 296–305
requiring two operations, 301–302
with variables on both sides, 302–303
Solving formulas for variables, 305
Solving inequalities, 313–314
Solving problems with proportions, 310
Special keys on calculators, 437–438
Sphere, 356
Spinner, 232
Spreadsheet formulas, 453
Spreadsheet graphs, 456
Spreadsheets, 452–456
Square centimeters, 372
Square function key, 442
Square inches, 372
Square roots, 176–178
estimating, 177–178
Squares, 351
perfect, 177
Squares of numbers, 169
Squaring triangles, 180
Standard form, converting from scientific notation to, 185–186
Statistics, 222–229
Stem-and-leaf plot, 208
Stems, 208
Straight angle, 345
Strip graph, 244
Subsets, 266
Subtraction
calculator keys for, 435
of decimals, 132–133
of fractions, 116–117
of integers, 93
of mixed numbers, 118–120
order of operations with, 82
writing expressions involving, 277–278
Sums, 277
of angles in triangle, 347
decimal, estimating, 133

Surface area, 378–380
- of cylinder, 380
- of pyramid, 379
- of rectangular prism, 378
- of triangular prism, 379

Surveys, 196–197
Symmetry
- lines of, 360
- reflection, 362

T

Tables, 202
Tally marks, 199
Tangent function key, 442
Tangent table, 400
Tangents of angles, 398–399
Ten, powers of, 173
Tenths in decimals, 126
Terminating decimals, 159
Terms, 276, 284
- constant, 284
- like, 289

Tetrahedron, 357
Theoretical probability, 241–242
Thousandths in decimals, 127
Time, 422
Transformations, 360
Translations, 364
Trapezoid, 352
- area of, 376

Tree diagrams, 232–234
Triangle inequality, 348
Triangles, 346–348, 353
- area of, 375
- classifying, 346–347
- squaring, 180

Triangular prism, surface area of, 379
True equations, 296
Turns, 363

U–V

Union of sets, 266–267
Values, absolute, of integers, 92
Variables, 276
- graphing equations with two, 319–321
- solving equations with variables on both sides, 302–303
- solving formulas for, 305

Venn diagrams, 267–268
Vertex of angle, 343
Vertical columns in spreadsheets, 452
Vertical lines, 322
- slope-intercept form and, 332
- slopes of, 327

Vertical number line, 316
Vertices of quadrilateral, 350
Volume, 382–385, 417
- of cone, 384–385
- of cube, 170
- of cylinder, 384
- of prism, 383
- of pyramid, 384–385

W

Weight, 420
Weighted averages, 229
Whole numbers, place value of, 74–76
Writing
- fractions in lowest terms, 106–107
- improper fractions, 109
- large numbers using scientific notation, 183
- mixed numbers, 109
- ordered pairs, 317
- products, 122
- small numbers using scientific notation, 184

Writing equations, 282
- of lines, 332–334
  - from two points, 333–334
- in slope-intercept form, 330–331

Writing expressions, 276–281
- involving addition, 276–277
- involving division, 279–280
- involving multiplication, 278–279
- involving subtraction, 277–278
- involving two operations, 280

X–Z

x-axis, 316
y-axis, 316
y-intercept, 328
- graphing lines using slopes and, 329

Years, leap, 422
Zero property
- of addition, 79
- of multiplication, 79

Zeros
- in division of decimals, 137
- multiplication of decimals with zeros in products, 135
- to right of decimals, 129